SYSTEMS BIOLOGY - THEORY, TECHNIQUES AND APPLICATIONS

RECENT ADVANCES IN SYSTEMS BIOLOGY RESEARCH

SYSTEMS BIOLOGY - THEORY, TECHNIQUES AND APPLICATIONS

Additional books in this series can be found on Nova's website under the Series tab.

Additional e-books in this series can be found on Nova's website under the e-books tab.

SYSTEMS BIOLOGY - THEORY, TECHNIQUES AND APPLICATIONS

RECENT ADVANCES IN SYSTEMS BIOLOGY RESEARCH

A. X. C. N. VALENTE
ABHIJIT SARKAR
AND
YUAN GAO
EDITORS

New York

For permission to use material from this book please contact us:
Telephone 631-231-7269; Fax 631-231-8175
Web Site: http://www.novapublishers.com

Library of Congress Cataloging-in-Publication Data

Recent advances in systems biology research / editors, Andri Xavier Valente (University of Coimbra and Biocant--Biotechnology Innovation Center, Portugal), Abhijit Sarkar (The Catholic University of America, USA), and Yuan Gao (Johns Hopkins University, USA).
pages cm
Includes index.
ISBN 978-1-62948-736-6 (hardcover)
1. Biological systems. 2. Systems biology. I. Valente, Andri Xavier.
QH313.R37 2014
571.7--dc23
2013046310

Published by Nova Science Publishers, Inc. † New York

Contents

Preface

Systems biology has emerged as an exciting and rapidly growing area of research. This book presents 14 original chapters written by experts in the field covering topics ranging from the definition and scope of systems biology, to the experimental and computational techniques driving the data-rich biology revolution, to applications of the systems paradigm to clinical problems. The first two chapters discuss the systems approach and contrast it with the prevailing reductionist techniques. What emerges is a clear realization that the two paradigms complement each other. Neither can be dispensed off in favor of the other; however, what we learn from either approach must be combined into an integrated description of physiological processes.

The following seven chapters deal with experimental, statistical, kinetic, and graph theoretic techniques for organizing and interpreting systems data. A wide variety of computational strategies are covered along with a comprehensive but concise description of mass spectrometry methods that (combined with deep sequencing) are the source of many of the large data sets driving systems biology.

The last five chapters present applications of the systems approach to clinical problems and are a specially valuable feature of the book. These chapters provide interesting perspectives and literature reviews that should be very useful to anyone considering how to translate systems biology into clinical practice.

The book is written at a level accessible to a wide readership and presents much useful information in a concise and accessible format. In addition, each chapter contains references to important research in this field. The reader will find the material presented here to be an up-to-date introduction to this emerging discipline.

This project would not have been completed without the support of our students, postdocs, and family members. We thank them all for contributing in one way or another to our understanding of this subject. We would also like to acknowledge the assistance of the publisher in carrying out the task of preparing the individual manuscripts for publication in a book format. And, most importantly, we would like to acknowledge the time and energy spent by the contributing authors in preparing their respective chapters. Clearly, without their effort, this book would not have been possible.

A. X. C. N. Valente
Abhijit Sarkar
Yuan Gao

In: Recent Advances in Systems Biology Research
Editors: A. X. C. N. Valente, A. Sarkar and Y. Gao
ISBN: 978-1-62948-736-6

Introduction

A. X. C. N. Valente[1,2,3], Yuan Gao[5] and Abhijit Sarkar[4]

[1]Systems Biology Group, Biocant - Biotechnology Innovation Center, Cantanhede, Portugal
[2]CNC - Center for Neuroscience and Cell Biology, University of Coimbra, Coimbra, Portugal
[3]Institute of Fundamental Medicine and Biology, Kazan Federal University, Kazan, Russia
[4]Physics Department and Vitreous State Laboratory, Catholic University of America, Washington DC, US
[5]Lieber Institute for Brain Development, Johns Hopkins Medical Campus, Baltimore, Maryland, US

Systems biology is an attempt to come to grips with the inherent complexity of biological systems. Taking the cell as an example, its physiological state is determined by the mutual interaction of a large number of proteins, polynucleotides, and small molecules. It is tempting to view the processes involved as linear cascades of chemical reactions initiated and maintained by internal and external signals. However, it has become increasing clear that a more appropriate model is that of a coupled network of protein-protein, protein-small molecule, and protein-DNA interactions. The role of chance fluctuations in determining cellular–level outcomes has also come to be recognized. In other words, to infer a cell's response to some change and to develop a predictive picture of how it responds to stimuli, studies of individual genetic processes have to be complemented by research focused on building a fuller picture of the mutual interdependence of these processes. Attention has to be equally given to how molecular-level events interact and become correlated, and how these mutual couplings respond to probabilistic variations resulting from the noisy environment inside cells. This perspective recognizes the importance, the centrality, of multi-scale and emergent phenomena in cells, and strives to incorporate them to form a more complete picture of cellular and higher-level biological systems.

Revolutionary advances in DNA and protein sequencing technologies have made possible just the sort of studies required to integrate and decipher the interdependence of individual biochemical reactions and the deterministic and chance variability in these links. The resulting perspective has come to be called systems biology, and this book brings together the research of a number of groups at the leading edge of this rapidly growing area of biology.

We start with two chapters that consider the question "What is systems biology?" As one should expect, definitive answers to this questions are not (and perhaps cannot be) provided. However, these two chapters serve to highlight some important differences between reductionist and systems approaches. On Chapter one, Bizzarri et al. start by describing the traditional view of biological systems as systems whose behavior can be fully specified at the level of genes and their interactions. Phenotypes can then be built up from a complete understanding of genetic processes and this is akin to identifying a chain of events and the participating molecules that starts at the gene level and ends up at the phenotype level. As the authors argue, this mode of explanation does not comport with the stochastic and non-linear environment inside the cell. The focus on individual genetic events, although valuable, has to be integrated within a framework that emphasizes functional and integrative explanations. For this purpose they propose the use of the concept of morphogenetic field and a system phase space to classify the multitude of cellular dynamical configurations that give rise to observed cellular responses.

On Chapter two, Bertolaso et al. begin by arguing that the central question of systems biology is to explain how order arises in biological systems. Indeed, they want to characterize the "inter-level couplings", the regulatory processes that bind the different organizational levels in the cell, through which the ordered phenomena emerge. After reviewing existing attempts to move away from purely reductionist explanations, for example using network topological concepts to characterize interactions at the whole-cell level, they argue for the incompleteness of both bottom-up and top-down approaches, the purely reductionist and the fully holistic approach. Instead, the authors present the merits of an intermediate approach, which they call the mesoscopic level, which places explanatory weight on how different organizational levels "communicate" with each other and lead them to cohere together into an integrated system. They develop their proposal in a variety of interesting ways and provide some guidance on how to select, for any given problem, the appropriate mesoscopic level that provides maximum explanatory leverage.

The next seven chapters discuss different modeling methodologies and experimental approaches that are useful for generating and analyzing large-scale data sets of the sort that are driving the systems approach. We start with the contribution of Valente et al. who present on Chapter three a novel methodology for analyzing genome-wide association study (GWAS) data. These datasets are generated by sequencing single nucleotide polymorphisms (SNPs) in both a control group and a group of subjects afflicted with a particular illness. The goal is to identify SNPs that differentially modulate the incidence of the condition. A key problem in these studies is how to convert existing biological insight about the condition into a more effective statistical search for the genetic risk factors. The authors present a so called hypothesis-rich approach, based on the construction of Rational Classes, sets of candidate risk factors that share an underlying common rationale. Of note, the authors point out that many insights and hypotheses that can straightforwardly be incorporated via Rational Classes, would be of no or difficult use under a classical Bayesian approach. The power of the method is illustrated by an application to a Parkinson's GWAS.

On Chapter four, Rietman et al. present a thermodynamic framework for analyzing the dynamics of molecular interaction networks consisting of mutating and replicating units. They show how a description of these dynamics can be formulated in terms of a stochastic kinetic model and its time evolution mapped by using concepts from irreversible thermodynamics. This wide-ranging chapter illustrates how the basic concepts of statistical mechanics and phase transition theory can be used to generate predictive models of the time course of interaction networks inside cells. An interesting example of the transition from a normal to a cancerous cell is presented as a case study of how the formalism can be used to develop a systemic perspective on the transition to cancer.

Chapter five by Tolga Can discusses an algorithmic way to indentify functional subunits in protein-protein interaction networks. This chapter begins to address the more general question of how to partition complex systems – systems consisting of many strongly interacting parts where the dynamics of the individual parts are well-characterized but cannot be aggregated in a simple way to furnish a predictive dynamical model for the system as whole – and partition it into subcomponents whose interactions are known and do provide a basis for constructing useful models at the system-wide level. After a brief but useful introduction to web-accessible interaction datasets, the author discusses general graph-theoretical ways to identify sub-graphs or modules defined according to some criteria. This is followed by a review of various algorithms developed specifically for studying protein-protein interaction datasets. Finally, the chapter turns to a discussion of model-based approaches to module identification. The key concept here is to utilize known signaling or interaction pathways to guide or bias the search for functional subunits in non-annotated networks. This chapter is a useful, non-technical review of the methods developed so far and some of the remaining challenges in the field.

Chapter six by Cheng et al. reviews advanced protein mass spectrometry (MS) methods, specifically quantified liquid chromatography (LC) MS/MS techniques, and how they are used to generate protein-protein interaction pathways. These techniques together with advances in so-called deep sequencing technologies for DNA have greatly advanced the availability of datasets that, when combined with appropriate analytical methods, are well-suited to provide a systemic perspective on cellular processes. After reviewing the steps involved in preparing samples for MS, the authors discuss how these techniques are used to identify the protein elements in a pathway. This is followed by a discussion of strategies for quantifying changes in protein abundance from MS data. Next, the authors provide a thorough discussion of how LC-MS/MS can be used to discover which proteins have been subject to post-translational modifications. The chapter finishes by describing how protein-protein interactions can be identified by LC-MS/MS.

The theory of reliability has been developed as a way to formalize and predict the effect of a component failure on the functioning of a network or system to which it belongs. Most applications have been confined to engineering analyses of the robustness of networks; however, this framework seems appropriate to describe the response of cells or even of the whole organism to dysfunctional elements within it. This is especially true if one considers that the design mechanism in biology – Natural Selection – would be particularly sensitive to reliability features of an imperfectly self-replicating system. This topic is taken up in Chapter seven by Koltover, who reviews how reliability concepts and formalisms can be used in the systems biology context. After discussing the basic theoretical terms, these concepts are applied to a study of enzymes to discover general principles of enzymatic design that may reflect reliability considerations. Next, an interesting application at the whole organismal level is presented. This is followed by a discussion of how differential reliability characteristics may show themselves in the birth-death processes of heterogeneous populations. The chapter ends with a discussion on reliability testing in biology.

Chapter eight by Moskon et al. presents a comprehensive review of various computational approaches that have been used to model the evolutions of molecular species in cells as well as interaction network dynamics. This wide-ranging chapter usefully summarizes a number of graph-theoretic techniques as well as concepts and strategies from deterministic and stochastic kinetic models. This is followed by a discussion of various statistical parameter estimation methods that may be used to develop estimates for rate and other constants that appear in these equations. Taking up a theme from the previous chapter on reliability and robustness, the authors present a number of criteria that may be used to quantify and evaluate these characteristics of biological systems. The chapter ends with a discussion of how techniques from dynamical systems theory, especially concepts associated with the stability of dynamical systems, can be applied in the systems biology context.

Chapter 9 by Davies and Husmeier discusses how Bayesian statistics can be used to generate a model for transcriptional regulation using infinite-parameter Gaussian Process (GP) priors. After defining GP priors, the authors present calculations describing how the posterior probability may be computed for models generated from stochastic kinetic differential equations using GPs as priors. This technical section is followed by a discussion of identifiability and noise assumptions. The authors present simulation testing of their model by considering the case of transcriptional dynamics under Michaelis-Menten kinetics. A brief section on extending the model to non-linear regulation rounds out the chapter.

On chapter 10, Blair et al. present a review of various strategies that may be used to model cancer at the systems-level. While focusing on one area – cancer biology – the techniques and approaches they present are equally applicable to the study of other biological processes. After describing the context of the chapter, the author review differential-equation-based kinetic models and well as graphical representations of systems data. This is followed by a review of databases available to provide inputs to these modeling approaches. This chapter is marked by the absence of technical discussions and should be useful to a wide range of readers.

We end with a group of four chapters presenting applications of the systems approach to clinical problems. Chapter 11 by Louridas and Louridas describes how a systems approach can provide a new perspective in cardiology. After describing the context of the discussion, the authors review how large scale genomic and proteomic – primarily transcriptomics and metabolomics – data sets have led to the possibility of individualized biomarkers for heart disease and related conditions. These results they have termed personalized cardiology. They end with a discussion of how clinical care can also be facilitated by the systems-level data they describe earlier.

Chapter 12 by Petrasek and Petrasek presents some thoughts on how a systems-level perspective may be relevant to the understanding of diabetes. After presenting their views on the reductionist versus systems paradigms, the authors provide an eclectic argument for their preferred approach. This conceptually wide ranging article describes the role of uncertainty in biology and how it should be handled. Examples of the DCCT and ACCORD trials are presented as case studies.

Chapter 13 by Hong et al. presents another perspective on cardiac systems biology. The authors begin by describing the various genome-wide association studies (GWAS) related to heart disease. They discuss –omics data, and then summarize relevant protein structural biology information. The rest of the chapter focuses on modeling and integration of these data streams and reviews studies in this direction.

Chapter 14 by Saidi and Kenari, the last chapter of the book, reviews systems and other approaches to the study of transplant immune tolerance. After defining tolerance in the clinical context, the authors describe the results of clinical tolerance studies. This is followed by a review of clinically-relevant biomarkers and system-type studies that have been undertaken to analyze immune tolerance.

In: Recent Advances in Systems Biology Research
Editors: A. X. C. N. Valente, A. Sarkar and Y. Gao
ISBN: 978-1-62948-736-6

Chapter 1

The Conceptual Foundations of Systems Biology: An Introduction

Mariano Bizzarri[1], Alessandra Cucina[2] and Alessandro Palombo[3]
[1]Department of Experimental Medicine, "Sapienza" University of Rome, Roma, Italy;
[2]Department of Surgery "P. Valdoni", "Sapienza" University of Rome, Roma, Italy.
[3]Department of Clinical and Molecular Medicine, "Sapienza" University of Rome, Roma, Italy

Abstract

Systems Biology is more than just a summary of different sciences, given that Systems Biology deals with 'systems', and it is concerned with the complex, emergent properties that arise from the relationship between molecules, cells and tissues. Functional properties are not yet in the 'molecules', instead they 'emerge' from a self-organized process, which geometrically shape the living structure into a system, characterized by hierarchical levels. Interaction among them leads to both top- and down-ward causation. Thus, biology is now facing a crisis — or is it an opportunity — reminiscent of the state of biology in pre-double-helix time. The time is gone for biology to "concern itself with the great 'non-reductionist' 19th century biological problems that molecular biology left untouched: e.g., the nature of the complex organization of the living matter. Precisely, Systems Biology affords that task.

Introduction: What We Do Mean by Systems Biology?

Despite the widespread use of the term 'Systems Biology' - a word coined by Mesarovic in 1968[1] - a shared and unified consensus about what Systems Biology is is still lacking. At a glance Systems Biology may be considered a kind on 'unusual' combination of biology with a great many other sciences, from physics to ecology, mathematics, medicine, chemistry and even sociology and linguistic. According to this 'minimalist' approach, Systems Biology has been viewed as the 'successor' of Molecular Biology[2], mainly characterized by the use of large-scale molecular interactions ('omic' framework) focused on building complex signalling networks by applying mathematical modelling, and thus showing how cells make decisions based on the information flow through those networks[3]. Eventually, this approach still relies on the molecular level thought as the privileged level of explanation and, despite its 'practical' usefulness, it has to be considered nothing more than a mere 'epicycle' of the existing, reductionistic theories[4].

Yet, several reports[5,6] have clearly showed that Systems Biology is more than just a sum up of different sciences, given that Systems Biology deals with 'systems', and it is concerned with the complex, emergent properties that arise from the relationship between molecules, cells and tissues. Functional properties are not yet in the 'molecules', instead they 'emerge' from a self-organized process, which shape geometrically the living structure into a system, characterized by hierarchical levels. Interaction among them lead to both top- and down-ward causation. Since the behaviour of such, complex systems is mainly ruled out by non-linear dynamics, it follows henceforth that the classical concept of linear causality, is meaningless in Biology[7]. Thus, *ad-hoc* approaches are clearly insufficient to cope with complexity, and classical physics as well as Molecular Biology are unfit to address such issues, given that "the existing philosophy of biology fails to address the rather profound issue of what distinguishes the living from the non living, except to say that something lives because its ancestors lived"[8]. Therefore, Systems Biology should have a unique and robust philosophical foundation and a different methodological approach in order to take up the challenge of understanding living organisms as wholes.

We need to build a reliable approach, allowing us to copy with the intrinsic 'disorder' of living processes, e.g., with the chaotic fluctuations underlying biological function[9]. What we are used to call 'biological noise', may be not really noise, as it actually reflects normal biological fluctuations. A great part of "relevant biological information is embedded in this part of the data: the small variations of small signals, which collectively could contain more information than the large variations of a handful of genes"[10]. Therefore, this is the challenge: how to include chaotic and non-linear, unpredictable, processes into our comprehension of Biology. Systems Biology addresses such issue, and should be defined as the study of phenomena in terms of *how the objects are related*, rather than what they are composed of. Systems Biology is "a science no more confined to simplified and idealised phenomenon: but a science facing itself with the complexity of the real world"[11]. We posit hence that a new approach, a shift from reductionism to an "integrated" perspective is by now required.

However, more than just a pronouncement of a new approach is required. What is needed is to provide a conceptual framework able to integrate some entrenched aspects, as complexity, hierarchical structured levels of observation, geometrical relationships, non-linear dynamics, network modelling, influence of biophysical constraints, operating on different scales, rather than solely focusing on building numerical mathematical or computer models[12]. Those aspects must be collectively considered in order to find organizing principles that exactly outline the evolution of systems in space and time[13].

The Newtonian Paradigm and the Emergence on Non-Linear Dynamics

Science is generally viewed as strongly grounded in the Newtonian paradigm of mechanics. According to that paradigm (to which Descartes proved a robust philosophical underpinning), every phenomenon we observed can be analytically reduced to its components. No matter the number of constituents are, you only need to take your analysis one step further, and look at their interactions. By continuing this subdivision long enough, you will end up with the smallest possible parts (atoms) or 'elementary particles', that are seen as 'separate' pieces of the same substance (matter). A fundamental property that distinguishes particles is their position in the space and its change along time. Any development or evolution is conceived as geometrical rearrangement due to the movement, and the movement is entirely governed by deterministic laws of cause and effect. By knowing the initial positions and the velocities of each system's particles, together with the forces acting on the system, you will be able (at least in principle) to predict the evolution of the system with complete certainty and accuracy, as claimed by the well known Laplace's aphorism[1]. The trajectories of the system are not only determined towards the future, but also towards the past. Therefore, you can theoretically 'reverse' the process to its past states: within that frame there is no room for the time's arrow[14]. The mathematical description of such kind of evolution can be depicted by a phase space diagram, in which changes are represented as phase transitions. A phase transition is described by a point along the line where the control parameter(s) runs. A phase transition is critical, i.e., it allows the system to acquire a different configuration, when some known (or eventually, hidden) observables diverge, thus leading to the appearance of a coherent structure. In Newtonian physics, the dynamics of the system within the phase space diagram is entirely described by linear mathematic equations providing a full knowledge of the object. Randomness ("noise") in such a framework is thought to be due to unknown ('hidden') variables, from whose interactions unpredictability comes. Therefore, according to that classical mechanistic view, variability in physical processes is due to a lack of *epistemic* knowledge: a physical phenomenon is "per se, a geodesic. This is a unique, optimal and 'critical' path, completely determined by the Hamiltonian, and may be computed as an optimum of a Lagrangian functional"[15]. As a consequence, such process possesses a unique basin of attraction. However, in the real world such 'beautiful' paradigm, rarely works, given that its validity is limited to systems behaving according to a linear dynamics. Indeed, most of the systems modelled by traditional science are linear. This means basically that effects are proportional to their causes and the system is thought to adopt only one configuration. Those systems are symmetrical. In some instances, the symmetry can be broken when one or more system's parameters fluctuate over a well-defined threshold value, reaching thereafter a bifurcation point where it experiences a symmetry breaking[16]. The system loses it homogeneity, and

[1] "We may regard the present state of the universe as the effect of its past and the cause of its future. An intellect which at a certain moment would know all forces that set nature in motion, and all positions of all items of which nature is composed, if this intellect were also vast enough to submit these data to analysis, it would embrace in a single formula the movements of the greatest bodies of the universe and those of the tiniest atom; for such an intellect nothing would be uncertain and the future just like the past would be present before its eyes". P.-S. Laplace, A Philosophical Essay, New York, 1902, p. 4.

can acquire an efficient notion of distance in space. The symmetry breaking discloses different solutions for the same parameters values (hysteresis and bistability), therefore opening the system evolution towards novelty and variability. The system drives along different trajectories, thus converging into one or more 'attractors'. An attractor is a stable solution to the set of mathematical equations that describe a dynamical system, representing the state of equilibrium to which the system will tend to move. Attractors are distributed along a complex landscape, in which stable (valleys), as well as metastable or unstable (hills) states are depicted [17]. By this way new configurations and self-organized structures arise. In non-linear systems randomness is an intrinsic property. A quite similar situation happen in Quantum Mechanics, where variables cannot always be associated to single points (quantum entanglement) and unpredictability is intrinsic, given that – according to Heinsenberg's principle and Schrödinger equation - the parameter value is expressed as a *probability*. In Quantum mechanics, some quanta interact and form one system in which they are 'entangled': this means that they are *non-separable* by measurements. By analogy, a complex, living system is characterized by non-separable, 'entangled' processes, behaving according to non-local rules, and henceforth requiring for a deeply understanding, a global, 'systemic', view.

Science is generally used to deal with (ideal) non-linear, isolated systems. Such models have played an important role in enabling us to conceptualise the rules governing their physico-chemical behaviour. Those systems are accurately described by equilibrium Thermodynamics, given that they behave like 'isolated' systems: they are insensitive to their environment. This 'insensitivity' implies that the state of equilibrium is stable (small perturbations are quickly damped and the system is driven back to the equilibrium), and external influences (like that exerted by gravitational or electromagnetic field) are small compared to thermal energy, and cannot trigger any significant effect. The state of equilibrium is characterized by maximum entropy, and there is no cooperativity within the system, except near a phase transition. Let us to consider a gas whose molecules have a masse m enclosed in a volume of length l, in a gravitational field of strength g: the maximum energy of interaction is therefore mgl. That energy will likely produce a density variation proportional to mgl/kT. As $mgl/kT \ll 1$, such effects of external fields on equilibrium systems can be ignored: they might be observed only when they are larger than the thermal fluctuations. However, for both non-equilibrium and equilibrium systems, there exist situations in which small external cues could cause large effects. Usually, for equilibrium systems, this happens when cooperative effects underwent near a symmetry breaking (bifurcation point), leading thereafter the system towards a phase-transition and, subsequently, allowing it to acquire a 'new' macroscopic order[18]. In non-equilibrium systems such phenomenon are more frequently encountered, given the richness of instabilities (due to non-linearity), and the consequent transition to very complex dissipative structures. By this way, sensitivity to the external field allows systems to become correlated whit the environmental cues. Keeping in mind the number of transitions that a non-equilibrium system can experience, such sensitivity becomes a basic and relevant feature, unlike in equilibrium systems in which phase transitions are isolated occurrences. By this way the system becomes complex. This is a remarkable and central aspect that permits to link the 'birth' of complexity to the emergence of different stable and unforeseen solutions. This is the background that has allowed Longo and Montévil to claim living, dissipative systems as characterized by extended critical transitions and symmetry changes: "The unifying theoretical framework in biology

[…] focuses […] on the permanent change of symmetries that per se modify the analysis of the internal and external processes of life [henceforth] …. variability may be considered as the main invariant of the living state of matter"[15].

Genes in Search of a Role

The aforementioned considerations leave no room to the classical theoretical framework, as expressed by Schrödinger[19], according to which in Biology "order comes from order".

The molecular biology paradigm posits that 'information' flows unidirectionally from genes to proteins, 'instructing' the system how to develop in time and space. Thus, cell differentiation and organism development are traditionally described in deterministic terms of program and design. The Central Dogma posits the DNA as a "digital information carrier", which means the information is highly stable and reproducible. However, the lack of a digital receiver in most biological and physiological systems prevents digital information processing: if the processing includes even only one analog step ''the perfection of the digital steps is wiped out. The processing then becomes effectively analog''[20]. This means that biological "information" is not embedded exclusively in DNA sequences, but is also contained in epigenetic and regulatory signal (both external and internal) networks. Moreover, it must be stressed that the adoption of the information theoretical paradigm failed to provide biology with a pertinent observable for understanding and measuring organization, and the concepts linked to information (code, program, signal) have eventually hindered the comprehension both of physical dynamics and of biological organization[21].

Indeed, genotypes and phenotype are interconnected in a non-univocal manner into a 3-dimensional structure: the same or very similar phenotypes may be associated with distinct patterns of gene expression[22] . Therefore, no simplistic linear correlations can be drawn from raw gene expression profiles to cellular phenotypes. The classical genetic analysis has taught us to expect every genotype to determine a single phenotype[23]. However, even in cases in which there is a single allelic difference between two strains, certain phenotypic traits cannot merely be considered as the result of the additive effect of independently acting genes[24]: even ''simple'' Mendelian disorders are complex traits[25], and ''a clinical disorder is not the product of the single gene that is disrupted, but the genetic disruption is embedded within the context of the individual entire genome and environmental experience''[26].

The implicit assumption underlying the DNA-based paradigm considers genes able to govern cellular functions through a linear and deterministic process: gene expression is "deterministically" triggered by "signals" leading to synthesis of proteins that henceforth instruct the system how to behave and function in a "reproducible" manner. However, a large amount of data suggests gene expression relies fundamentally on a stochastic mechanism[27]. Yet, the molecular realm is dominated by random, aleatory processes that eventually drive the system towards ordered macroscopic structures[28], and statistical physics gives many such examples. Stochasticity has been proven to underlie the behaviors even of genes[29], isogenic macromolecules[30], cells[31,32] and organisms as well[33]. In turn, stochasticity coupled to non-linear dynamics is at the core of self-organizing processes, which lead the system towards ordered macroscopic structures[34]. Even if deterministic laws rule out the process, non-linearity allows the system to reach some few stable states (attractors), each of them characterized by a different likelihood. Thus, a deterministic process leads to different stochastic outputs. Causality is "distributed" along that pathway, in which several cues (both internal as well as external to the system) interact each other. By this way non-linear dynamics produces novelty and acquire unpredictability.

According to that model biophysical as well as biochemical interactions between cells and the surrounding microenvironment (stroma, extracellular matrix) act by stabilizing the cellular phenotype to coordinate of it and addressing their differentiating fate[35]. Indeed, self-organizing complex systems are dynamically interacting with the gene-regulatory networks: from those interactions hence, emerges the direction to which system's evolution moves. There are no essences, only processes, governed by non-linear dynamics. This implies that development is subject to sensitivity to initial conditions, symmetry breaking, extended critical transitions and so forth. For that the gene-driven "causative" role in biology cannot be separated from the context in which they actually are thought to "operate". According to Denis Noble, genes are deemed to be 'physiological prisoners': "they are trapped in huge colonies, locked inside highly intelligent beings, moulded by the outside world", and the system from which they belong " allows their code to be read"[36]. Thereby, the system itself is the playmaker that permits, controls and modulates the ultimate gene expression pattern. Thereby, causality in biology cannot be "extrapolated" from the context, and it should be hypothesized that a relevant role for genes emerges only when the systems becomes unstable after experiencing a phase-transition (i.e., a symmetry-breaking).

The System

Molecular biology takes for exhaustive an implicit definition of 'system' as well as of 'level of observation', by "reducing" both of them to genes. As a consequence, within this framework "organisms became nothing but the *vehicles for genes*"[37]. On the contrary, Systems Biology needs to preliminary conceptualize again these notions, given that the aforementioned considerations evidence how neither the genome nor eukariotic cells may be considered a system on its own. In Biology we are dealing with the non-linear interplay in between different structures – intra-cellular organelles, epithelial and stromal cells, extracellular matrix components – and consequently this implies that the "observable" parameters could not be longer "reduced" to "gene expression" or intracellular biochemical pathways only. Explanations in biology should be pursued through an explicit search for a proper biological observable(s), caught at the right level of organization: this imply biologists should embrace the complexity and the unity of the organism, and seek explanations at or nearby the mesoscopic level[38]. This is especially true for development and carcinogenesis, as particular instances of biological dynamics and organization, where multi-level phenomena take place [39,40].

Indeed, some complex biological functions – like differentiation or pathological states - take place within tissues: it is therefore mandatory to consider the integrated interplay between cells and stroma as the proper level of investigation. Hence, we posit the observable "system" must include the cell and its relationships with the surrounding tissue, in a three-dimensional geometric perspective. Overall, these factors determine the shape (or form) the system acquires. Such a complex system is thermodynamically open, and it is characterized by a non-linear dynamics. Such system has a history: this means that the *present* behaviour of the system is in part determined by its *past behaviour;* a complex system displays both sensitivity and resilience (robustness), in respect to the perturbations exerted by internal and/or external stimuli, weak or strong they would be. In addition, complex living systems are characterized by both local and long-range interactions (non-locality), as well as by complex interactions in between molecules and structures, that makes their determination "non-separable" (i.e., "entangled")[41]. That model implies a hierarchy of levels, interacting each other according to both a bottom-up and a top-down complex causal chain, distributed . across different scales. However, separating a hierarchical level from others is not possible in general; the relations with different hierarchical levels are not necessarily represented by appropriate parameters. We encounter a fundamental difficulty in relating micro-objects and macro-objects when a simple summation (or integral) does not work adequately: it is the difficulty of *non-linearity*. A summation of the different interacting elements cannot be done when they are entangled into organized structures or structured pathways, like in living organisms: those levels are generally tightly linked each other, even if they may be thought as 'autonomous'[15, 42]. In addition, higher levels of matter aggregation display astonishing, "emerging" properties that cannot be explained by "fundamental laws" nor by analyzing the single parts. It must be stressed that "each level is both characterized and governed by emergent laws that do not appear at the lower levels of organization. This implies that [...] we require a theory that operates at the corresponding hierarchical level"[43]. In turn, hierarchical organization in between different levels creates 'downward causation'[44].

Biophysical Constraints

Diffusible chemical factors alone, as well as genes products, are not sufficient to fully explain cell fate regulation, and even gradients of morphogenetic molecules cannot entirely explain morphogenesis as firstly proposed by Turing[45,46]. Morphological and functional plasticity, like that experienced by an organism during growth and development, rather than being the result of a mere genetic "adaptation", reflects deeply the influence of external physico-chemical parameters on any material system and is therefore an inherent, inevitable property of organisms[47]. Physical, electromagnetic and gravitational forces affect both cell behaviour, gene expression and phenotypic traits (shape and differentiation)[48,49,50]. Cells dynamically adapt to force (shear and tensile stress, compressive forces, hydrostatic pressure) by modifying their behaviour and remodelling their shape; cells can in turn exert a reciprocal influence on their microenvironment (mechano-reciprocity) through actomyosin- and cytoskeletal-dependent modification[51]. By their own, cytoskeleton and other structural elements of the cell form an interconnected tensionally integrated system, which is able to collectively coordinate gene expression profiles and complex biological functions in response to structural and mechanical perturbations, via mechanotrasduction[52]. Given the pluripotent role of force in tissue function, it is not surprising that several diseases, including cancer, are characterized by compromised tensional homeostasis[53]. On the contrary, normalizing the tissue tensional state of a tumor, cells can be reverted toward a non-malignant phenotype[54].

However, biophysical factors (electromagnetic and gravitational fields, stiffness, surface tension, shear stress and so forth) must not be viewed as "signals", coming up from the field, by analogy with the "instructive" model proposed by molecular biology. Biophysical factors are not substitutes for genes, and should not be considered as mere causative agents linearly unchained into a causative biological framework. They act as "constraints", this means that their "causal weight" cannot be extrapolated from the complex non-linear dynamics into which several factors are entangled and superimposed into themselves (Figure 1). This is a key point, implying a distributed and interconnected chain of causal factors in biological processes, and scientists are not epistemologically trained nor they have the habit to copy with. We are generally used to think that complete disorder or chaos is the only "usual" state, as learned from the thermodynamics of open system. In fact, the combination of the random motion and the homogeneous (structureless) field is a trivial equilibrium solution, leading to heat death. Yet, there are a variety of structures and examples of evolution in the universe. For instance, the beautiful spiral structures of galaxies: they are displaying self-organized order in the collective motion of many stars and gases interacting by gravity. Such a system is able to find a self-consistent "solution" without any program or a priori aim.

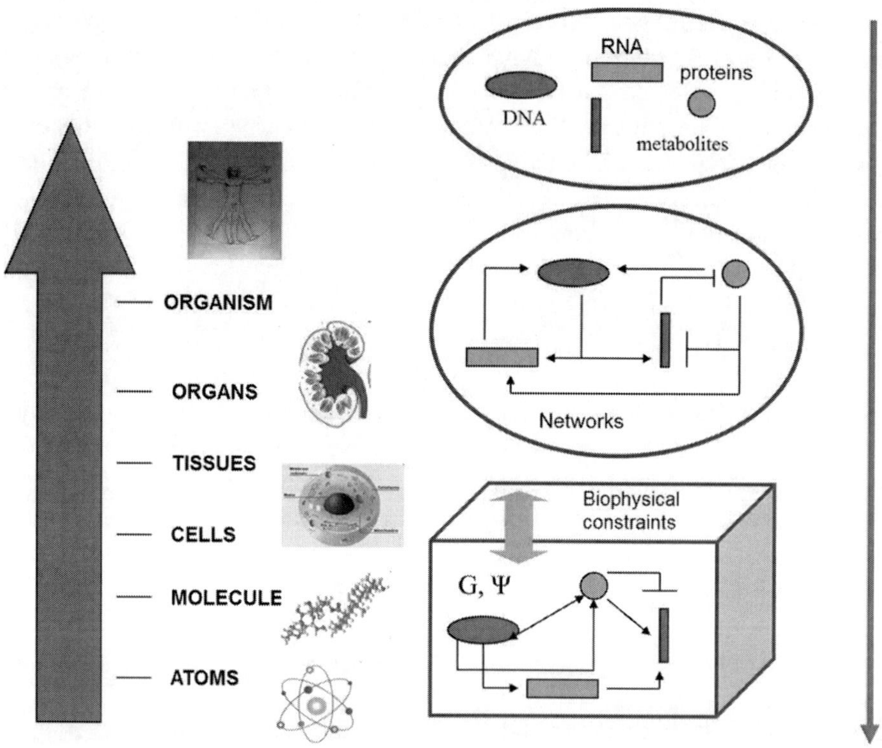

Figure 1. A biological system is constituted by a set of elementary entities and processes. System's elements interact according to specific relationships, and following a non-linear dynamics. They are furthermore constrained by biophysical forces and specific boundary conditions (environmental and surrounding factors). Overall, these factors determine the shape (or form) the system acquires. Entanglement among different levels and self-organizing processes lead the system to acquire increased complexity, as we proceed from cells to tissues, organs and organism.

Order from Disorder

Cell behaviour and differentiation are complex functions that cannot be mechanistically explained according to a linear gene-based program, and could be viewed as self-organized process emerging from different actors participating within the so-called morphogenetic field[55]. Indeed, a biological process is dominated by a *field* - a function of space-time producing *force - which is a common rule,* and, at the same time, a common product of a group of elements. In other words, the system organizes a recurrent relation such that the motion creates a field and the field rules the motion: interactions between particles produce the field, and, in turn, some characteristics of the single particle are transferred into the field. The aforementioned considerations, encompassing a cyclic notion of cause very different from the linear one of molecular biology, are of foremost importance, when we are dealing with a special kind of field: the 'morphogenetic field'[56]. A morphogenetic field is "a system of order such that the positions taken up by unstable entities in one part of the system bear a definite relation to the positions taken up by other unstable entities in other parts of the system [...] A field is bound to a particular substratum, from which a dynamic pattern arises [...] and can, like a magnetic field, maintain its pattern when its mass is either reduced or increased"[57]. Such framework in any way denies the relevance of genetic factors: but their effects are significantly amplified, modulated or hindered by the field in which they are operating[58]. Changes in these fields, then changes the ways that tissues, organisms or animals develop (Figure2).

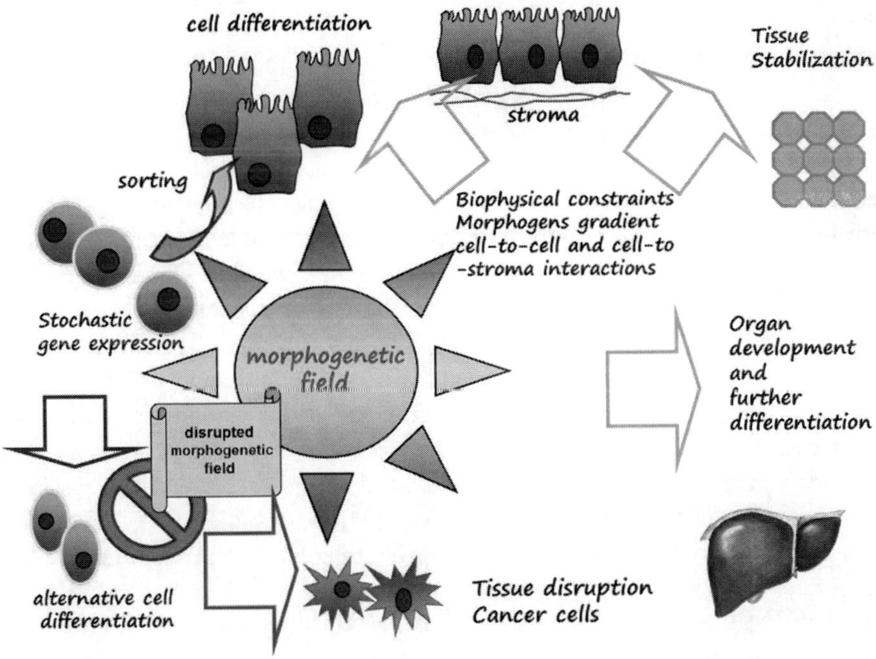

Figure 2. Cell behaviour and differentiation are complex functions that cannot be mechanistically explained according to a linear gene-based "program", and should be viewed as self-organized process emerging from different actors participating within the morphogenetic field. The field integrates different biophysical and biochemical cues, driving the cell evolution towards selected differentiating fates. Disruption of the morphogenetic field may eventually lead the system into pathological states, like cancer.

The aforementioned considerations allow us to properly address the question "from where order comes in Biology?".

Reductionism, that was the methodology to search for the "order" (arbitrarily identified with the simplest description of the process under study) has progressively became a strategy to bundle off the "disorder" (e.g., the "complexity") to the periphery of science. Now, the exiled "complexity" tries to reverse its position by shaking the structure of the traditional science that holds "order" in the centre.

The gene-centric paradigm has been proven to be inadequate in explaining the emergence of complex structures[55], and now we take for granted that the interplay between biophysical forces and gene activity, both governed by non-linear dynamics, lead the biological system to acquire organized structures and increased 'complexity'. How this happens? Self-organization emerging from several cues interacting according to a non-linear dynamics has been addressed by Prigogine[59], who provides compelling evidence about how dissipative structures self-organize through fluctuations and instabilities, which lead to irreversible bifurcations and new stable states. This approach is the only able to solve an old paradox: how to explain the increasing ordering that occurs in developmental biology (where self-organizing processes 'decrease' the system's disorder) when entropy (chaos) must be increased. A solution may be obtained by coupling a process producing order (negentropy) with another leading to increased disorder (entropy): close to the equilibrium, the dependence of the one process rate on the driving force of the other process should equal the dependence of the other process rate on the former driving force[60]. These approaches have contributed to the non-equilibrium thermodynamics development[61,62] and "constituted a prelude to systems biology [...] in that they 1) dealt with integration quantitatively and 2) aimed to discover general principle rather than just being descriptive"[63].

A linear system can be fully described by linear differential equations that are thought to give us a reliable prediction about the position the system will have in the future. On the contrary, a system like the Bénard cells, after experiencing a thermal fluctuation, will becomes non-linear, then attaining different stable and unstable states (hysteresis and bistability). This happen when a defined threshold value of a system's parameter (like temperature) has been crossed and the system undergo a *symmetry breaking*. The rupture of the symmetry – at a bifurcation point – allows the system to acquire different as well as unpredictable solutions, given that the systems drives towards one or more stable states (attractors)[64]. A cell state could be ''lifted up'' from its attractor and moved to another attractor as a consequence of a wide range of efficient factors (biophysical and biochemical cues). In normal development, we can envisage that the nature of the landscape is such that the cell states would most probably ''move'' in a single direction, but the possibility must also exist of moving in the reverse or alternative directions, resulting in the reversal of a cell-fate decision, or trans-determination and trans-differentiation. Even mild, gradual variation in a single control parameter can significantly affect processes characterized by non linear-dynamics and bistability, thus switching cells between distinct phenotypes. This behaviour is reminiscent of phase transitions observed in physical systems and it is a profound concept to explain cell differentiation, since it allows us to understand what it is unconceivable according to the classical molecular paradigm: cell with identical genotypes, exposed to identical environments, can exhibit extremely different phenotypes as a function of the history of the system, or in response to minimal fluctuations in some (weak) biophysical forces[65]. The ability of attractors to integrate distributed signals could explain how physical perturbations - such as cell shape distortion – can trigger a particular cell fate, such as switching between proliferation, apoptosis, differentiation or neoplastic transformation[66]. Such perturbations carry no specific information encoded by a molecular structure and hence do not operate a single stimulus-specific receptor, but instead, impose a *distributed set of stimuli* on structural elements, such as the cytoskeleton, that simultaneously influence numerous signalling pathways inside the cell[67]. This regulation model explains how several hundreds of genes can be collectively and coherently involved in phase-transitions from one attractor to another[68] and claims in favour of a non univocal genotype-phenotype relationship, evidencing that the 'robustness' of the phenotypic state cannot be linearly ascribed to the gene's configuration[69].

The rupture of the symmetry gives the system a historical dimension, a sort of memory of the past event that took place at a critical moment and which will affect the next evolution. Let now to highlight some relevant consequences linked to the rupture of the symmetry, generally addressed by the physics of criticality[70]. Transition to the critical point permits to acquire relevant features, as long-range correlations and scale invariance. Long range correlations means that the determination of the system is thereafter global and not local. . Scale invariance (i.e., the system presents the same behaviour at each scale) allows the system to acquire a 'structure of coherence', that is to say that local processes are 'globally' determined. In addition, scale invariance confers fractal geometry to the system[71]. The attractors the system is thought to reach during its evolution share those characteristics. In turn, changes in fractal dimensions, indicate that system's parameters have overreached a threshold value and the system is undergoing a transition beyond the critical point, i.e., experiencing a symmetry breaking. Therefore, fractal analysis promises to be of strategic relevance in analyzing system's behaviour[72], given that "cell shape can be considered as "the most critical determinant of cell function (…) and cell shape per se appears to govern how individual cells will respond to chemical signals in their local microenvironment"[73]. As a consequence, measurable parameters describing shape could be considered as system's descriptors of the specific level of observation represented by the cell–stroma system[40].

How to Describe a Complex, Non-Linear System?

A complex system can be described by means of a phase-space diagram, described by parameters ("observables"), that still has to be defined, since only few attempts has been performed to carefully recognize them[74]. This model enable us to move from 'local' properties to more 'global' complex networks, as firstly guessed by Waddington[75], when he proposed the concept of 'epigenetic landscape', thought as a metaphor for the trajectory that a complex biological system might be travelling in response to genetic, physical, and environmental cues.

According to that model, functional states, depicted as attractors, have been usually conceived as mainly specified by the gene-regulatory networks, embedded into interconnected and complex networks, and they operate according to a non-linear dynamics[76,77]. That task requires a huge amount of data from which statistical analysis can be inferred and high-throughput techniques (functional genomic, metabolomic, proteomic and fractal analysis) are currently needed to obtain a reliable and understandable picture from unsupervised data in order to allow further simulation by means of *in silico* models[78,79]. However, even such an approach is likely to be not enough, keeping in mind that "the big question to understand in biology is not the regulatory linkage but the nature of biological systems that allows them to be linked together in many non-lethal and even useful combinations"[80]. Indeed, the stability of functional states clearly also (mainly) depends on external cues and the system's dynamics in the phase space cannot be "reduced" either to a genetic wiring diagram or, even to the integrated functioning of a genome–proteome–metabolome network. To give an answer to this question we need to adopt an approach by which changes in shape and functions may be ascribed to the overall system and not to a single component, as important it should be[5]. We and others[81,82] posit that cell types cluster in a multidimensional state space in which each axis define a value of a physiological (system) parameter, among which Fractal Dimension, Metabolomic Fingerprint and the Morphogenetic Field Strength are of paramount relevance. Undoubtedly, additional axes could be added, representing more variables (including the patterns of gene expression, electric trans-membrane potential, and so forth), distributed along separate state space ensembles (Figure 3).

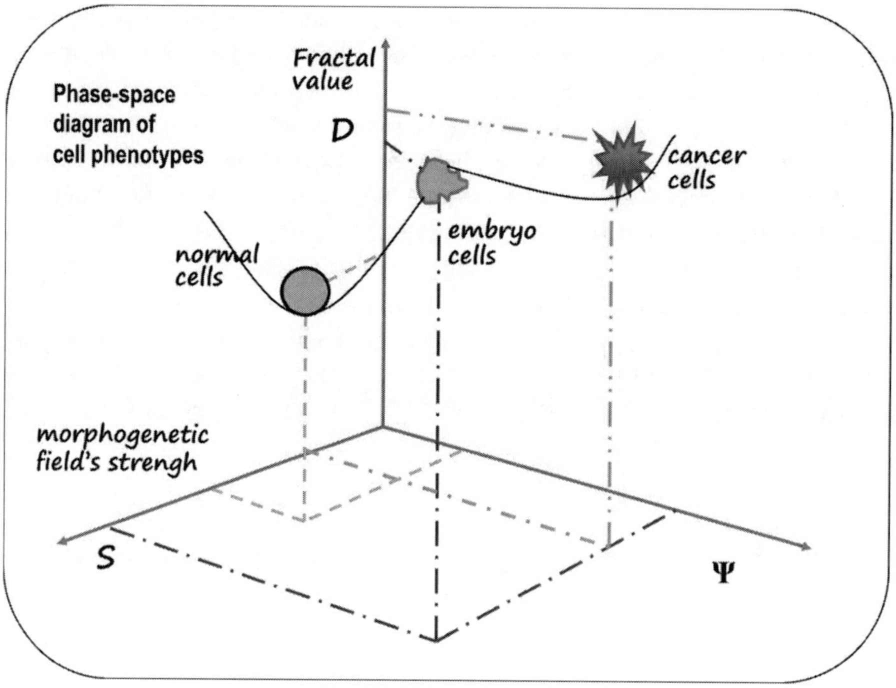

Figure 3. A complex system can be described by means of a phase-space diagram (remnant of Waddington's epigenetic landscape), described by state-parameters ("observables"). Cell types cluster in a multidimensional state space in which each axis define a value of a physiological parameter, among which Fractal Dimension, Metabolomic Fingerprint (exemplified by the dissipative function ψ) and the Morphogenetic Field Strength are of paramount relevance.

Once appropriate data are gathered, cells could be displaced from their "attractor" and moved to a different stable state by different types of stimuli (physical, biochemical, pharmacological), shifting them along each axis, toward a different attractor within the state space. This model highlights how cells are sensitive to a wide array of different unspecified cues (molecular as well as biophysical). The model posit that phenotypic fluctuations may be a general feature of living cells: the cellular microenvironment contributes actively in shaping the overall system, and in generating new state (phenotypic) transitions. As a result the cell phenotype is determined by the joint action of cell-intrinsic, stochastic fluctuations and by collective cell-to-cell interactions, involving both genes and biophysical forces[83]. Furthermore, the model easily fit with experimental data evidencing that cells can evolve into a more differentiated phenotype or they can be reversed into less differentiated states and may even acquire pathological features. Namely, that framework could explain the astonishing property of some morphological fields in reverting neoplastic cells into well-differentiated, "normal" cells[84,85,86,87].

Conclusion

The natural world consists of hierarchical levels of organization that range from subatomic particles to molecules, ecosystems and beyond. Each level is both characterized and governed by emergent laws that do not appear at the lower levels of organization[43]. This implies that, in order to explain the features and behaviour of a whole system, we require a theory that operates at the corresponding hierarchical level. For instance, emergent phenomena that occur at the level of the organism cannot be fully explained by theories that describe events at the level of cells or macromolecules. As forecasted by Paul Weiss: "There is no phenomenon in a living system that is not molecular, but there is none that is only molecular, either [...the molecule-based approach] ignores the relevance of morphological form and morphogenesis, which are of salient importance for investigations on cancer, which after all is also a phenomenon of development at the tissue level"[88].

Thus, nowadays Biology is facing a crisis — or is it an opportunity — reminiscent of the state of biology in pre-double-helix time[89]. As a matter of fact, time is gone for Biology to "concern itself with the great 'non-reductionist' 19th century biological problems that molecular biology left untouched"[90]: all of these problems are different aspects of a more general question, e.g., the nature of the complex organization of the living matter. Precisely, Systems Biology affords that task.

Whatever the press releases might say, molecular biology, the acme of structural reductionism, has improved diagnostic methodology but has done almost nothing to improve patient care[91,92], while imposing huge cost burdens[93]. The gene-centric paradigm posits that knowing all about the genome we could be able to find a cure for every disease. Many observers honestly believed that a pot of golden medicine would be found at the end of that structural rainbow - that we would be able to find the part and fix the patients. But structural reductionism has clearly not delivered on the second part of this equation[94]. Diseases are "complex context-dependent entities to which our genes make a necessary but only partial contribution"[95]. Thus, the attempt to "develop personalized medicine solely on the genetic level is a profound conceptual mistake; and it is predictable that it will fail simply because it fails reality"[94].

The organism (and its development) was considered until now an epiphenomenona of its genes[96]. However, we are now at the point where the bottom-up approach is meeting the top-down approach. Whole organisms and their environmental interactions are becoming understandable, and gene expression patterns are being seen as being controlled both from the bottom-up and from the top-down. As a consequence, a profound rethinking of the biological paradigm is now underway and it is likely that such a process will lead to a 'conceptual revolution'[97] emerging "from the ashes of reductionism"[98].

Therefore, by no means Systems Biology could be considered a 'simple' 'gradual' extension of Molecular Biology, despite the efforts to do so made by many scientists[99].

Systems Biology is today thought to promote an integration of different kinds of knowledge, not a simple collation of disciplines, but a true multidisciplinary synergy. There is no doubt that this challenging task needs a new epistemology and scientific methodology for the III millennium[100].

References

[1] Mesarovic, MD. Systems theory and biology – view of a theorician. In: Mesarovic MD, editor. Systems *Theory and Biology*. New York: Springer; 1968.

[2] Kellenberger, E. The evolution of molecular biology. EMBO Rep, 2004, 5, 546-549.

[3] Brent, R. A partnership between biology and engineering. *Nature Biotechnol.,* 2004, 22, 1211-1214.

[4] O'Malley, MA; Dupré, J. Fundamental issues in Systems Biology. *BioEssays*, 2005, 27, 1270-1276.

[5] Kitano, H. Systems Biology: a brief overview. *Science*, 2002, 295, 1662-1664. Saetzler, K; Sonnenschein, C; Soto, AM. Systems Biology beyond networks: generating order from disorder through self-organization. *Sem. Cancer Biol*., 2011, 21, 165-174.

[6] Dinicola, S; D'Anselmi, F; Pasqualato, A; Proietti, S; Lisi, E; Cucina, A; Bizzarri, M. A systems biology approach to cancer: fractals, attractors, and nonlinear dynamics. *OMICS*, 2011, 15, 93-104.

[7] Soto, AM; Sonnenschein, C; Miquel, PA. On physicalism and downward causation in developmental cancer biology. *Acta Biotheor.,* 2008, 56, 257-274.

[8] Boogerd, FC; Bruggeman, FJ; Hofmeyr, J-HS; Westerhoff, V. Systems Biology: Phylosophical Foundations. Amsterdam: *Elsevier*; 2007.

[9] Bak, P. How Nature works: the Science of Self-organized criticality. Berlin: Springer; 1996.

[10] Auffray, C; Imbeaud, S; Roux-Rouque, M; Hood L. Self-organized systems: conjunction of a stable organization with chaotic fluctuations in biological space-time. *Phil. Trans. R. Soc. Lond.* A, 2003, 361, 1125-1139.

[11] Prigogine, I. La fin des certitudes. Paris: Jacob O, editor; 1996, 15-16.

[12] Auffray, C; Nottale, L. Scale relativity theory and integrative systems biology: 1. Founding principles and scale laws. *Prog. Biophys. Mol. Biol.,* 2008, 97, 79-114.

[13] Mesarovic, MD; Sreenath, SN; Keene, JD. Search for organising principles: understanding in systems biology. *Syst. Biol.,* 2004, 1, 19-27.

[14] Prigogine, I; Stengers I. La Nouvelle Alliance, Paris, 1979.

[15] Longo, G; Montévil, M. From physics to biology by extending criticality and symmetry breaking. *Progr. Biophys. Mol. Biol*., 2011, 106, 340-347.

[16] Nicolis, G; Prigogine, I. Exploring complexity. New York: Freeman; 1989.

[17] Huang, S; Ingber, DE. A non-genetic basis for cancer progression and metastasis: self-organizing attractors in cell regulatory networks. *Breast Dis*., 2007, 26, 27–54.

[18] Nicolis, G; Prigogine, I. Symmetry breaking and pattern selection in far from equilibrium systems. *Proc. Nat. Acad. Sci. USA*, 1981, 78, 659-663.

[19] Schrödinger, E. What is Life? *The Physical Aspect of the Living Cell.* Cambridge: Cambridge University Press; 1944.

[20] Laird, PW. Cancer epigenetics. *Hum. Mol. Genet.,* 2005, 14, R65–R76.
[21] Longo, G; Miquel, PA; Sonnenschein, C; Soto, AM. Is information a proper observable for biological organization? *Prog. Biophys. Mol. Biol.*, 2012, 109, 108-114.
[22] [1] Waliszewski, P; Molski, M; Konarski, J. On the holistic approach in cellular and cancer biology: non-linearity, complexity, and quasi-determinism of the dynamic cellular network. *J. Surg. Oncol.*, 1998, 68, 70–78.
[23] Kacser, H; Small, JR. How may phenotypes from one genotype? The case of prion diseases. *J. Theor. Biol.*, 1996, 182, 209–218.
[24] . Weatherall, DJ. Phenotype-genotype relationships in monogenic disease: lessons from the thalassaemias. *Nat. Rev. Genet.,* 2001, 2, 245–255.
[25] Scriver, CR; Waters, PJ. Monogenic traits are not simple. *Trends Genet.,* 1999, 15, 267–272.
[26] Dipple, KM; McCabe, ERB. Phenotypes of patients with ''simple'' mendelian disorders are complex traits: thresholds, modifiers and systems dynamics. *Am. J. Hum. Genet.*, 2000, 66, 1729–1735.
[27] Elowitz, MB; Levine, AJ; Siggia, ED; Swain, PS. Stochastic gene expression in a single cell. *Science*, 2002, 297, 1183-1186.
[28] Paldi, A. Stochastic gene expression during cell differentiation: order from disorder? *Cell Mol. Life Sci.,* 2003, 60, 1775-1778.
[29] Laforge, B; Guez, D; Martinez, M; Kupiec, JJ. Modelling embryogenesis and cancer: an approach based on an equilibrium between the autostabilization of stochastic gene expression and the interdependence of cells for proliferation. *Prog. Biophys. Mol. Biol.,* 2005, 89, 93–120.
[30] Xie, XS; Lu, HP. Single-molecule enzymology. *J. Biol. Chem.,* 1999, 274, 15967–15970.
[31] Hume, DA. Probability in transcriptional regulation and its implications for leukocyte differentiation and inducible gene expression. *Blood*, 2000, 96, 2323–2328.
[32] Blake, WJ; KA, M; Cantor, CR; Collins, JJ. Noise in eukaryotic gene expression. *Nature*, 2003, 422, 633–637.
[33] Herndon, LA; Schmeissner, PJ; Dudaronek, JM; Brown, PA; Listner, KM; Sakano, Y; Paupard, MC; Hall, DH; Driscoll, M. Stochastic and genetic factors influence tissue-specific decline in ageing C. elegans. *Nature*, 2002, 419, 808–814.
[34] Camazine, S; Deneubourg, J-L; Franks, NR; Sneyd, J; Theraulaz, G; Bonabeau, E. Self-Organization in Biological Systems. Princeton: Princeton University Press; 2003.
[35] Till, JE. Cellular diversity in the blood-forming system. *Am. Sci.,* 1981, 69, 522–527.
[36] Noble D. The music of life. New York: Oxford University Press; 2006.
[37] Noble D. Biophysics and Systems Biology. *Phil. Trans. R. Soc.* A, 2010, 368, 1125-1139.
[38] Laughlin, RB; Pines, D; Schmalian, J; Stojkovic, BP; Wolynes, P. The middle way. *Proc. Natl. Acad. Sci. USA*, 2000, 97, 32–37.
[39] Bizzarri, M; Cucina, A; Conti, F; D'Anselmi, F. Beyond the oncogene paradigm: understanding complexity in cancerogenesis. *Acta Biotheor.,* 2008, 56, 173-196.
[40] Bizzarri, M; Giuliani, A; Cucina, A; D'Anselmi, F; Soto, AM; Sonnenschein, C. Fractal analysis in a systems biology approach to cancer. *Semin. Cancer Biol.,* 2011, 21, 175-182.

[41] Bailly, F; Longo, G. Biological organization and anti-entropy. *J. Biol. Syst.,* 2009, 17, 63-96.
[42] Soto, AM; Sonnenschein, C. The tissue organization field theory of cancer: a testable replacement for the somatic mutation theory. *BioEssays*, 2011, 33, 332-340.
[43] Mazzocchi, F. Complexity in Biology. *EMBO Rep*., 2008, 9, 10-14.
[44] De Haan, J. How emergence arises. Ecol Compl, 2006, 3, 293-301.
[45] Turing, AM. The Chemical Basis of Morphogenesis. Philosophical Transactions of the Royal Society of London. *Biol. Sci.,* 1952, 237, 37-72.
[46] Honda, H. Genes do not determine biological shape directly. Forma, 1999, 14, 287-293.
[47] Newman, SA; Forgas, G; Muller, GB. Before programs: The physical origination of multicellular forms. *Int. J. Dev. Biol*., 2006, 50, 289-299.
[48] Kirson, ED; Dbalý, V; Tovarys, F; Vymazal, J; Soustiel, JF; Itzhaki, A; Mordechovich, D; Steinberg-Shapira, S; Gurvich, Z; Schneiderman, R; Wasserman, Y; Salzberg, M; Ryffel, B; Goldsher, D; Dekel, E; Palti, Y. Alternating electric fields arrest cell proliferation in animal tumour models and human brain tumours. *Proc. Nat. Acad. Sci. USA*, 2007, 104, 10152-10157.
[49] Hammond, TG; Benes, E; O'Reilly, KC; Wolf, DA; Linnehan, RM; Taher, A; Kaysen, JH; Allen, PL; Goodwin, TJ. Mechanical culture conditions affect gene expression: gravity-induced changes on the space shuttle. *Physiol. Genomics*, 2000, 3, 163–173.
[50] Levin, M. Bioelectromagnetics in morphogenesis. *Bioelectromagnetics*, 2003, 24, 295-315.
[51] Butcher, DT; Alliston, T; Weaver, VM. A tense situation: forcing tumour progression. *Nat. Rev. Cancer*, 2009, 9, 108–122.
[52] Ingber, DE. Tensegrity: the architectural basis of cellular mechanotrasduction. *Annu. Rev. Physiol*., 1997, 59, 575-599.
[53] Tracqui, P. Biophysical model of tumor growth. *Rep. Prog. Phys.,* 2009, 72, 1-30.
[54] Paszek, MJ; Zahir, N; Johnson, KR; Lakins, JN; Rozenberg, GI; Gefen, A; Reinhart-King, CA; Margulies, SS; Dembo, M; Boettiger, D; Hammer, DA; Weaver, VM. Tensional homeostasis and the malignant phenotype. *Cancer Cell*, 2005, 8, 241–254.
[55] Levi, M. Morphogenetic fields in embryogenesis, regeneration and cancer: non-local control of complex patterning. *Biosystems* 2012, 109, 243-261.
[56] Beloussov, L.V. Morphogenetic fields: outlining the alternatives and enlarging the context. *Riv. Biol/Biol. Forum*, 2001, 94, 219-235.
[57] Needham, J.. Biochemistry and Morphogenesis. Cambridge: Cambridge University Press; 1950.
[58] Gilbert, SF; Opitz, JM; Raff, RA. Resynthesizing evolutionary and developmental biology. *Dev. Biol.,* 1996, 173, 357-372.
[59] Glansdorff, P; Prigogine, I. Thermodynamic theory of Structure, Stability and Fluctuations. New York: Wiley; 1971.
[60] Onsager, L. Reciprocal relations in irreversibile processes. *Phys. Rev.,* 1931, 37, 405-426.
[61] Rottenberg, H; Caplan, SR; Essig, A. Stoichiometry and coupling: theories of oxidative phosphorylation. *Nature*, 1967, 216, 610-611.
[62] Westerhoff, HV; Van Dam, K. Thermodynamics and control of Biological Free-Energy Transduction. Amsterdam: *Elsevier*; 1987.

[63] Westerhoff, HV; Palsson, BO. The evolution of molecular biology into systems biology. *Nat. Biotechnol.,* 2004, 22, 1249–1252.

[64] Huang, S; Ingber, DE. A non-genetic basis for cancer progression and metastasis: self-organizing attractors in cell regulatory networks. *Breast Dis.,* 2007, 26, 27–54.

[65] Kauffman, S. Origins of Order: *Self-Organization and Selection in Evolution.* New York: Oxford University Press; 1993.

[66] Huang, S; Ernberg, I; Kauffman, S. Cancer attractors: a systems view of tumours from a gene network dynamics and developmental perspective. *Semin. Cell Dev. Biol.* 2009, 20, 869–876.

[67] Ingber, DE. Mechanical control of tissue growth: function follows form. *Proc. Natl. Acad. Sci. USA*, 2005, 102, 11571–11572.

[68] Reuveni, E; Giuliani, A. Emergent properties of gene evolution: species as attractors in phenotypic space. Physica A, 2011, 391, 1172-1178.

[69] Kacser, H; Small, JR. How may phenotypes from one genotype? The case of prion diseases. *J. Theor. Biol.,* 1996, 182, 209–218.

[70] Binney, J; Dowrick, NJ; Fisher, AJ; Newman, MEJ. The Theory of Critical Phenomena: An Introduction to the Renormalization Group. New York: Oxford University Press; 1992.

[71] Spillman, WB; Robertson, JL; Huckle, WR; Govindan, BS; Meissner, KE. Complexity, fractals, disease time, and cancer. *Phys. Rev. E.*, 2004, 70, 1–12.

[72] Bizzarri, M; Pasqualato, A; Cucina, A; Pasta, V. Physical forces and non linear dynamics mould fractal cell shape. *Histol. Histopathol.* 2012, In press.

[73] Ingber, DE. How cells (might) sense microgravity. *FASEB*, 1999, 13, S3-S15.

[74] Guo, Y; Eichler, GS; Feng, Y; Ingber, DE; Huang, S. Towards a holistic, yet gene-centered analysis of gene expression profiles: a case study of human lung cancers. *J. Biomed. Biotechnol.*, 2006, 2006, 69141.

[75] Waddington, CH. The strategy of the genes. London: Geo, Allen & Unwin; 1957.

[76] Chang, HH; Hemberg, M; Barahona, M; Ingberg, DE; Huang, S. Transcriptome wide noise controls lineage choice in mammalian progenitor cells. *Nature*, 2008, 453, 544–547.

[77] Enver, T; Pera, M; Peterson, C; Andrews, PW Stem cell states, fates and the rules of attraction. *Cell Stem. Cell.*, 2009, 4, 387-397.

[78] Mar, JC; Quackenbush, J. Decomposition of gene expression state space trajectories *PLoS. Comput. Biol.,* 2009, 5, 1-10.

[79] Joyce, AR; Palsson, BØ. The model organism as a system: integrating 'omics' data sets. *Nature Rev.,* 2006, 7, 198-210.

[80] Kirschner, MW. The meaning of Systems Biology. *Cell,* 2005, 121, 503-504.

[81] Levin, M. Molecular bioelectricity in developmental biology: new tools and recent discoveries. *Bioessays*, 2012, 34, 205-217.

[82] Bizzarri, M; Giuliani, A. Representing cancer cell trajectories in a phase-space diagram: switching cellular states by biological phase transitions. In: Dehmer M; Emmert-Streib F; Graber A; Salvador A, editors. Applied Statistics for Network Biology: Methods in Systems Biology. New York: Wiley; 2011.

[83] Stockholm, D; Edom-Vovard, F; Coutant, S; Sanatine, P; Yamagata, Y; Corre, G; Le Guillou, L; Neildez-Nguyen, TM; Pàldi, A. Bistable cell fate specification as a result of stochastic fluctuations and collective spatial cell behaviour. *PLoS. One*, 2010, 5, 1-12.

[84] Pierce, GB; Lewis, SH; Miller GJ; Moritz, E; Miller, P. Tumorigenicity of embryonal carcinoma as an assay to study control of malignancy by the murine blastocyst. *Proc. Natl. Acad. Sci. USA*, 1979, 76, 6649–6651.
[85] Krause, S; Maffini, MV; Soto, AM; Sonnenschein, C. The microenvironment determines the breast cancer cell's phenotype: organization of MCF7 cells in 3D cultures. *BMC Cancer*, 2010, 10, 263.
[86] Kenny, PA; Bissell, MJ. Tumor reversion: correction of malignant behavior by microenvironmental cues. *Int. J. Cancer*, 2003, 107, 688–695.
[87] Bizzarri, M; Cucina, A; Biava, PM; Proietti, S; D'Anselmi, F; Dinicola, S; Pasqualato, A; Lisi, E. Embryonic morphogenetic field induces phenotypic reversion in cancer cells. *Curr. Pharm. Biotechnol.* 2011, 12, 243-253.
[88] Drack, M; Wolkenhauser, O. System approaches of Weiss and Bertalanffy and their relevance for Systems Biology today. *Semin. Cancer Biol.,* 2011, 21, 150-155.
[89] Strohman, RC. The coming Kuhnian revolution in biology. *Nat. Biotechnol.,* 1997, 15, 194–200.
[90] Woese, CR. A new biology for a new century. *Microbiol. Mol Biol. Rev.*, 2004, 68, 173-186.
[91] Horrobin, DJ. Innovation in the pharmaceutical industry. *J. Royal Soc. Med.,* 2000, 93, 341–345.
[92] Ness, RB. Fear of failure: why American science is not winning the war on cancer. *Ann. Epidemiol.*, 2010, 20, 89-91.
[93] Hawkes, N. High cost of cancer treatment doesn't reflect benefits, say specialists. *BMJ*, 2011, 343, d6220.
[94] Le Fanu, J. The rise and fall of modern medicine. London: Abacus; 1999.
[95] Rossenbloich, B. Outline of a concept for organismic systems biology. *Sem. Cancer Biol.*, 2011, 21, 156-164.
[96] Gilbert, SF; Sarkar, S. Embracing complexity: organicism for the twenty-first century. *Dev. Dyn.,* 2000, 219, 1–9.
[97] Ehrenberg, M; Elf, J; Aurell, E; Sandberg, R; Ttegner, J. Systems Biology is taking off. *Genome Res.*, 2003, 13, 2377-2380.
[98] Van Regenmortel, MHV. Biological complexity emerges from the ashes of reductionist. J Mol Recognit, 2004, 17, 145-148.
[99] De Backer, P; De Waele, D; Van Speybroeck, L. Ins and Out of Systems Biology vis-à-vis Molecular Biology: continuation or clear cut? *Acta Biotheor.,* 2010, 58, 15-49.
[100] Conti, F; Valerio, MC; Zbilut, JP; Giuliani, A. Will systems biology offer new holistic paradigms to life sciences? *Syst. Synth. Biol.,* 2007, 1, 161-165.

In: Recent Advances in Systems Biology Research
Editors: A. X. C. N. Valente, A. Sarkar and Y. Gao
ISBN: 978-1-62948-736-6

Chapter 2

The Mesoscopic Level and Its Epistemological Relevance in Systems Biology

***Marta Bertolaso*[1], *Alessandro Giuliani*[2] and *Simonetta Filippi*[3]**

[1]Institute of Philosophy of Scientific and Technological Practice, Faculty of Engineering, University Campus Bio-Medico of Rome, Italy
[2]Environment and Health Department, Istituto Superiore di Sanitá, Viale Regina Elena 299, Roma, Italy
[3]Laboratory of Non Linear Physics and Mathematical Modeling, Faculty of Engineering, University Campus Bio-Medico of Rome, Italy

Abstract

The approach advocated in Systems Biology has been defined as a system-level understanding of biological phenomena. The notion of 'mesosystem' seems to meet this explanatory requirement. However, its epistemological presuppositions have not been clarified yet. In this paper we analyse how the notion of mesoscopic level enters the explanatory accounts of biological processes such as cell differentiation and cancer development. We thus suggest that the epistemological interest of the notion of mesoscopic level is both related to the specific dynamics of collective behaviours and the enterprise of the identification of an adequate level of analysis. The heuristic and explanatory relevance of the mesoscopic level is eventually discussed in terms of Systemic Approach, that refers to the wider epistemological framework required to account for the process of identification and analysis of the regulatory levels in complex biological behaviours.

Keywords: Systems, levels, explanation, biological complexity, systemic approach, behaviours

1. Introduction

Systems Biology is often viewed as an advanced method, which, by integrating a multiplicity of data sources, is able to grasp the complexity of biological processes. It has also been seen as a reaction to mere bottom-up reductionist approaches in biology, consisting in the attempt to explain biological dynamics in terms of molecular parts and their local interactions. Both accounts of Systems Biology, however, have a common root in the challenge of explaining how biological processes are regulated at different levels of biological organisation and how their functional structure is maintained. This means that in practical science the aim is to provide an explanation of some robust phenomena, where robustness is the ability of a system of maintaining a system state in response to perturbations. From a philosophical perspective, this means to understand the multilevel phenomenology of complex biological behaviors and how it is related the hierarchical control of its organization.

In this chapter instead of discussing the scientific and philosophical relevance of the above-mentioned views on Systems Biology, that an extensive international literature has already addressed (Boogerd et al., 2007, O'Malley and Duprè 2005), we adopt a different strategy. We focus on how science deals in practice with regulatory phenomena and on the kind of scientific questions that call for a Systemic Approach. Our claim is that the reflection on Systems Biology program has thus an epistemological relevance for science at large.

In the next section we analyse a conceptual framework that seems to meet scientists' practice and their way of thinking when trying to explain the origin of the remarkable stability features of the dynamics relative to biological systems. In the following sections, instead, we spell out the problems that forced towards such systemic perspectives, and present some worked examples taken from cell biology to characterize the conceptual framework of a Systemic Approach disentangling the epistemological dimensions that it implies and that can shed light on the conceptual foundations of Systems Biology as well.

2. The Stones and the Bridge

Marco Polo describes a bridge, brick by brick. "But which is the stone that sustains the bridge?" Kublai Kahn asks. "The bridge is not sustained by this or that brick, Marco replies, but by the line of the bridge that they form." Kublai Kahn keeps silent, thinking. Then he adds: "Why do you then talk about the stones? It is only the arch that interests me". Polo replies: without bricks, there is no arch. (I. Calvino, Invisible Cities). This dialogue is often quoted in scientific literature when trying to explain biological dynamics (e.g., in Bizzarri et al., 2011, Vitiello 2001). The core issue of systems biology research programs eventually deals with inter-levels regulatory processes, and with the "generation of order" in nature, in terms of regular patterns of behaviour. Models span from the protein folding to the development of cancer and the origin of the high heterogeneity of forms in nature. We will thus focus on some epistemic issues that link the bridge metaphor with the scientific practice, so to identify the level of generalization at which a given explanation is performed. The explanatory enterprise and the adopted methodology are not autonomous from a specific epistemological point of view. This relationship has been already addressed, although, from different perspectives elsewhere and is commonly identified as the problem of the "system level understanding" (Kitano 2002, Richardson and Stephan 2007).

2.1. Beta Cells

Let us start with a brief analysis of the explanatory terms that characterize a study of a collective behavior. Cell collective behaviors are, in fact, paradigmatic among phenomena for which a systemic explanation is required. As a relevant case study we will deal with pancreas organization. Islets of Langerhans, the basic functional units of pancreas, have a complex anatomy including a core mainly consisting of insulin producing beta-cells, which are tightly interconnected by gap junctions and surrounded by a mantle of other endocrine cells secreting different hormones. The integrity of the interactions and the 3D architecture among beta-cell populations in pancreatic islets is critical for proper biosynthesis, storage and release of insulin, and is progressively destroyed by autoimmune progression towards type 1 diabetes. Experiments and simulations have been performed to evaluate the effect on beta-cells electrophysiological signaling of progressive lymphocytic islet cell infiltration by modeling the disruption of pancreatic islet's anatomy.

The activity of beta-cells has been investigated embedded into a regular 3D lattice with defined neighborhood relations and altered glucose concentrations. Many studies (Jonkers et al., 1999, Sherman et al., 1991), showed the contemporary presence of irregular spiking behavior of isolated beta-cells and the highly recurrent and regular collective bursting dynamics of coupled cells so pointing to the emergence of order from the cell interactions. In the reported simulation study (Portuesi et al., 2012) progressive damage of the cell-cell interaction pattern was modeled at different glucose concentrations, correspondent to the autoimmune progression towards diabetes. In Figure 1 some results are shown. It is worth noting we used a minimalistic model of cell-cell interaction derived by experimental data at both single cell and cell cluster level.

The numerical results highlight the relationship between the progressive beta-cell disruption in the pancreatic cluster and the system's electrophysiological response to the damage. In the final conditions (bottom of the picture) insulin pulsatility is not expected to occur. The most significant information emerging from this modeling study, applied to the pathogenesis of type 1 diabetes, is that 1) the islet tissue is biophysically robust enough to compensate high rates of cellular loss (robustness arises from the particular structure of the underlying wiring diagram between the component parts); 2) there are threshold effects so that is possible to explain why hyperglycaemia does not occur even in the presence of few functioning beta-cells; 3) stochastic ion channels opening and closing is the key mechanism to account for the difference between the single beta cell and the bursting behavior of coupled beta cells in a cluster.

These evidences show that through the effective coupling due to a channel sharing dynamics, noise intensity is reduced by gap junctions electrical connections and that beta-cells ensemble behavior creates an emergent temporal pattern at a large scale represented by another set of variables at a hierarchically higher level of structural organization. Overall, the complex behavior is synchronized throughout the islet so that the islet works as a functional syncytium. On the other hand it is important to note how the order arising at the cell-cell interaction scale influences the calcium dynamics at the single cell level by 'top-down' causation: not only the order 'emerges' from the interaction of many cells, but this 'higher scale' behavior influences the single cell dynamics.

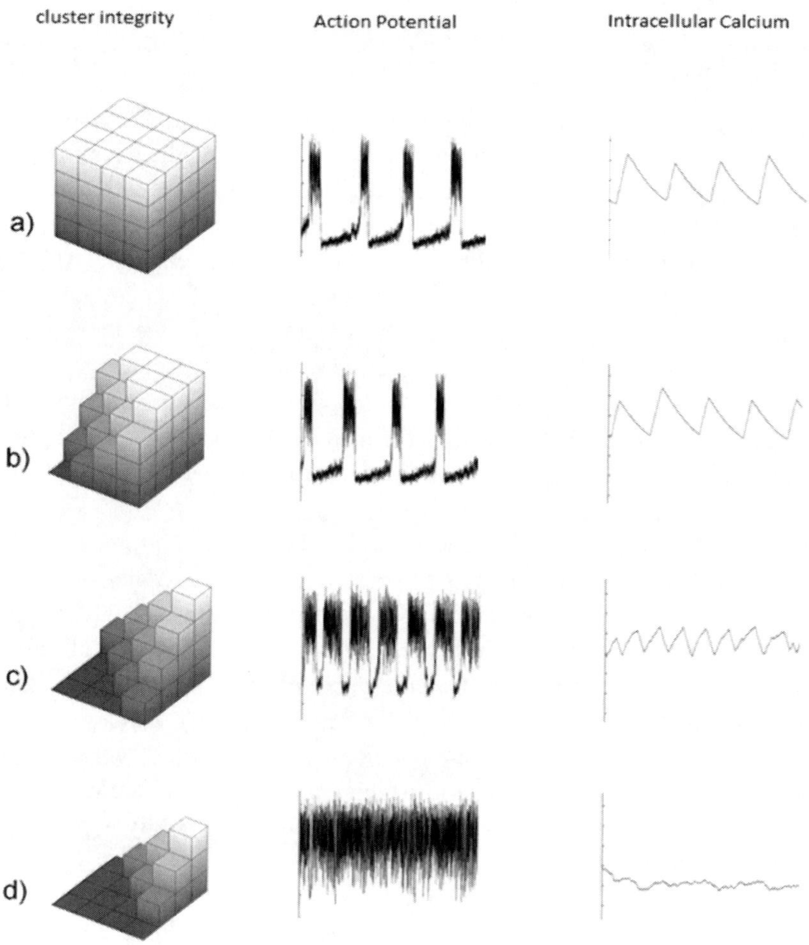

Figure 1. Left panel series: going from the top to the bottom of the figure we can appreciate the progressive damage of a cluster (initially consisting of 64 beta-cells) of beta-cells. Central panel series: the cluster membrane potential is registered. Right panel series: the related intracellular [Ca ++] for a generic beta-cell in the cluster during 80 simulated seconds is reported. It is worth noting how both the collective (central series) and individual (right series) behavior undergoes a progressive disorder and loss of temporal correlation. This fact suggests the onset of a 'top-down' causation process in which the single element behavior is modified according to the global system it is part of.

This model thus suggests that complex organisms either exploit stochastic cell population heterogeneity to achieve a smooth, macroscopically continuous tissue behavior, by coordinating cell behavior through structural and/or chemical cell-cell coupling, as is also the case in the pumping of the heart. This blend of microscopic stochasticity and macroscopic determinism is the kind of dynamics that Systems Biology tries to capture.

Collective behaviors can be thus understood as spatial configurations adopted by agents interacting through corresponding different structures over time (Minati et al., 2012). Two key concepts emerge here: dynamic order and long-range interactions creating a link between structure and function and making sense of the bridge-stones metaphor. The beta-cells are the stones, Islets of Langerhans are the arch, i.e., the level intermediate between the single cell and the entire organ: what we intend as a proper mesoscopic level. The general pancreatic function that is hold by the collective behavior of the beta cells, and which implies a dynamic order through long-range interactions, is the bridge. The new idea is that this hierarchy can be traversed also on the other way round and that the properties of the most elementary elements (beta-cells) depend on the kind of organization they are embedded into. This bi-directional causative fluxes can be rationalized in terms of the concept of functional field.

2.2. Cancer

An analogous notion of functional field arises from the effort to deal with the dynamic features of cancer and with the heterogeneity that tumor cells introduce in a tissue. Accumulating evidences coming from tissue microenvironment research, bioenergetics, epigenetics and physical chemistry tend to converge in the definition of cancer as essentially a genetically non-deterministic disease. Cancer is characterized by progressive dis-organization taking place at a variety of organizational levels, from the genome and metabolic networks, to tissue integrity (Sonnenschein and Soto 1999, Huang S, Ingber 2000, Nelson and Bissel 2006).

Cancer can be thus primarily understood as a disease of tissue organization. In explanatory models of cancer this new dimension of analysis is characterized through the notion of fields. They are generally defined as groups of cells from which specific morphological structures (tissues and organs) develop through the mediation of biophysical and biochemical cues, mainly through epigenetic changes. In cancer, these epigenetic changes are aberrant. It has been thus used the term of cancerization fields (Ushijima 2007). The common idea is that in cancer the functional stability of a field is compromised, and that this notion of field concerns functional definition of parts and parts' stability as well.

Moreover, recently theoretical models are proposed that suggest that impaired coherence dynamics of collective phenomena could lead to cancer, thus combining biological and physical approaches to understanding this complex pathology (Del Giudice 2012). Coherence of molecular dynamics is, in fact, increasingly recognized as an organizing principle linked to the long-range coordination of biological processes, which can explain the remarkable efficiency of biological systems (coherence is applied here as the coupling of cells within a 3D architecture like in the previous example).

System's dynamics in a state space cannot be thus "reduced" either to a molecular wiring diagram or, even to the integrated functioning of parts. Additional influences (i.e., those of the intracellular topology) must be taken into consideration in order to give a more reliable definition of functional field.

Self-organization does not only pertain to topological order, but it also has its temporal manifestation, i.e., synchronization. In this way, explanatory systems are not characterized in terms of molecules and cells, but cancer is rather explained in terms of a functional state of the system, identified by defining a field and describing it in organizational terms.

The link between topology and dynamics (deeply explored in protein folding studies (Dokholyan et al., 2000) is constitutive of a graph-like structure, i.e., of a system made by elements (nodes) and mutual interactions (arches) connecting them (no matter what kind of interaction we are speaking about), implies that the system (network) has some preferred states corresponding to specific patterns of values (positions, expression levels, concentrations) taken by the nodes, consequently each displacement from a stable configuration of the system generates a relaxation dynamics, whose features are dictated by the wiring structure of the corresponding network. This feature is so general that is at the basis of spectroscopic measurements like NMR that identify the structure of the investigated 'pieces of matter' in terms of relaxation dynamics of the investigated system following a perturbation stimulus (Callaghan 1991). This is why "Understanding the complexity of cancer is not only a matter of alternative theories, but requires first and foremost, a methodological approach that reflects a new epistemological view, in light of the fact that cancer is increasingly held to be a systems biology disease" (Rew 2000).

3. The Mesoscopic Level

The above sketched examples are a proof-of-concept of the possibility of an easily communicable and understandable systemic explanation of a biological object of inquiry in terms of functional fields. The contemporary presence of both a 'bottom-up' (the bridge is made by stones, Islets of Langherans are made by beta-cells) and a 'top-down' (the stones are taken together by arch geometry, a single beta-cell into an islet has a different behavior with respect to an isolated beta-cell) causative way is indicated with the term 'middle-out' and, in our opinion, is the most genuine peculiarity of Systemic Approach allowing to overcome the classical (and in many cases sterile) 'reductionism' vs. 'holism' opposition. This 'middle-out' approach acquires the concept of 'functional field' as epistemological foundation: like in usual physical fields (think of electromagnetic or gravitational fields) the properties of the field (i.e., the particular energy landscape of a portion of space) are both generated by the constituting elements (bottom-up) and influence the elements behavior (top-down).

This style of reasoning has the name 'mesoscopic' because it concentrates on the between elements relation structure that is considered as the channel along which the microscopic and macroscopic views are related. It implies the necessity to take into consideration, even when we concentrate on a single element (e.g., the lethal character of a specific mutation), the general frame in which the element is inserted. "The mesoscopic level is where organizational principles act on the elementary biological units that will become altered, or constrained, by both their mutual interaction and the interaction with the surrounding environment. In this way and in this place is where general organization behaviour emerges and where we expect to meet the elusive concept of complexity" (Bizzarri et al., 2011).

Framing the picture from this point of view, and explaining functional states in terms of dynamic order and long-range interactions, avoids unfair debates about reductionism in scientific practice (Bertolaso 2013). In our view, and at this level of analysis, it might be enough to acknowledge that science works through progressive analysis and synthesis of the decomposed subsystems, where the notion of analysis and synthesis have to be understood in their procedural meaning (a different meaning these terms, in fact, have in the philosophical literature). Taken in this way, the challenge we aim to contribute to is well introduced by the philosopher Woodwards who says: "philosophers have often misdescribed or oversimplified the content of the causal and explanatory claims they have hoped to reduce. We need more careful description of just what such claims say (…). Only after this has been done should we investigate what sorts of reductionisms are possible" (Woodwards 2003, 22).

What is at stake is, thus, understanding how the phenomenology of biological systems and their dynamics are equally relevant for the explanation of the dynamic generation of order in nature. In philosophical terms this concern meets the question posed by O'Malley and Dupré when saying: "The issues that are most fundamental are how the systems that are the focus of Systems Biology are defined, and how those definitions affect the research agenda that arise from earlier scientific legacies" (O'Malley and Dupré 2005).

Systems defined in terms of functional fields, in fact, both entail different levels of order and can be considered themselves as peculiar ordered layers because they present unitary properties at each level that emerge not only from the properties of the components but from the peculiar relationships among them. A Systemic Approach (SA) is thus defined by the kind of scientific question addressed while being not committed to any specific kind of explanation (molecular, organizational, topological).

SA mainly focuses on the study of a system's behavior in terms of the emerging robustness of a new entity (the system object of inquiry) from the between elements interaction pattern. It is worth noting that when we speak of 'interaction' we must not only intend the local links of a specific 'node' of a network, but the general wiring structure emerging from both the global architecture and eventual general fields the system as a whole is embedded into.

4. Old Problems and New Concepts of Systems Biology

To better understand the epistemological relevance of the mesoscopic level and the SA that endorse it, we need to spell out some difficulties that have been encountered in science and that forced scientist towards systemic views and methodologies. They are summarized in Table 1.

Table 1.

Traditional approach	However ….
The classical form in which biological systems are described corresponds to a set of basic elements of the described system	Graphs became larger and larger and asked for some form of global analysis in order to get rid of their wild multiplicity
Cell biology since its birth relied on population data coming from millions of cells growing on plates but sketched its models by means of node and arrows cartoons implicitly posited in the interior of a single cell	Discovery of single highly noisy and un predictable behavior and of the presence of collective order parameters spanning several order of magnitudes forces cell biologists to a complete change of paradigm
Substantial equivalence between all the cells inside a population	Self sustained oscillations in cell ensembles points to a sort of 'structuring' of the intercellular variance with the consequent departure of the cell cultures from ergodicity and the need to consider a sort of 'ecology-in-a-plate'.

The classical form in which biological systems are described (be they metabolic charts, gene expression regulation pathways, protein-protein interaction maps, food webs and so forth) corresponds to a set of basic elements (genes, proteins, metabolites and so forth) and the edges connecting them some rules of the kind 'is transformed into' or 'is increased (decreased) by'. However, with the development of high throughput methodologies, these graphs became larger and larger and asked for some form of global analysis in order to get rid of their wild multiplicity.

Cell biology, since its birth, relied on population data coming from millions of cells growing on plates but sketched its models by means of node and arrows cartoons implicitly posited in the interior of a single cell. However, the contemporary discovery of single cell highly noisy and unpredictable behavior and of the presence of collective order parameters spanning several order of magnitudes of, for example, expression intensity involving thousands of genes simultaneously at the cell cultures level forces cell biologists to a complete change of paradigm.

The presence of self sustained oscillations in cell ensembles points to a sort of 'structuring' of the intercellular heterogeneity with the consequent departure of the cell cultures from ergodicity and the need to consider a sort of 'ecology-in-a-plate' and higher level constructs linked to cell-cell communication with respect to classical cell-centered molecular biology models. This point is particularly cogent if we take into consideration a second aspect of the ergodic assumption: namely the substantial equivalence between all the cells inside a population. This is a very important point because it is strictly related to the onset of non-genetic heterogeneity between cells.

All in all this case has shown the existence of a mesoscopic level whose behavior cannot be simply derived by the knowledge of the constituting elements while, in an apparently counterintuitive manner with respect to the reductionist paradigm, influencing the microscopic level. "A good question to ask, therefore, is whether Systems Biology sets itself apart from earlier genomics because of its object ('systems') or because of the way it does things ('systematically')" (O'Malley and Dupré 2005). A clarification of this relationship might help us to overcome some aspects of the reductionism-anti-reductionism debate and to move forward in developing epistemological tools that are still needed to explain the hierarchical control of biological systems as well.

4.1. Networks

We have seen that the system are defined through a process of analysis and synthesis of the explanatory relevant elements identified through the concept of functional field and the establishment of systems (often also called emergence, although we will avoid this term for the different meaning that has in the scientific and philosophical literature) is related to the concept of interactions. Such interactions are often represented through pathways that are implicitly considered as linear causative chains in which a signal starting from a molecular perturbation, after a sequence of if-then events, emerge as a biological end-point (Tun et al., 2011). Normally these processes are referred as 'cascades' provoking a progressive amplification of the initial stimulus (MacFarlane 1964). On the contrary, biological effectors (being them genes, proteins, hormones, etc.), with only few exceptions, work in networks, and this fact implies a completely different form of biological regulation with respect to the 'cascade' model: the entire network has, thanks to its wiring structure, few preferred modes corresponding to the stable configuration of the network itself (Tun et al., 2011, MacFarlane 1964, Kauffman 1993) , any perturbation, being it pharmacologically induced or coming from a mutation in a crucial gene, ends up into one or the other of these allowed states without any simple relation with the features of the applied perturbation (Tun et al., 2011).

The approach considering the graph as a system of differential equations in which an entering stimulus, correspondent to a modification of a peripheral node of the network, is progressively processed according to the wiring architecture and kinetics constraints of the network itself, while being the most potentially exhaustive avenue of research is severely hampered, in the case of biological systems, by a lot of problems. First of all the practical impossibility to attach to the whole set of edges reliable kinetic-like weights for quantifying the entity of the between elements correlation. Only in the case of very small networks can this be done by means of statistical estimation of the parameters from experimental data, but it is well known that in physiological settings these weights can vary of orders of magnitude (Laughlin 2000). Moreover in many cases we cannot rely on the complete knowledge of the wiring diagram of the network. For these (and other) reasons many authors preferred a purely topological approach to the analysis of biological networks. It implies considering the presence of a link between two nodes as a pure yes/no binary relation and limiting themselves to statistical descriptions making use of the so called graph-invariants, i.e., a collection of indexes that enable the analyst to identify crucial elements of the network or to highlight specific features of the entire network architecture responsible for some aspects of the studied system behaviour. This is the case of the so called 'scale-free' architecture that was demonstrated to be at the basis of the huge resilience of biological systems (Watts and Strogatz 2004).

The consideration of biological systems at the coarse-grain level of the graph topological approach is an important first step for the development of a sort of biological statistical mechanics in which the actual behaviour of the global system can be predicted by a convenient statistics over its constituent parts (Giuliani 2010). The common experience of any experimentalist dealing with microarray data is a good example to understand some aspects of this picture.

Any two independent samples, in fact, of the same cell kind when correlated over the expression of more than twenty thousand different gene products display a near to unity correlation (see Figure 2).

This marked invariance is normally 'taken for granted' by biologists that historically focused on the (small) deviations from the native tissue profile in gene expression space looking for specific altered genes without having any perception of the 'elephant-in-the-room' correspondent to the invariance. Systems Biology is starting to look at the nature and origin of this elephant not only for knowledge reasons, but for incredibly urgent and dramatic practical motivations.

The same basic idea of network stable states allows us to understand what happened: the only 'simple targets' whose modification is expected to give rise to a macroscopic, organism-scale observable effect are those located at the extreme periphery of the interaction network, while the modification of a node located in the internal position of the network is immediately buffered by the feedback relations so that the system cannot be modified by pharmacological intervention (Tun et al., 2011).

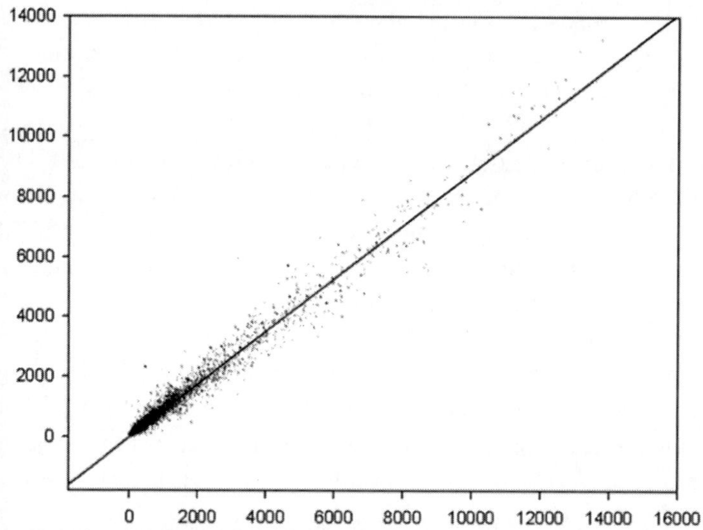

Figure 2. The X and Y axes of the figure correspond to two independent samples of the same tissue type (in this case macrophages). The approximately 23000 vector points correspond to different gene expressions coming from a microarray experiment. Notwithstanding the fact the graph spans four order of magnitudes there is a remarkable order of gene expression level corresponding to a Pearson correlation coefficient r = 0.99 between the profiles. This invariance comes from the fact each tissue is an attractor in the gene expression space, the (relatively small) scattering around the identity line corresponds to the motions 'inside the attractor', these motions are analogous to the dynamics of a protein molecule around its native state and are the only ones eventually affected by disease states or pharmacological perturbations (Giuliani 2010).

4.2. Attractors

The attractor view, that often is used to highlight the dynamic relevance of networks allows scientists to eliminate the need to impose the presence of 'intelligent agents' in order to get rid of specific and finely tuned behaviours: the system 'lives' in a non-uniform state space (the ensemble of all the possible system configurations) characterized by a so called 'rugged landscape' where the energy minima (valleys of the landscape, quasi-equilibrium configurations) correspond to attractor states (Frauenfelder et al., 1991). Thus the specific differences in mechanism of action of different effectors are not so relevant in terms of the resulting effect that is largely dependent on the affected system modes. A very important consequence of the presence of an attractor-like regulation is the impossibility to maintain the classical discrimination of house-keeping vs. specifically regulated genes and the importance of low-variance genes with the consequent need to re-cast the idea of what a 'pathway' (or a 'gene signature') is (Tsuchiya et al., 2010, Venet et al., 2011).

Clearly, as above stated, this is a pure phenomenological proposal that does not enter the molecular mechanisms supporting it, the hard problem to solve being the establishment of an accurate spatial compartmentalization of the system allowing different ordered phenomena to be maintained, while a classical diffusive regiment cannot sustain such highly structured and specific processes, in any case we can speculate cytoskeleton organization (Ingber 1999) or confinement by phase transitions (Hyman et al., 2012) allowing the selection of specific pathway in complex microenvironments are very plausible candidates.

Limiting ourselves to data analysis coming from actual biological experimentation, it is sufficient to imagine these discrete states corresponding to 'allowed positions in the transcriptome space', coming from the presence of a still unknown origin field sensed by the entire genome and driving its collective behaviour (Huang et al., 2005).

It is worth noting that a very basic 'toolbox' made of principal component analysis (principal components being the coordinated fluxes of variations of many different genes), network invariant descriptors (with the assignment to each node a set of measures related to its role in the network wiring), cluster analysis (very dense clusters in the phase space correspond to attractors) are sufficient to undergo such 'biological dynamics' avenue of research (Huang et al., 2005, Huang 2009, Di Paola et al., 2013). A very recent work describing the architecture of whole genome regulation as emerging from results coming from ENCODE project (Gerstein et al., 2012) is consistent with the view of a dense interconnected network working as a whole and thus asking for system-level description of gene expression dynamics.

4.3. Complementing the Macroscopic and Microscopic Levels

The interaction of a multiplicity of scales in both time and space is thus a fundamental feature of biological systems. The complementation of macroscopic (entire organism) and microscopic (molecular biology) views with a mesoscopic level of analysis able to connect the different planes of investigation is urgently needed. This will allow to both obtain a general frame of reference for rationalizing the burden of data coming from high throughput technologies and to derive effective operational views on biological systems.

What is important to reiterate here is that the network paradigm introduces a unique synthesis between reductionist (all is in the molecules) and holistic (all is in the whole) approaches. A clear example of the efficiency of this 'graph-based' reasoning more linked to Systems Biology problems is the prediction of lethal mutants in yeast by the graph analysis of metabolic network (Palumbo et al., 2005, Palumbo et al., 2007).

From a purely topological point of view, each node of a network is uniquely defined by its position in the graph. Obviously, when dealing with experimentally derived and not abstract networks, each node has a name (a particular gene, protein, metabolite) pointing to a rich basin of knowledge and evoking cognition resonance to the specialist mind and the same is true for the edges. However, if we are interested in discovering what can be inferred solely from topological information (so acquiring a Boltzmann-like statistical attitude sacrificing the unique personality of the element to the search of a mesoscopic principle), we should try and predict some relevant features of the studied system without relying on the particular 'nature' of nodes and edges, but only taking into consideration their connectivity pattern.

In other terms all the properties relative to each node (edge) must be derived only by its pattern of relations and thus by its peculiar location in the complete graph. In (Palumbo et al., 2005, Palumbo et al., 2007) the authors checked for the possibility to derive, from purely topological information on the metabolic network of yeast (*Saccharomyces Cerevisiae*), the lethal character of genetic mutations. The metabolic network of microorganisms is very well understood: it can be considered as a graph having enzymatic reactions as edges and metabolites as nodes. Since an enzymatic reaction is catalysed by one or more enzymes, an edge can also represent the enzymes involved in the reaction. If it is possible to pick up a connectivity descriptor able to unequivocally define essential enzymes (those enzymes whose lack provoke the yeast death) we can safely assume the biological relevance of the metabolism 'wiring structure', irrespective of the specific nature of the involved enzymes, and consequently deriving a mesoscopic biological principle (Giuliani 2010).

It is relatively easy to extract from graphs many descriptors located at local (single nodes), global (entire network) and mesoscale (clusters of nodes, optimal paths) levels. Thus we can compute the degree of each node (how many links are attached to a given node) that is a local, microscopic characteristic by which we can in principle locate the most important elements in a complex system (bottom-up approach) or we can compute the so called 'average shortest path' or 'characteristic length' of a graph corresponding to the average length of minimal paths connecting all the node pairs (this corresponds to a mesoscopic feature of the system) or the general connectivity of the network (this allows for a top-down study of the network as a whole) (Watts et al., 2004, Di Paola et al., 2013). Finally, it is important to stress all these different views are strictly intermingled among them, given they derive from the same basic representation (the graph) so that any view influences (and in turn is influenced) by all the others. The necessary (and natural) interaction of different level views is called 'middle-out' approach to stress the fact the interest is focused on the mesoscopic level, i.e., on the pattern of between elements relation and not on the fundamental features of the constituting elements (Csermely et al., 2005)[1].

5. A Systemic Approach: Essentiality-By-Location

The understanding of the process of establishment (technically emergence) of systems based on interactions among components detected by an observer equipped with a sufficiently sophisticated cognitive model, implies a conceptual framework of crucial importance in scientific disciplinary applications and theories. This process of identification of the mesoscopic level is also known as the 'essentiality-by-location' mesoscopic principle. In the above mentioned example it consists in equating the lethal character of a mutation to the lack of an alternative path in the network, as was confirmed (Palumbo et al., 2007) demonstrating that a double mutation involving two enzymes, which per se are not essential, acquires essentiality and then causes the death of the organism, if the double knock-out provokes the 'lack of alternative path' condition. This illustrates the emergent character of the 'essentiality by location' principle: the arising of lethality by the summation of two non lethal events derives from the existence of a global metabolism architecture and thus cannot be inferred by going in depth into the nature of the two enzymes, in other words is a collective emergent property of the network system (Giuliani 2010). Moreover, all in all this case showed the existence of a mesoscopic level (the network) whose behaviour cannot be simply derived by the knowledge of the constituting elements while, in an apparently counterintuitive manner with respect to the reductionist paradigm, influencing the microscopic level (Bertolaso 2011).

Going back to cancer, we have historically been called to the study of the neoplastic processes through an epistemological and methodological approach, which breaks cancer down into simple elements, identified as the main causes that, once isolated, can be understood immediately. Molecular biology in this way has provided a suitable experimental

[1] For the relevance and implication in chemistry of these issues see Di Paola et al. 2013.

platform for this kind of scientific work. As mentioned earlier, however, empirical evidence forced the revival and resurgence of a more systemic, comprehensive understanding of the phenomena (Bertolaso 2009), which implies a synthetic approach[2] besides the analytic one that by default characterize the methodology of the experimental design but does not exhaustively represent the scientific practice as a rational enterprise.

There are therefore two epistemic dimensions that guide the study of complex phenomena and that call for a systemic perspective: one has to do with the identification of the explanatory level or level of generalization, the other one with the characterization of the parts-whole system that holds the dynamic behaviour. The correct choice of explanatory and analytical level is, in fact, not self-evident. Moreover, pragmatic reasons in scientific practice are embedded of the effort to identify mesoscopic levels where external (objective) and internal (subjective) dimensions of science meet.

Thus, "depending on the details of the case, description or causal explanation can be either inappropriately broad or general, including irrelevant details, or overly narrow, failing to include relevant detail. Which level is most appropriate will be in large part an empirical matter" (Woodward 2010). However, from a systemic perspective what an "empirical matter" is should be understood in a wider sense. It is not just a pragmatist problem (what is possible here and now) but a rational problem that start with a scientific question (why this and not that). This leaves open a philosophical reflection about the correspondence between the world and the way we know it through science. A compositional and pluralistic view of the scientific enterprise is better suited to explain why and how science works, but the philosophical implication of what is a common practice in science does not seem to be still at hand. Molecules and collective dynamics are equally involved but this does not necessarily imply that they are equally explanatory relevant, in the sense that –this is our suggestion- they play a different explanatory role in the process of the understanding.

The scientific question and the related choice of the level at which to work influences both the results obtained and the form in which the explanation presents itself. A well-posed scientific question determines the focus of the research, together with the theoretical elaborations of the experimental data. From this point of view, the analytical and synthetic approaches converges into an integrative systemic approach, for brief SA, that is capable of integrating experimental data with the theoretical working hypothesis formulated at the level of biological complexity that has the highest degree of explanatory coherence. It is at this mesoscopic level that the properties of the parts emerge in a system of collective behaviour that can be fully studied by means of Systems Biology.

5.1. Non-Trivial Determinism: A Statistical Answer to the Quest for the Optimal Mesoscopic View

How can we identify the most 'convenient' (in terms of explanatory power) mesoscopic level for a given scientific enterprise? A very intriguing proposal is the one made by statistical methodology: the most convenient mesoscopic level is that level that maximizes the

[2] As already mentioned, in this context, synthetic means the procedural aspect that completes the analytical one and we use it with the same restriction that was first made for the notion of the analytic approach. That is, we refer to the process by which a data series is taken together with their specific relationships, allowing for considerations of the phenomenon at a different level, with any unnecessary details stripped away.

'non trivial determinism' (Pascual and Levine 1999). This proposal stems from a very straightforward line of reasoning: the equations of a deterministic model express a causal and, therefore, totally predictable relationship between past and future: given exact initial conditions, they completely specify the future state of the system (of course this is a purely ideal and never fully accomplished condition in practice given measurement and dynamical noise are never absent). When explicit equations are not available (like in the quasi-totality of biological enterprises), the degree of determinism can be evaluated from data, through the prediction accuracy of a given algorithm; think for example to the percentage of correct predictions of receptor binding ability of a set of molecules inferred by structure-activity statistical prediction algorithms (Bender and Glen 2005) or to any other machine-learning procedure. The determinism estimated by prediction accuracy must in any case be 'nontrivial': the prediction of the number of different amino-acid species constituting a protein is trivial because ANY protein is made by the same 20 amino-acid species. Thus the 'nontrivial' character of a prediction has to do with the 'original variance' of the system at hand, i.e., of the variety of *a priori* possible different outcomes, this opens the way to a straightforward definition of nontrivial determinism as:

$$\text{Nontrivial Determinism } R^2 = 1 - E^2/\sigma^2 \quad (1)$$

In (1) E^2 is the square of mean prediction error (mean difference between estimated and actual outcomes), σ^2 is the natural variance of the system. Thus the formula is forced to have a maximum unitary value (no prediction error) and a zero value in the case of no explanatory power at all (prediction error equates natural variance), and is formally consistent with the usual R^2 statistics routinely used to compute the goodness of fit of a given model. This simple index allows us to identify in a non-ambiguous way the correct level of analysis: e.g., we could imagine to look for the variation of model accuracy (R^2) in the prediction of the outcome at time = t (X(t)) of a time series by changing the 'window size' (L) of the past outcomes used for model construction (namely the values X(t-1), X(t-2)...X(t-L)). In general we will reach a maximum of prediction accuracy at a given value of L, from that window size onward (i.e., adding other variables to the model going further into the past), prediction accuracy will not increase, thus indicating *L* as the time scale maximizing the nontrivial determinism or, in other words, the best 'mesoscopic' view of the entire process. This methodology can be generalized to other situations by defining 'windows' in both space and time: under this heading the Islets of Langherans we introduced before is a very convenient mesoscopic scale because it maximizes the determinism of burst occurrence.

5.2. Reconstructing Dynamics

This simple, purely syntactical, rule, avoids the risk of making methodological recommendations about the ontological restrictions, as done in epistemological or ontological reductionism, while also avoiding deriving methodological indications from holistic principles. The synthetic perspective is closely correlated with the natural processes that involve the appearance of functional characteristics and the reorganization of actions through new interactions. As biology shows, different levels of complexity interact and operate in a synergistic manner, requiring a more comprehensive explanation.

In experimental work, the analytical and synthetic approaches imply mutual interaction and assume a systemic perspective. It is the latter approach that is capable of capturing the explicative level that gives the most essential explanation of a specific phenomenon. This, however, does not mean that we can find a single level (being it the lower, the central, or the higher) from which all the explanations automatically derive. Looking for the level maximizing nontrivial determinism must be intended as a proposal for looking at the place 'where to start the investigation' and its primacy is in terms of the descriptive enterprise and not in terms of the particular piece of world, by the way we are convinced a physical reality exists out there independently of our models. On the contrary, the regularity and functionality of a level may be dependent in different ways on those at a lower or higher level. This is clearly the case in many biological morphogenetic phenomena of cellular differentiation or pathogenesis, such as cancer. These processes show that in nature there are relational factors that play a crucial and directional role from a systemic perspective. This is the reason why biological theories often appear as "collections of prototypical inter-level causal models largely qualitative that can be extrapolated by analogy to different organisms" (Schaffner 2007) and it is also the reason for which an analytical approach furnishes important information about biological systems and their organization and operation.

From this, however, the possibility or even desirability of an inter- or intra-theoretical reduction cannot be logically derived, but only the possibility and reality of their epistemological and methodological relationship can be inferred. The discussion of the terms, which are involved in this framework, goes beyond this contribution and requires a more detailed analysis of the properties that define the levels of explanatory and causal attribution. The data presented so far, however, allow the rejection of easy simplifications of epistemological issues in biological sciences.

6. Beyond the Dichotomies: From Model Choice to the Framing of the Problem

The contribution of Systems Biology in contemporary science is thus mainly related to its epistemological assumptions, although often implicit, that have interesting impacts on methodological aspects as well. These epistemological assumptions depend on the organizational principles that Systems Biology seeks to explain. A clarification of this relationship might help us to overcome some aspects of the reductionism-anti-reductionism debate and to move forward in developing epistemological tools that are still needed to explain the hierarchical control of biological systems. When a SA underlies the explanatory model, even if used only implicitly by researchers, its use is, rather, a sort of self-correction. Fallibility of the procedures may become an epistemic advantage, as correctly noted by Wimsatt (Wimsatt 2007). The question of the mesoscopic level is fundamental in experimental research and its epistemological status is the real challenge for the philosophy of systems. It will contribute to a clarification of the organizational principles of biological systems and to an explanation of the philosophical implications thereof.

The ultimate test of the adequacy of a SA is whether it implements and demonstrates the validity, through the developing of experimental models, of different methodological approaches. When asked then what the most appropriate model is, the answer is, "it depends".

This diversity of approaches, however, does not mean relativism of response, because the rationality of the explanation does not belong to the model itself, but to the model as representative of a phenomenon adequately identified. As a matter of fact, the possibility to obtain a straightforward and completely data-driven quantification of the amount of nontrivial determinism allows us to give a strong factual basis to our choices. Nevertheless the optimal mesoscopic scale (in terms of maximum nontrivial determinism) can be found at different organization layers: it is the integrative dimension of the SA, as presented above, that argues for this pluralism[3].

These considerations allow us to appreciate the fundamental epistemological importance of systemic explanations, which integrate an analytical and synthetic approach. They allow us to understand distinctly and isolate from the rest and therefore consider scientifically as an object of investigation, the level of organization that in reality is integrated into a hierarchical organization. This is also why those who maintain an anti-reductionist epistemology consider Systems Biology to be predicated on the insufficiency of mechanistic epistemology to cover and explain emergent functional properties and behaviours of living systems. To pretend that technological innovations will enable the understanding of these complex phenomena may just be wishful thinking. Instead, we need a novel way of thinking about these problems that takes all levels of biological organization into consideration.

References

Bender A., Glen R.C. (2005) A Discussion of Measurement of Enrichment in Virtual Screening: Comparing the Information Content of Descriptors with Increasing Levels of Sophistication. *J.Chem.Inf.Model.* (45): 1369-1375.

Benigni R. and Giuliani A. (1994) Quantitative modeling and biology: The multivariate approach. *American Journ. of Physiol.* (266) (35), R1697-R1704.

Bertolaso M. (2009) Towards an Integrated View of the Neoplastic Phenomena in Cancer Research, *Hist. Phil. Life Sci.,* 31:79-98.

Bertolaso M. (2011), "Hierarchies and Causal Relationships in the Interpretative Models of Cancer", in Special Issue "Causation and Disease in the Post-Genomic Era", *Hist. Phil. Life Sci.*, vol. 33(4): 389-396.

Bertolaso M. (2013) La Dimensione Non-Riduzionista del Riduzionismo nella Ricerca Sperimentale Dai Modelli Molecolari a Quelli Sistemici nella Ricerca sul Cancro. Rivista Di Filosofia Neo-Scolastica.

Bizzarri M., Giuliani A., Cucina A., D'Anselmi F., Soto AM., Sonnenschein C. (2011) Fractal Analysis in a Systems Biology Approach to Cancer, "*Seminars in Cancer Biology*, 21:175-82, 176.

Boogerd F. C., Bruggerman, F. J., Hofmeyr, J. S., Westerhoff, H. V. et. al. (2007) Systems biology: *Philosophical foundations*. New York, 123 – 144.

Callaghan P.T. (1991) *Principles of Nuclear Magnetic Resonance Microscopy*. Clarendon Press Oxford).

[3] Mitchell (2009) has offered a fruitful analysis of pluralism in scientific research and we believe that the data we have presented here as well as their epistemological discussion can easily seen in accordance with it.

Csermely P., Agoston V., Pongor S. (2005) The efficiency of multi-target drugs: the network approach might help drug design. TRENDS in *Phram.Sci.* (26): 178-182.

Di Paola L., De Ruvo M., Paci P., Santoni D., Giuliani A. (2013) Protein Contact Networks: an Emerging Paradigm in Chemistry. *Chemical Reviews* DOI: 10.1021/cr3002356, in press.

Dokholyan N.V., Buldyrev S.V., Stanley E.H., Shakhnovic E.I. (2000) Identifying the Protein Folding Nucleus Using Molecular Dynamics. *J. Mol.Biol.* 296: 1183-1188.

Frauenfelder H., Sligar S.G., Wolynes P. (1991) The energy landscapes and motions of proteins. *Science* (254): 1598-1603.

Gerstein M.B., Kundaje A., Hariharan M. et al., (2012) Architecture of the human regulatory network derived from ENCODE data. N*ature*, (489): 91-100.

Giuliani A. (2010) Collective motions and specific effectors: a statistical mechanics perspective on biological regulation. *BMC Genomics* 11 (suppl. 1): S2.

Huang S., Eichler G., Bar-Yam Y. et al.,. (2005) Cell fates as high-dimensional attractor states of a complex gene regulatory network. *Phys. Rev. Lett.* (94): 128701-5.

Huang S., Ingber D.E. (2000) Shape-dependent control of cell growth, differentiation, and apoptosis: switching between attractors in cell regulatory networks. *Exp. Cell Res.* 261: 91-103.

Huang S. (2009) Reprogramming Cell Fates: reconciling rarity with robustness. *BioEssays* 31: 546-560.

Hyman A.A., Simons K. (2012) Beyond Oil and Water: Phase Transitions in cells.*Science*, (337): 1047-49.

Jonkers F.C., Jonas J.C., Gilon P., Henquin J.C. (1999) Influence of cell number on the characteristics and synchrony of Ca2+ oscillations in clusters of mouse pancreatic islet cells. *J. Physiol.* 520.3:839-849.

Kauffman S.A. (1993) The Origins of Order. New York: Oxford University Press.

Kitano, H. (2002) Systems biology: A brief overview. *Science* 295: 1662-1664.

Laughlin R.B., Pines D. et al., (2000) The middle way. *Proc.Natl.Acad. Sci. USA* (97): 32-37.

MacFarlane R.G. (1964) An Enzyme Cascade in the Blood Clotting Mechanism, and its Function as a Biochemical Amplifier. *Nature* 202: 498-499.

Minati G. and Licata I. (2012) Meta-structural properties in collective behaviours, *International Journal of General Systems,* DOI:10.1080/03081079.2011.651136.

Mitchell SD. *Unsimple truths. Sciene, complexity and policy*. The University of Chicago, 2009.

Nelson C.M., Bissell M.J. (2006) Of extracellular matrix, scaffolds, and signaling: tissue architecture regulates development, homeostasis, and cancer. *Annu. Rev. Cell Dev. Biol.* 22: 287-309.

O'Malley M., J. Duprè (2005) Fundamental issues in systems biology, *Bioessays*, 27:1270-1276.

Palumbo M.C., Colosimo A., Giuliani A., Farina L. (2005) Functional Essentiality from Topology features in metabolic networks: a case study in yeast. *FEBS Letters.* (579): 4642-4646.

Palumbo M.C., Colosimo A., et al., (2007) Essentiality is an emergent property of metabolic network wiring. *FEBS Letters* (581) (13): 2485-2489.

Pascual M., Levine S.A. (1999) From Individuals to Population densities: Searching for the Intermediate Scale of Nontrivial Determinism. *Ecology* 80(7): 2225-2236).

Plankar M., Del Giudice E., Tedeschi A., Jerman I. (2012) *The Role of Coherence in a Systems View of Cancer Development*, vol. 2 (in press).

Portuesi R., Cherubini C,. Gizzi A., Buzzetti R., Pozzilli P., Filippi S. (2012) A stochastic mathematical model to study the autoimmune progression towards type 1 diabetes". *Diabetes Metab Res Rev*. Dec 10. doi: 10.1002/dmrr.2382. [Epub ahead of print].

Rew DA (2000) Modelling in surgical oncology. Part III: massive data sets and complex systems. *Eur. J. Surg. Oncol*. 26: 805-809.

Richardson, R. C., and Stephan, A. (2007) Mechanism and mechanical explanation in systems biology. In Boogerd F. C., Bruggerman, F. J., Hofmeyr, J. S., Westerhoff, H. V. et. al. (2007) *Systems biology: Philosophical foundations*. New York, 123 – 144.

Schaffner K.F.2007 Theories, models and equations in systems biology. In Systems Biology. Philosophical foundations, Boogerd F.C., Bruggeman F.J., Hofmeyer J. H. S., Westerhoff H.V. (Eds) *Elsevier*, p. 145.

Sherman A., Rinzel J. (1991) Model for synchronization of pancreatic beta cells by gap junction coupling. *Biophys. J.* 59:547-559.

Sonnenschein C., Soto A.M. (1999). *The Society of Cells: Cancer and Control of Cell Proliferation*. Springer-Verlag Inc, New York.

Tsuchiya M., Piras V., Giuliani A. et al.,.(2010) Collective dynamics of specific gene ensembles crucial for neutrophil differentiation: the existence of genome vehicles revealed. *PLoS ONE*, 5(8): e12116.

Tun K., Menghini M., D'Andrea L., Tanaka H., Dhar P., Giuliani A. (2011) Why so few drug targets: a mathematical explanation ? *Current Computer Assisted Drug Design* 7/3: 206-213.

Ushijima T. (2007) Epigenetic Field for Cancerization. *Journal of Biochemistry and Molecular Biology* 40: 142-150.

Venet D., Dumont J.E., Detours V. (2011) Most random gene expression signatures are significantly associated with breast cancer outcome. *PLoS. Comp. Biol.,* 7(10): e1002240.

Vitiello G. (2001) *My Double Unveiled, Advances in Consciousness Research*, John Benjamins Publishing Company, Amsterdam/Philadelphia.

Watts D.J., Strogatz S.H. (2004) Collective dynamics of 'small world' networks. *Nature Rev. Genet.* (5): 101-113.

Wimsatt W.C. (2007) Re-Engeneering Phoilosphy for Limited Beings: Piecewise Approximations to Reality. Cambridge: Harvard University Press, ch. 6.

Woodward J. (2003), *Making things happen – A theory of causal explanation*. Oxford University Press, p. 22.

Woodward J. (2010). Causation in Biology: Stability, Specificity, and the Choice of levels of Explanation. *Biology and Philosophy,* 25 (3).

In: Recent Advances in Systems Biology Research
Editors: A. X. C. N. Valente, A. Sarkar and Y. Gao
ISBN: 978-1-62948-736-6

Chapter 3

Mathematical Framework for Analyzing Genome-Wide Association Study Data with Rational Classes

A. X. C. N. Valente[1,2,3,*], Abhijit Sarkar[4,#], Joo Heon Shin[5], Jie Wu[5] and Yuan Gao[5]

[1]Systems Biology Group, Biocant - Biotechnology Innovation Center, Cantanhede, Portugal
[2]CNC - Center for Neuroscience and Cell Biology, University of Coimbra, Coimbra, Portugal
[3] Institute of Fundamental Medicine and Biology, Kazan Federal University, Kazan, Russia
[4]Physics Department and Vitreous State Laboratory, Catholic University of America, Washington DC, US
[5]Lieber Institute for Brain Development, Johns Hopkins Medical Campus, Baltimore, Maryland, US

Abstract

Genome-wide association studies can statistically expose genes and biological processes involved in disease mechanics. Development of algorithms that increase the statistical power of genome-wide association studies remains an important research challenge. We present a mathematical approach to analyze a genome-wide association study. The method enables the use of Rational Classes to biologically direct the search for biomarkers and thus increase statistical power. Ranking of candidate biomarkers avoids the need for a Bonferroni-type correction. A modified q-value is introduced to provide a false-discovery rate type of measure in this setting.

[*] Corresponding author email - andre.valente@biocant.pt
[#] Co-Corresponding author email - SARKAR@cua.edu

Introduction

A major approach for the identification of genetic risk factors to disease is the genome-wide association study (GWAS) [1]. Analytical methods that increase the statistical power of the GWAS are essential as therein lies a fundamental limitation of the approach [2]. A number of research avenues are being pursued in this regard. Prior biological knowledge [3,4], for example in the form of gene expression [5,6], biological pathway [7,8], or protein interaction data [9,10], is being exploited. There are proposals for multilocus association tests, that combine data from physically close SNPs in linkage disequilibrium [11]. There is ongoing work on improving measures of statistical significance in the GWAS context [12,13,14,15,16]. Finally, as an alternative to p-value frequentist methods, research is active on Bayesian schemes [17,18].

We have recently introduced the concept of Rational Class to translate biological thought into increased statistical analytical power [19], in particular in the gene expression analysis context [20]. A Rational Class is a set of candidate biomarkers that share a common underlying rationale.

In here, we present a GWAS data analysis method that enables the use of Rational Classes. This permits introduction of biological considerations to guide the search for biomarkers, thus increasing statistical power. An unbiased assessment is obtained from a proper mathematical treatment of the associated multiple hypotheses testing. Overall, the method allows a targeted search with an unbiased assessment. The approach has been already successfully applied to analyze a Parkinson's disease GWAS in Valente et al., [21]. Specific Rational Classes and their construction are described in that article. In here, the focus is instead on presenting the underlying mathematical approach. In particular, in this article we show how the method: i) enables testing of multiple Rational Classes, ii) renders a Bonferroni-type correction unnecessary via appropriate interpretation of the ranking of candidate biomarkers and iii) provides a modified q-value false-discovery rate type of measure.

Analysis

The GWAS

In a GWAS, a cohort of healthy and diseased individuals (or of any relevant distinct phenotypes) is genotyped at a dense set of SNP sites across the genome. Mathematically, this assigns the individuals to locations within a high-dimensional SNP space (Figure 1). The objective is then to associate regions of the SNP space with a differential disease susceptibility.

Large enough differential susceptibilities have practical prognostic value. So far though, most differentials found have been too small to be relevant in this sense [22]. However, such statistical associations can still be invaluable as indicators of the potential involvement of particular genes or biological processes in the disease mechanics.

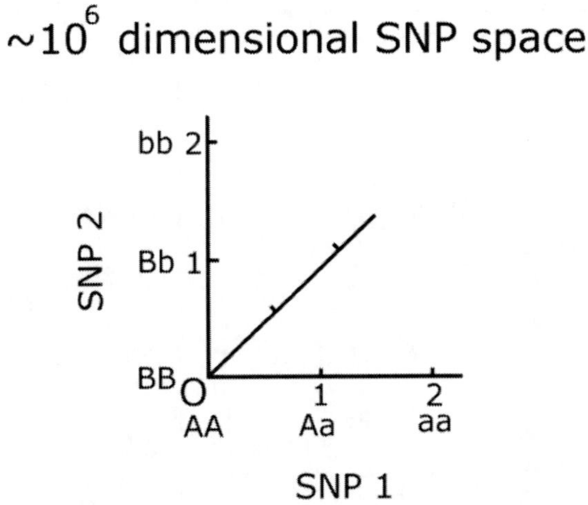

Figure 1. SNP space. In a genome-wide association study (GWAS), individuals are vectors in SNP space. The GWAS objective is to associate regions of the SNP space with a differential disease risk. Depicted is one sensible coordinate system for SNP space (capital letters represent the major allele, lower case letters the minor allele; to each SNP corresponds an axis with 3 admissible values, 0, 1 and 2).

Caveats of the GWAS Statistical Association

A statistical association does not guarantee direct causality between genotype and phenotype. A major source of non directly causal associations is population heterogeneity in the analyzed cohort [23]. A number of mathematical techniques exist to mitigate the problem [24]. For instance, we recently proposed homogenizing a cohort by knocking-down the statistical weight of select cohort individuals [25]. In here, we assume that either the population has been well-mixed over recent generations, or the sampled cohort has been homogenized as suggested above. Even then, other indirect explanations for a statistical association cannot be ruled out. An elementary example is a haplotype that increases propensity for a certain behavior which in turn alters the risk of disease.

Mathematical Problem Setup

The differential disease risk between two SNP space regions can be quantified via the Odds Ratio (OR) [26]. In this setting, the OR is the ratio of the odds of disease for individuals in one region, to the odds of disease for individuals in the other region. Note that swapping the regions inverts the value of the OR. Hereafter, we follow the convention of always employing the equal or larger than unity value of an OR. In a set of candidate region pairs (splits, for shorthand) the objective is to identify those with a biologically noteworthy non-unity OR (i.e., an OR of 1.0001 is unlikely to be interesting biologically) (Figure 2). We label these splits as G ('good') and label the remaining splits as B ('bad').

Now, it is not possible to measure the actual OR of a split. Only an *estimated OR* is available, based on the cohort sampled from the population. The finite cohort size introduces a stochasticity in the OR estimates.

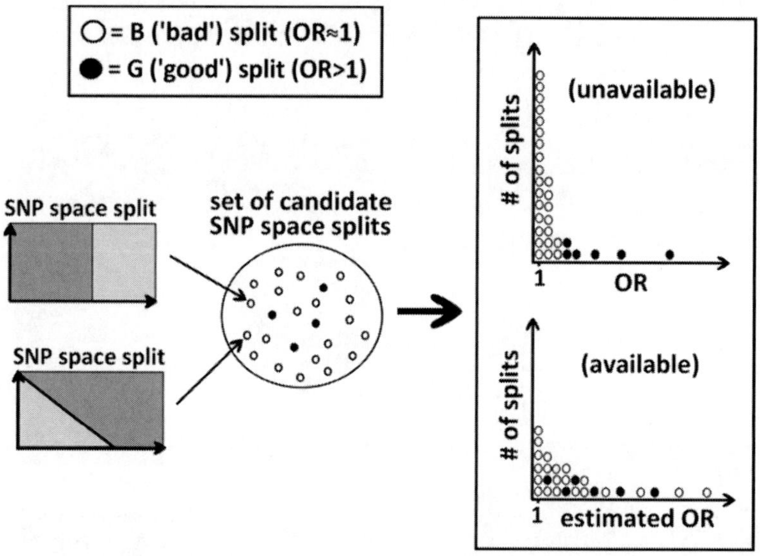

Figure 2. Mathematical problem set up. A SNP space region pair (SNP space split) has an associated Odds Ratio (OR) that quantifies the differential disease risk between the two regions. We use the convention of always utilizing the equal or larger than unity value of an OR. In a set of candidate splits, the objective is to identify the G ('good') splits, with a biologically noteworthy non-unity OR, amongst the B ('bad') splits, with an approximately unity OR. However, stochasticity in the OR estimates due to the finite cohort size, may lead to the top ranked splits by estimated OR being Bs.

The stochasticity is further increased by genotype and phenotype assignment errors (although these are usually not overly significant in a GWAS [27]). Altogether, because of stochasticity, given a large enough number of Bs in the set of candidate splits, the largest estimated OR splits will be Bs. Herein, lies the difficulty in positively identifying G splits in a GWAS.

Mixing Level

One approach to the problem is replacing the OR estimate with a better metric at separating Gs from Bs. In here we adopt a standard choice, the Fisher exact test p-value [28]. This associates with a split the probability (p-value) that, assuming the OR is unity, a cohort (preserving the present net number of sampled subjects in each SNP space region and of each phenotype) would produce an OR estimate equal or larger than the observed one. In Figure 3, we depict -log(p-value), which is monotonically related to the p-value metric, but conveniently visually inverts the p-value scale and expands the relevant low p-value range. In this figure, the B split with the highest -log(p-value) score sets the mixing level. Above the mixing level, all splits are Gs. At or below the mixing level, Bs can also be found. Thus, only above the mixing level is a split guaranteed to be a G. Any selected {set of candidate splits, ranking metric} pair produces a mixing level, defining these two distinct regions (convention: the Gs-only region is the one *above* the mixing level). With the OR estimate as the ranking metric, a B split containing by chance a single cohort individual in one of its SNP space regions would have infinity for its OR estimate, thus placing the mixing level also at infinity.

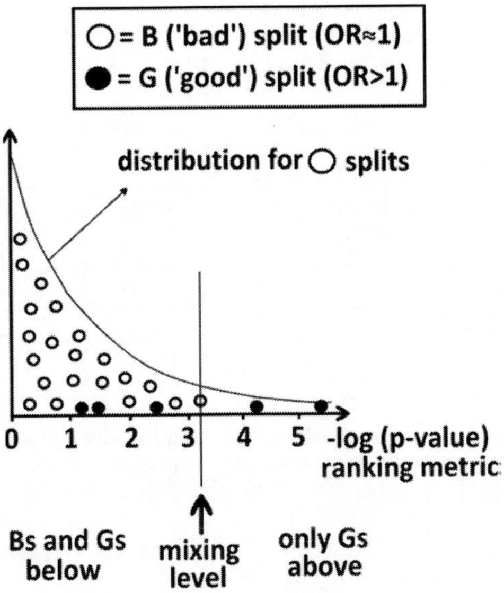

Figure 3. Mixing level. Each split in the set of candidate splits is assigned a Fisher exact test p-value, based on the cohort data. The sketch shows -log(p-value) on the x-axis, which is monotonically related to p-value, but visually conveniently inverts the p-value scale and expands the low p-value range. The B split with the highest score sets the mixing level. Above the mixing level, all splits are Gs. At or below the mixing level, both G and B splits may be found. The B split distribution on this -log(p-value) scale follows an exponential decay. However, the correlations amongst the Bs make it difficult to estimate theoretically the location of the mixing level.

This problem is avoided with the p-value as the ranking metric. Note that larger cohort sizes tend to yield smaller p-values for the Gs, thus in the Figure 3 depiction raising the Gs, rather than affecting the mixing level.

Split Ref-P Assignment

We would like to estimate the chance that a given split is at or below the mixing level. The split Šidák adjusted p-value, i.e., $1-(1-\text{p-value})^k$, where k is the total number of splits in the set of candidate splits, effectively provides such an estimate (shorthand: Šidák p-value) [29]. A small Šidák p-value is well approximated by the more common Bonferroni adjusted p-value, i.e., p-value*k. Use of the Šidák p-value relies on the following three approximations. First, all Bs are assumed to have exactly unity OR. This is valid as, by continuity, there is no practical difference between a unity or a marginally larger than unity OR. Second, it is assumed that all splits in the set of candidate splits are Bs. This assumption is conservative in the sense that it leads to overestimating the height of the mixing level. It is a good approximation for the typical set of candidate splits that contains many orders of magnitude more Bs than Gs. Third, it is assumed, similarly conservatively, that the p-values of the Bs are independent. This last approximation however can be significantly off, due to linkage disequilibria introduced in the past and perpetuated by SNPs in close proximity within a chromosome tending to travel together as they pass from parent to offspring [30]. A Šidák p-value may therefore considerably overestimate the height of the mixing level.

○ = B split (OR≈1)
● = G split (OR>1)

Top-B → (mixing level): rank 3

Rank	Split	p1-value	p2-value	ref-p
1	●	$2*10^{-6}$	$7*10^{-4}$	$7*10^{-4}$
2	●	$5*10^{-5}$	$1*10^{-4}$	$7*10^{-4}$
3	○	$1*10^{-4}$	0.2	0.2
4	○	$3*10^{-4}$	0.01	0.2
5	○	$5*10^{-4}$	0.5	0.5
6	●	$9*10^{-4}$	0.02	0.5

⋮
etc.

Figure 4. Split ref-p assignment. The original cohort is divided into two cohort sets. Splits are ranked by p1-value, the p-value based on the first cohort set. The highest ranked B split (top-B) sets the mixing level. Splits are then assigned a second p-value (p2-value), based on the second cohort set. The reference p-value (ref-p) of a split is the largest p2-value amongst the p2-value assigned to it and the p2-values assigned to splits ranked above it. The split ref-p is used as a basis to reject (or not) the null hypothesis that the split is at or below the mixing level. Values shown are for illustrative purposes.

In here, we suggest an alternative approach. It associates with every split a p-value (denoted below as ref-p, for distinction) to serve as a basis for rejecting an assumed null hypothesis that the split is at or below the mixing level (Figure 4). We partition the original population cohort into two cohort sets. The first set is utilized to generate the p-value ranking of the splits. In this ranked list the mixing level is set by one specific split, the highest ranked B (top-B). Based on the second cohort set, a second p-value (p2-value) is assigned to each ranked split. In particular, this assigns a specific p2-value to top-B. Now, the null hypothesis that a split X is at or below the mixing level can equally be stated as top-B being X or a split ranked above X. If top-B is X or a split ranked above X, then the p2-value of top-B can be at most as large as the largest p2-value amongst those assigned to X and the splits ranked above X. We call this largest p2-value the reference p-value (ref-p) for split X. Thus, ref-p provides a basis for rejecting (or not) the null hypothesis. It remains safely conservative, given that it is only an upper bound on the p2-value of top-B (under the null hypothesis).

Smaller cohorts tend to increase the p-value of a G split. In the first stage, a larger p-value decreases ranking, possibly pushing a G below the mixing level. In the second stage, for a G above the mixing level, a large p2-value prevents ruling out that the G is a B and hence that the mixing level is at that height. Note that in lieu of the elementary partition of the original cohort into two disjoint cohort sets, a resampling-recursive implementation akin to the one implemented in the Rational Classes analysis of gene expression data could be considered [19,20]. However, this would have to be balanced against the high computational time demand created by the typically large sets of candidate splits in a GWAS [21]. A fundamental advantage of the ref-p rank-based approach is that it obviates the need for a Šidák/Bonferroni correction.

Rational Classes (RCs)

The choice of the {set of candidate splits, ranking metric} pair determines the mixing level and the G splits that surpass it. We have adopted the standard p-value as the split ranking metric. We now discuss the set of candidate splits. The unbiased, hypothesis-free set of all mathematically possible splits of SNP space is clearly a bad choice as, even though by definition it contains every G split, the much larger number of B splits will place those G splits below the mixing level. Therefore, a more restricted, biologically minded set of candidate splits must be chosen. Now, different biological rationales and hypotheses can lead to different sets of candidate splits, or Rational Classes (RCs) [19]. It is advantageous to analyze these RCs separately because their distinct underlying rationale ensures likely distinct mixing levels/sets of Gs above the mixing level (Figure 5). Conversely, a merging of the RCs could result in the highest individual RC mixing level drowning G splits that were above the mixing level of their original RC.

Note that an insightful biological rationale for dividing a RC into two separate RCs can be valuable even if no a priori preference is assigned to one of the new RCs over the other. It is enough that the biological rationale likely results in distinct mixing levels/sets of Gs above the mixing level for the now separate RCs. This is so, since one of these new RCs may have Gs above its mixing level that would be below the mixing level of the other (and hence below the mixing level of a joint RC).

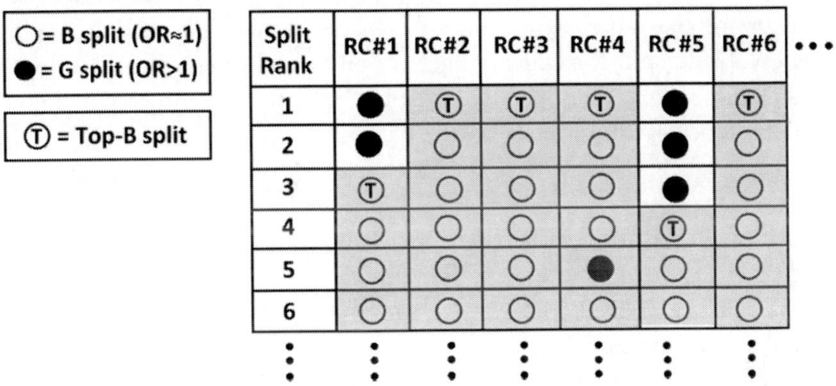

Figure 5. Rational Classes (RCs). Different Rational Classes (RCs), i.e., sets of candidate splits, are kept separate to prevent the highest individual RC mixing level from covering G splits that may otherwise be above the mixing level of their RCs.

Split Q-Value Assignment

The analysis of multiple RCs constitutes a form of multiple testing. Therefore, when analyzing multiple RCs, the significance of a split cannot be assessed solely based on its ref-p value. In this section, we adapt the q-value approach [16] to make it applicable to our framework. A split's *adapted q-value* (henceforth: q-value) provides a false discovery rate when calling that split significant, in the sense we now explain. The q-value of a split X is defined and calculated as follows. Place the ref-p threshold for significance at the ref-p value of split X. Then, over all the RCs, the splits with a ref-p value equal or smaller than the ref-p of X are considered significant. Let *# of significant RCs* be the number of RCs containing at least one (presumed) significant split. Let *# of B significant RCs* be the estimated number of these RCs where at least one B split was (incorrectly presumed) significant. Then we define the temporary q-value of X as *# of B-significant RCs / # of significant RCs*. Finally, the q-value of X is defined as the smallest temporary q-value amongst the temporary q-value of X and the temporary q-values of splits with a ref-p larger than the ref-p of X. This last step ensures that the ranking of splits based on q-values matches the ranking based on ref-p values. It remains to provide an estimate for *# of B-significant RCs*. Over all the RCs, the number of top-B splits with a p2-value equal or smaller than the ref-p of X can be estimated as *total number of RCs * ref-p of X*. This is also a conservative overestimate for the *# of B-significant RCs*. To see this, note that a RC that contain at least one B significant split, must contain its top-B split as significant and for that top-B to be significant, its p2-value must be below ref-p of X. It is an overestimate because a top-B split having a p2-value below ref-p of X, does not ensure that top-B was deemed significant. Namely, the ref-p of that top-B split could still be larger than the ref-p of X due to a G split ranked higher in the RC than that top-B having a larger p2-value than the ref-p of X.

We emphasize that the q-value of X, i.e., the estimated *# of B-significant RCs / # of significant RCs* when calling X significant, is not the probability that X is a B split, or even the probability that X is in a RC containing a B significant split. In particular, note that q-value calculation is based on X having the largest ref-p value amongst the splits being called significant. Therefore the probability of X being a B split may be substantially higher than that probability for the other significant splits. Finally, a liberal choice was made when defining the q-value of split X as the minimum of all temporary q-values of splits with higher ref-p than X. These limitations parallel exactly those found with the traditional q-value measure and discussed by Storey et al., [16].

Assignment of Q-Values Under Ranked Rational Classes

In practice, RCs believed to have the most potential are generally tested early on. Arguably, it may thus be desirable to have a non-uniform ref-p threshold for significance, more accommodating in earlier tested RCs and progressively stricter as more and more RCs are proposed and tested. In this section, we show how to assign q-values under such a scenario. Let the RCs be ranked, reflecting the order in which they have been proposed and tested so far. Let the permitted, gradually stricter, ref-p significance threshold curve have the form C/r, where r is the rank of the RC and C is the constant that allows setting of the threshold curve at the desired stringency. We calculate the q-value for a split X in a RC ranked r_x, as follows. For split X to be considered significant, the ref-p significance threshold curve at r_x must be at least ref-p of X. This sets $C=r_x$*ref-p of X*, fixing the ref-p significance threshold curve for the q-value of X calculation. At another RC ranked r_y, the ref-p threshold is now C/r_y. All splits above the ref-p threshold at their RC are considered significant. The *# of significant RCs* is thus determined. The *# of B-significant RCs* is in turn given by the sum of all the RC thresholds, by an argument parallel to the one used in the previous section. As before, the split temporary q-value is then the ratio *# of B-significant RCs / # of significant RCs*. The final split q-value is defined as the smallest temporary q-value amongst the temporary q-value of X and the temporary q-values of all the splits requiring a less stringent (i.e., with a larger C constant) ref-p threshold curve for significance than split X requires.

Application to a Parkinson's Disease GWAS

We have applied our approach to the analysis of a Parkinson's disease (PD) GWAS in Valente et al., [21]. The analyzed NGRC cohort of Hamza et al., [31] consisted of 2000 sporadic late-onset PD patients and 1986 control subjects. All individuals were North Americans of self-reported European descent. As a pre-processing step, cohort heterogeneity (see *Caveats of the GWAS statistical association* Section) was further decreased via subject weight knock-down [25]. This reduce the dataset to a net weight of 1904 PD patients and 1802 controls (a 7% reduction). A total of twenty three RCs were created. The construction of the splits and RCs is described in Valente et al., [21]. The RCs were analyzed in ordered fashion with a gradually stricter split significance threshold proportional to 1/r, where r is the rank of the RC, as described in the previous Section. The analysis yielded the following potential genetic associations with PD: 19 SNPs in the 17q21.31 'MAPT' region (q-values 0.08 to 0.11), 3 SNPs in the 4q22.1 'SNCA' region (q-values 0.07 to 0.08), 1 SNP in the 6p21.3 'HLA-DRA' region (q-value 0.08) and 1 coding non-synonymous SNP in the 13q32.1 DZIP1 gene (q-value 0.07). Note that the number of SNPs associating a chromosomal region with PD likely depends on the degree of correlation between SNPs in that region. Thus, it should not be equated with degree of evidence for the association. The q-values herein reported are not available in Valente et al., [21], as we had not yet adapted the q-value approach to our framework at the time (instead, a 'reference probability' consisting of the split ref-p value multiplied by the RC rank is reported in that article). The association of the MAPT, HLA-DRA and SNCA regions with PD had been previously reported. The novel finding was the increased susceptibility to PD conferred by the rare (0.7% allele frequency in the cohort), coding, non-synonymous SNP variant in the DZIP1 gene. Besides its potential as a prognostic marker, biologically this association highlighted for the first time in a genetic association study a possible connection between stem-cell regulation and PD [20].

Discussion

In this article we presented a mathematical approach to analyze a Genome-Wide Association Study (GWAS). We began by mathematically framing the objective of a GWAS, in terms of identifying SNP space splits with a biologically noteworthy non-unity Odds Ratio (OR). The challenge lies in identifying such G 'good' splits amongst a large pool of B 'bad' splits. The notion of mixing level was introduced to capture the statistical dynamics presented by this problem. The pair {set of candidate splits, metric for ranking the splits} sets the mixing level and hence the ability to positively identify G splits. The Fisher test p-value provides a standard valid metric for ranking the splits. We showed how to associate with every split a ref-p value. The ref-p value of a split provides a basis for rejecting (or not) an assumed null hypothesis that the split is at or below the mixing level. We explained why an hypothesis-free type of approach is not viable and why biological knowledge and insights are essential. In our approach, biological thought is introduced into the mathematical analysis via the Rational Class (RC), a set of candidate splits with an underlying common biological rationale. We further described how utilizing multiple RCs, rather than a single joint one, can increase the statistical ability to detect G splits. The use of several RCs introduces again a problem of multiple testing. The q-value approach [16] was modified to assess significance of findings under this setting. Namely, we defined the q-value of a split X as the ratio of the estimated number of RCs where at least one B split is incorrectly declared significant to the number of RCs containing at least one declared significant split, assuming every split with a lower ref-p than X is declared significant (see the *Split q-value assignment* Section for a minor adjustment to this definition that ensures split ranking by ref-p and by q-value is identical). This proposed q-value therefore provides a false discovery type of assessment for findings under our GWAS analysis framework. We also proposed a modification to this q-value calculation that enforces a non-uniform significance threshold across RCs. This is motivated by arguably in practice more promising RCs tending to be tested early on. With the RCs ranked by testing order, we thus let the ref-p significance threshold stringency increase with RC rank. Finally, q-values calculated using this non-uniform threshold were reported for the main findings in a Parkinson's disease GWAS analyzed under this framework [21].

Given this framework, a successful GWAS analysis then hinges on the quality of the proposed RCs, i.e., on the biological knowledge and insight that goes into constructing them. We thus expect most interesting developments in future applications of the approach to be centered on the proposed RCs. This article instead focused exclusively on presenting the underlying framework. In this regard, development of computationally tractable improved resampling/ recursive methods for partitioning of the available cohort remains a relevant open research problem.

Acknowledgments

A.V. would like to acknowledge the support of the Center for Neuroscience and Cell Biology (project PEst-C/SAU/LA0001/2013-2014) and of Portuguese national funds via the programs FEDER and COMPETE and via Fundação para a Ciência e a Tecnologia.

References

[1] Manolio TA (2010) Genomewide association studies and assessment of the risk of disease. *New Engl. J. Med.* 363 (2): 166-176.

[2] Wang WYS, Barratt BJ, Clayton DG, Todd JA (2005) Genome-wide association studies: theoretical and practical concerns. *Nat. Rev. Genet.* 6: 109-118.

[3] Li MX, Sham PC, Cherny SS, Song YQ (2010) A knowledge-based weighting framework to boost the power of genome-wide association studies. *PLoS One* 5 (12): e14480.

[4] Chasman DI (2008) On the utility of gene set methods in genomewide association studies of quantitative traits. *Genet Epidemiol.* 32 (7): 658-668.

[5] Hsu YH, Zillikens MC, Wilson SG, Farber CR, Demissie S (2010) An Integration of Genome-Wide Association Study and Gene Expression Profiling to Prioritize the Discovery of Novel Susceptibility Loci for Osteoporosis-Related Traits. *PLoS Genet* 6(6): e1000977.

[6] Xiong Q, Ancona N, Hauser ER, Mukherjee S, Furey TS (2011) Integrating genetic and gene expression evidence into genome-wide association analysis of gene sets. *Genome Res:* Epub ahead of print.

[7] Bakir-Gungor B, Sezerman OU (2010) A new methodology to associate SNPs with human diseases according to their pathway related context. *PLoS One* 6 (10): e26277.

[8] Wang K, Li M, Hakonarson H (2010) Analysing biological pathways in genome-wide association studies. *Nat. Rev. Genet.* 11: 843-854.

[9] Pan W. Network-based model weighting to detect multiple loci influencing complex diseases (2008) *Hum. Genet.* 124: 225-234.

[10] Jia P, Zheng S, Long J, Zheng W, Zhao Z. dmGWAS: dense module searching for genome-wide association studies in protein–protein interaction networks (2011) *Bioinformatics* 27(1): 95-102.

[11] Gauderman WJ, Murcray C, Gilliland F, Conti DV (2007) Testing association between disease and multiple SNPs in a candidate gene. *Genet Epidemiol.* 31(5): 383-395.

[12] Johnson RC, Nelson GW, Troyer JL, Lautenberger JA, Kessing BD (2010) Accounting for multiple comparisons in a genome-wide association study (GWAS). *BMC Genomics*. 11: 724.

[13] Galwey NW (2009) A new measure of the effective number of tests, a practical tool for comparing families of non-independent significance tests. *Genet Epidemiol.* 33(7): 559-68.

[14] Han B, Kang HM, Eskin E (2009) Rapid and accurate multiple testing correction and power estimation for millions of correlated markers. *PLoS Genet* 5(4): e1000456.

[15] Benjamini Y, Hochberg Y (1995) Controlling the false discovery rate: a practical and powerful approach to multiple testing. *J. R. Statist. Soc. B.* 57 (1): 289-300.

[16] Storey JD, Tibshirani R (2003) Statistical significance for genomewide studies. *Proc. Natl. Acad. Sci. USA* 100 (16): 9440-9445.

[17] Stephens M, Balding DJ (2009) Bayesian statistical methods for genetic association studies. *Nat. Rev. Genet.* 10: 681-690.

[18] Wakefield J (2007) A bayesian measure of the probability of false discovery in genetic epidemiology studies. *Am. J. Hum. Genet.* 81: 208-227.

[19] Valente AXCN (2011) Prediction in the hypothesis-rich regime. In: Valente AXCN, editor. Science and engineering in high-throughput biology including a theory on Parkinson's disease. *Lulu Books*. pp. 23-38.

[20] Valente AXCN, Sousa JAB, Outeiro TF, Ferreira L (2011) A stem-cell ageing hypothesis on the origin of Parkinson's disease. In: Valente AXCN, editor. Science and engineering in high-throughput biology including a theory on Parkinson's disease. *Lulu Books*. pp. 43-73.

[21] Valente AXCN, Shin JH, Sarkar A, Gao Y (2012) Rare coding SNP in DZIP1 gene associated with late-onset sporadic Parkinson's disease. *Sci. Rep.* 2: art. no. 256.

[22] Hindorff LA, Sethupathy P, Junkins HA, Ramos EM, Metha JP, et al., (2009) Potential etiologic and functional implications of genome-wide association loci for human diseases and traits. *Proc. Natl. Acad. Sci. USA* 106 (23): 9362-9367.

[23] Marchini J, Cardon LR, Phillips MS, Donnelly P (2004) The effects of human population structure on large genetic association studies. *Nat. Genet.* 36: 512 - 517.

[24] Tiwaria HK, Barnholtz-Sloan J, Wineinger N, Padilla MA, Vaughan LK, et al., (2008) Review and evaluation of methods correcting for population stratification with a focus on underlying statistical principles. *Hum. Hered.* 66 (2): 67-86.

[25] Valente AXCN, Zischkau J, Shin J H, Gao Y, Sarkar A (2012). Genome-Wide Association Study Heterogeneous Cohort Homogenization via Subject Weight Knock-Down. PLoS ONE 7(10): e48653.

[26] Edwards AWF (1963) The measure of association in a 2x2 table. *J. R. Stat. Soc. Ser. A-G* 126 (1): 109-114.

[27] Miller MB, Schwander K, Rao DC (2008) Genotyping errors and their impact on genetic analysis. *Adv. Genet.* 60: 141-152.

[28] Fisher RA (1954) Statistical Methods for Research Workers. Oliver and Boyd.

[29] Abdi H (2007) Bonferroni and Šidák corrections for multiple comparisons. In: Salkind NJ, editor. *Encyclopedia of Measurement and Statistics.* Sage Publications.

[30] Reich DE, Cargill M, Bolk S, Ireland J, Sabeti PC, et al., (2001) Linkage disequilibrium in the human genome. *Nature* 411: 199-204.

[31] Hamza TH, Zabetian CP, Tenesa A, Laederach A, Montimurro J, et al., (2010) Common genetic variation in the HLA region is associated with late-onset sporadic Parkinson's disease. *Nat. Genet.* 42: 781–785.

In: Recent Advances in Systems Biology Research
Editors: A. X. C. N. Valente, A. Sarkar and Y. Gao
ISBN: 978-1-62948-736-6

Chapter 4

A Review and Perspective on Thermodynamics Applied to Molecular Systems and Cancer Biology

Edward A. Rietman, Philip Hahnfeldt and Lynn Hlatky
Center of Cancer Systems Biology, GeneSys Research Institute,
Tufts University School of Medicine, Boston, MA, US

"Ask not what physics can do for biology, ask what biology can do for physics."
Stanislaw Ulam

Abstract

We review the basics of thermodynamics showing that biological systems are robust dynamical systems able to rapidly process high-energy and dissipate it as heat, a lower-energy. This process takes place on an energy manifold and the steeper the slope, the more rapid the processing. We also show that evolution of complexity is a thermodynamic enabled process governed by the second law of thermodynamics. Finally, we describe how cancer is a thermodynamic driven process.

Introduction and Motivation

There is an obvious evolution of complexity from protons to atoms to molecules and from self-replicating molecular networks to metazoans (Chaisson, 2001; Jantsch, 1992; Bonner, 1988). The progression to complexity is strictly dictated by free energy degradation and entropy production (Demetrius, 1997; Schneider and Sagan, 2005). This results in what is called self-organization, or in essence structures formed in space and time, via symmetry breaking (Muller, 2007, Rietman, Karp, Tuszynski, 2011). The self-organized structures look like patterns in the space where the dynamics takes place and are able to channel free energy more rapidly than if they didn't exist (Schneider and Sagan, 2005). An example is Bernard cells and Taylor vortices (Nicolis and Prigogine, 1989) that increase the flow of free energy from their surroundings.

As the number of these self-organized structures increases and as the number of hierarchies increases the number of ways to distribute the energy increases (increasing the rate of entropy production and further breaking of symmetry). At the molecular level, this increase in number of ways to distribute energy, results in more types of molecules and more molecules of each type. At the microbiological level this results in a larger number of types of microbes (e.g., different bacterial and fungal species). Scaling this up we find at the ecological level, thermodynamics is a key factor associated with ecosystem formation and evolution (Demetrius, 2000; Brooks and Wiley, 1988; Gillespie, 1991; Schneider and Sagan, 2005). We argue that these concepts, combined with others, can be integrated to not only describe molecular and cellular biology more effectively, but that these concepts can be used to develop a sub-branch of condensed matter physics. Living systems are dynamical condensed matter structures. In keeping with Ulam's quote above, we feel that biology can become the foundation for a condensed matter field theory. In this review/perspective chapter we suggest that molecular network thermodynamics will be an important topic in this research area, and potentially contribute to both physics and clinical medicine. This review will always be from the perspective of molecular systems biology and will consider genome stability and cancer.

Biological systems are complex chemical networks with hundreds to thousands of molecular components interacting with each other in a timed sequence. The molecular network comprising a cell consists of protein, DNA, RNA, peptide, small molecule, lipid, and other molecular interactions. Associated with each of these molecules is a potential energy surface which contributes to a larger energetic network. When this network is perturbed with, for example a sugar molecule being transported into the cell, it will trigger a cascade of molecular reactions to bring the overall network back to an equilibrium, which in this case will likely be a complex limit cycle like a strange attractor (Hilborn, 1994; Gurel and Gurel, 1983). The driving force to push the reaction back to the limit cycle is a perturbation of Gibbs free energy expressed as concentration of sugar.

In order for the molecular network to evolve there must be a mutation in one or more of the molecular species. This mutation manifests as a change in potential energy and thus a perturbation of the overall molecular network energy landscape. If this "new" molecular network is able to utilize the Gibbs free energy more rapidly, it will be producing entropy at an increased velocity. This is analogous to Bernard cells being able to channel free energy across a thermal gradient more rapidly than a random configuration. *A symmetry break provides this acceleration.* Likewise if the mutation in the original molecular network now provides an *additional* molecular species without deleterious effects on the overall limit cycle it will essentially be an increase in the number of ways to dissipate free energy.

The drive to complexity thus comes about from symmetry breaking and an acceleration of entropy production across an energy gradient. That is the main thesis of this Perspective paper. To support these arguments we will describe analogies between molecular network thermodynamics and electronics. This is an advantageous marriage because extensive simulation and design tools for electronics may then be exploited to potentially design new molecular network systems. Prior to that discussion we will review kinetics and thermodynamics of simple polymer replicating networks, linear and circular chains of interacting molecular species (Kuppers, 1983; Eigen, 1971) and molecular network evolution (Demetrius (1997). Finally we will discuss genome stability from an energy and symmetry breaking perspective.

Note, because of the wide ranging subtopics reviewed we reuse symbols and redefine as needed.

Molecular Networks: Thermodynamics and Kinetics

Much has been written about molecular networks. Palsson (2006) in particular has done some very significant research on flux balance in biochemical and biological networks. His approach has been to use systems of differential and/or difference equations and solve the system on a computer.

Networks of this type are necessarily not very large because of computational resource limitations. Palsson's approach to circumvent this problem is to use linear transforms and extensive use of stoichiometry-connectivity matrices. His networks consist of a few hundred to a thousand chemical species and the computer code is used to look for convex surfaces in the reaction space. Beard and Qiun et al., (2004) describe an approach using mathematical objects known as matroids to describe biochemical reaction networks. As we will discuss, the two approaches are complementary.

Since our goal, in this chapter, is to present a thermodynamic perspective on molecular networks, we start with a simple differential equation system of chemical reaction networks, in particular one introduced by Eigen (1971) and Kuppers (1983). We follow a review by Casti (1992) and Rietman (1993) and will introduce only enough to make some points concerning thermodynamics. We start by looking at self-replicating information carriers, or quasi-species. This approach can be applied to cancer development by drawing on the work of Sole (2003), Sole and Deisboeck (2004) and Pastor-Santorras and Sole (2001). These authors show a phase transition, or error catastrophe, in mutation leading to aneuploidy.

Information is said to be the foundation of both life science (Avery, 2003; and Yockey, 2005) and physical science (von Baeyer, 2003; Seife, 2006; Muller, 2007; Davies and Gregersen, 2010).

Because information theory is probabilistic and Bayesian, this ontological assumption, of reality being based on information, implies that either probabilities have an objective reality, as essentially stated by quantum mechanics (Nielsen and Chuang, 2000; Leff and Rex, 1990; Zurek, 1990) or that our brains are inherently Bayesian (Coombes and Bressloff, 2005; Doya et al., 2007). Here we simply make the assumption that information carrier polymers (or genomes) are the key replicators associated with biological systems and we leave the debate to others (Fuchs, 2011).

Consider a very primitive autocatalytic molecular network consisting of $i = (1, \ldots, N)$ self-replicating entities (information carriers or genomes) of various types of quasi-species that are competing for resources in a semipermeable reaction vessel. The fraction of each type of information carrier (i.e., each quasi species or genome) is given by the vector,

$x=(x_1,...,x_N)$. Assume a function expressing the *fitness* of the molecular information carries exists, $f(x)\doteq(f_1,...,f_N)$. We normalize as follows

$$0\le x_i\le 1,\quad \sum_{i=1}^{N}x_i=1$$

so the flow describing the change of various quasi-species exists on a unit simplex S^N in $\Re^N$.

The rate of increase in molecular species *i*, is proportional to the current fraction of x_i and the fitness of $f_i(x)$ as compared to other molecular species, we can now write the rate equation as function of the fitness of that species minus the average population fitness.

$$\dot{x}_i=x_i\left[f_i(x)-\Phi(x)\right]$$

where

$$\Phi(x)=\sum_{j=1}^{N}x_jf_j(x)$$

is the average fitness of the population.

Let the fitness function *f*(*x*) be represented by only the linear terms, implying linear interactions only.

$$f_i(x)=(Ax)_i=\sum_{j=1}^{N}a_{ij}x_j$$

The terms in the matrix $A=\left[a_{ij}\right]$ provide interaction between molecular species *i* and *j*.

If we include rate of feedstock input (blood supply in the case of cancer cells) to the reaction vessel or tissue, rate of output, rate of utilization, rate of decay, mutation probability, selection pressure and assume constant population, we get the rate equation (Eigen, 1971; and Kuppers, 1983)

$$\dot{x}_i=\left(W_i^{cp}-\overline{E}\right)x_i+\sum_{i\neq j}a_{ij}x_i$$

where W_i^{cp} represents a rate function and the average excess production is given by

$$\bar{E}(t) = \frac{\sum_{i}^{N} E_i x_i}{\sum_{i}^{N} x_i}$$

This is similarly described by Sole (2003), Sole and Deisboeck (2004) and Pastor-Santorras and Sole (2001) in terms of cancer development.

Let the interaction be zero. We now rewrite the rate equation as a difference equation $x_{n+1} = ((W_i^{cp} - \bar{E})t + 1)x_n$, and assume we have five quasi species competing with the following initial conditions: W_1=1.0; W_2=4.0; W_3=6.0; W_4=9.0; W_5=10.0; E_1=1.0; E_2=4.0; E_3=6.0; E_4=9.0; E_5=10.0; $(X_1,\ldots,X_5)$=0.25; $\bar{E} = 6.0$, solving the system of equations we get Figure 1 showing the quasi species competition. Under these conditions, which are not too unrealistic, we see all but one of the quasi species vanishes. Those that vanish remove their genomic information from the mix. This is strictly based on kinetics. As we can see from the graphics, three of the qiasi species fail to compete very well and their lineage quickly dies off. A fourth quasi species tends to compete well but it also eventually dies off. The fifth species competes the best by consuming all the available molecular food supply. We emphasize that this system is simply rate limiting term W_i^{cp} being set to a constant. The quasi species are simply competing for resources, but there is no parasitic cross-reaction between species *i* and *j*.

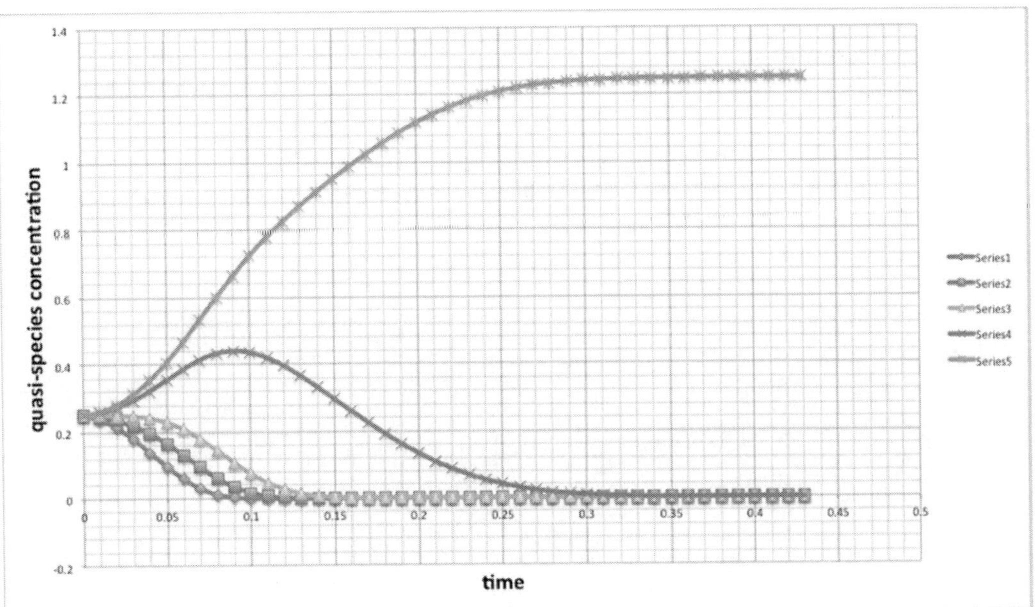

Figure 1. Competition between five molecular replicators. Only one wins the competition.

If we included mutation, we effectively modify the dynamics to become stochastic. A Markovian process would further improve the dynamics (Bharcucha-Reid, 1988). Instead of doing that we will focus on the theoretical aspects of hypercycles where there are beneficial interactions between the quasi species. There is some evidence that cancer cells can cooperate with each other (Axelrod et al., 2006). The following paragraphs will describe autocatalytic molecular cooperation such as partial polymer A participating in extending the polymer B construction.

We start with the most general differential equation for the concentration of species x_i,

$$\dot{x}_i = \Gamma_i\left(\vec{X},\vec{K},\vec{X}(0)\right) - \phi x_i$$

where Γ_i is the growth function and ϕ is a dilution parameter related to global flux with the now redefined symbol Φ_0 by the relation

$$\phi = \frac{\Phi_0}{\sum_j x_j}$$

The arguments for the growth function depend on $\vec{X},\vec{K},\vec{X}(0)$. These column vectors represent a concentration parameter, kinetic parameter and an initial condition respectively, with *j* as the index variable. The total concentration over time (i.e., rate) is given by

$$\dot{c} = \sum_i \dot{x}_i = \sum_i \Gamma_i\left(\vec{X},\vec{K},\vec{X}(0)\right) - \Phi_0$$

Under constant population (cp) conditions we get

$$\phi_0 \equiv \sum_i \Gamma_i\left(\vec{X},\vec{K},\vec{X}(0)\right)$$

Now the variables x_i are given by the equations

$$\dot{x}_i = \sum_i \Gamma_i\left(\vec{X},\vec{K},\vec{X}(0)\right) - \frac{x_i}{c_0}\sum_j \Gamma_j\left(\vec{X},\vec{K},\vec{X}(0)\right)$$

$$(i = 1,\ldots\ldots, N)$$

where c_o is the stationary concentration of all information carriers at cp.These are the selection equations and can be evaluated for any given *Ansatz* of Γ_i. Specifically we can write the growth function's first three terms

$$\Gamma_i = k_i + \sum_j k_{ij} x_j + \sum_j \sum_k k_{ijk} x_j x_k + ...$$

The composition at any time t is given by

$$\vec{X}(t) = \begin{pmatrix} x_1(t) \\ \vdots \\ x_N(t) \end{pmatrix}$$

which can be represented as a point in N-dimensional concentration space X^N similar to the approach used by Palsson (2006). By normalizing the concentration

$$\zeta_i = \frac{x_i}{c(t)}$$

we now have an N-1 space. Points in this space are a unit simplex. Fixed points in this space are invariant over time.

Above we saw the results of a quasi species competition in Figure 1. Only one species won the competition. In a non-coupled systems this will always be the case that one species wins the competition.

Mutant species will compete with the original and may or may not supplant the initial species. This type of dynamics limits the amount of information that can be transmitted from one generation to the next. By coupling the reactions we can get far more complicated dynamics. Here we will examine linear chains and ring circuits constructed from these competitions.

For a linear chain of interacting components $I_1 \rightarrow I_2 \rightarrow \cdots \rightarrow I_N$ the growth of each information carrier at i promotes growth at i+1. Only the first element in the chain is not promoted and the last element in the chain does not have a successor. To define the terms first consider a group of independent replicators. The linear growth function is $\Gamma_i = k_i x_i$ and the relation for change in concentration is

$$\dot{x}_i = k_i x_i - \frac{x_i}{c_0} \sum_{j=1}^{N} k_j x_j$$

We define the normalized average rate constant as

$$\bar{k} = \frac{1}{c_0} \sum_j k_j x_j$$

so the simplified change in concentration is

$$\dot{x}_i = \left(k_i - \bar{k}\right)x_i$$

Now in the case of a linear chain, we assume the linear rate law for the first $\Gamma_1 = k_1 x_1$ element, the others have the rate function

$$\Gamma_i = k_i x_i + k_i^{'} x_i x_{i-1}$$
$$(i = 2, 3, ..., N)$$

The relevant set of differential equations is

$$\dot{x}_1 = k_1 x_1 - \frac{x_1}{c_0}\left[k_1 x_1 + \sum_{j=2}^{N}\left(k_j x_j + k_j^{'} x_j x_{j-1}\right)\right]$$

and

$$\dot{x}_i = k_i x_i + k_i^{'} x_i x_{i-1} - \frac{x_i}{c_0}\left[k_1 x_1 + \sum_{j=2}^{N}\left(k_j x_j + k_j^{'} x_j x_{j-1}\right)\right]$$
$$(i = 2, ..., N)$$

This scheme can be extended to branched chains and include mutation. As an aside, we note that at each replication if there is a probability of mutation leading to one normal replicator and a mutate replicator, then the process can be described by a Markov chain (Demetrius et al., 1985) where we can write entropy H of the process in terms of elements in the Markov matrix $P(= p_{ij})$ as

$$H = -\sum_{ij} p_{ij} \log p_{ij} \,.$$

If we now close the loop and connect the last element to the first to form a cycle, the growth factors for elements 2 through N are the same. The overall hypercycle growth function now becomes

$$\Gamma_i = k_i x_i + k_i^{'} x_i x_j$$
$$\begin{bmatrix} j = i - 1 + N\delta_{i1} \\ i = 1, ..., N \end{bmatrix}$$

and the change in concentration is given as

$$\dot{x}_i = k_i x_i + k_i' x_i x_j - \frac{x_i}{c_0} \sum_{l=2}^{N} \left(k_l x_l + k_l' x_l x_{l-1} \right)$$

$$\begin{bmatrix} j = i - 1 + N\delta_{i1} \\ k = l - 1 + N\delta_{l1} \\ i = 1, \ldots, N \end{bmatrix}$$

Hypercycles like this provide the machinery to produce stable information carriers over generations of replication; and could likely be extended for modeling cooperating cancer genotypes. Of course these replicator networks are comprised of networks of smaller molecules and oligomers in a complex interacting network. The thermodynamics and kinetics of those networks are important for our continued analysis.

Thermodynamics dictates the possible. Kinetics dictates the probable. Thermodynamics is primary and kinetics is secondary. The kinetics of replicators needs to be such that self-replicating catalytic networks are able to rapidly utilize free energy. They are exploiting thermodynamic gradients that exist in the environment (Schneider and Sagan, 2005; Schuster 1986). The basic reactions for replicators without mutation and with mutation are:

$$(A) + I_i \xrightarrow{f_k} 2I_i + (A)$$
$$(A) + I_i \xrightarrow{f_m} I_i + I_j + (A)$$

where I represents the molecular species, (A) represents the catalysts (e.g., enzyme) or molecular machines, and f_k, f_m represents the rate constant for the reactions. The first reaction tells us that the molecular species I_i is faithfully replicated to create two copies. The second reaction tells us that during the replication only one copy of I_i was produced and a mutant variation I_j was also produced. In order to bring thermodynamics into the discussion, we start with the kinetics reaction for the faithful replicator and the mutation reactions:

$$\frac{[I_i]^2 [(A)]}{[(A)][I_i]} = \exp\left(-\frac{\mu_{I_i}^0 - \mu_{(A)}^0}{k_B T} \right)$$

$$\frac{[I_i][I_j][(A)]}{[(A)][I_i]} = \exp\left(-\frac{\mu_{I_j}^0}{k_B T} \right)$$

The species written in square brackets is conventional chemical physics representation for concentration of the specific molecular species. We have explicitly written these equations to show temperature, T and the chemical potential, μ_0^k for the relevant species. The sum of the chemical potentials is related to Gibbs free energy and thus energy gradients. Also note, the concentration of the catalyst (enzyme), or the number of molecular machines available, influences the rate of the reaction.

The next step in understanding cellular molecular networks is to describe the relationship between generation time and temperature, which is implied in the above equation. The focus will now be on entropy. This part of the review will follow Demetrius (1974, 1997). For a more recent review see Demetrius (2013). Entropy is conventionally given by:

$$S = k_B \ln W$$
$$S = -k_B \sum_i p_i \log p_i$$

The first equation is by Boltzmann and says that the natural logarithm of the number of *ways* to distribute energy over different energy levels, W is related to the entropy by a constant (Boltzmann's constant). The second equation, due to Gibbs, denotes this in terms of probability of energy state i.

The central parameters related to entropy are free energy F, the mean energy U and the temperature T, related through the equation: $F = U - ST$. Temperature is best thought of as the kinetic energy of the molecules in the system. The mean kinetic energy of the molecules is given by the amount of heat that must be added to increase the entropy one unit. This is the well-known relation

$$\frac{1}{T} = \frac{\Delta S}{\Delta U}$$

Evolutionary theory, including cancer evolution, is concerned with dynamical behavior of populations (i.e., numbers of types of quasi-species, or in this case numbers of types of cancer genomes).

The main parameters for us to consider are the growth rate r, the reproductive potential θ, the generation time τ, the age of the molecular replicator, and population entropy $\tilde{S}$, an age dependent process. The evolutionary entropy H, is given by the usual probability arguments

$$H = \frac{\tilde{S}}{\tau} = -\frac{\sum_j p_j \log p_j}{\sum_j j p_j}$$

In this expression, evolutionary entropy is normalized according to age class of the replicator. This is an entropy measure of the change in the population from one steady-state to the next as a result of mutation and genetic drift. The reproductive potential is a measure of the ability of a quasi-species, or a cancer genotype, to reproduce itself under ideal conditions and its normalized relation is given by

$$\theta = -\frac{\sum_j p_j \log \Lambda_j}{\sum_j j p_j}$$

where the Λ represents the fraction of individuals surviving to age j. In both cases p_j represents the probability that a randomly chosen replicator is in class j. We can now write the rate as $r = \theta + \tilde{S} / \tau$, and derive generation time of the replicator in terms of population entropy

$$\tau = \frac{\Delta \tilde{S}}{-\Delta \theta}.$$

Molecular and cellular networks are sensitive to chemical and thermal, or energy, gradients. For example, the rate constant ν for some enzymatic reaction in terms of Gibbs free energy (Liberles, 1968) is

$$\upsilon = \exp\left(-\frac{\Delta G^0}{k_B T}\right)$$

Chemical Flux and Potential

In the this subsection we review some ideas by Qian and Beard (2005) and Qian et al., (2003). Many biological chemical reactions are reversible and le Chatelier's principle can be applied (Atkins and de Paula, 2006).

For example, the reaction can be driven forward by adding extra reactants or driven in reverse by adding extra products. The chemical equilibrium expression for a simple transformation of A into B is:

$$J_+ \rightarrow A \underset{k_{-1}}{\overset{k_1}{\rightleftarrows}} B \rightarrow J_-$$

$$\frac{[B]}{[A]} = \frac{k_1}{k_{-1}} = \exp\left(-\frac{\mu_B^0 - \mu_A^0}{k_b T}\right) \qquad \text{Eq. [1]}$$

where k_1 and k_{-1} are the forward and reverse rate constants; k_B is Boltzmann's constant; and μ_0^B, μ_0^A are the chemical potential of the two molecular species at equilibrium. This equation is important because it takes us into the topic of energy gradients and will connect to entropy. The concentration of A and B, c_A, c_B, is, of course related to the flux and the rates, and is given by:

$$c_A = \frac{k_{-1}C_T + J}{k_1 + k_{-1}}, \quad c_B = \frac{k_1 C_T - J}{k_1 + k_{-1}}$$

This is not a common representation but is utilized often by biochemists and is the representation given by Qian and Beard.

Adding zero to the rate of change in concentration allows us to derive this representation. For example, for the above reaction system we can write using conventional chemical engineering notation of square brackets to represent concentration,

$$d[A]/dt = k_{-1}[B] - k_1[A] - J_+.$$

We now add zero in the form $k_{-1}[A] - k_{-1}[A]$, assume steady state, regroup and get, $k_{-1}([A]+[B]) - (k_{-1}+k_1)[A] - J_+ = 0$. After rearranging and substituting $[A]+[B]=C_T$ and $[A]=c_A$, we get the above form in agreement with Qian and Beard.

The difference in chemical potential of the two species, in terms of the equilibrium potential is

$$\Delta\mu = \mu_B - \mu_B = \mu_B^0 - \mu_A^0 + k_B T \ln\frac{c_B}{c_A} = -k_B T \ln\frac{J_+}{J_-}$$

where the forward and reveres fluxes are given by $J_+ = k_1 c_A, J_- = k_2 c_B$ respectively and $J = J_+ - J_-$. The flux will, of course be zero when $J = \Delta\mu = 0$. We can now write

$$-J\Delta\mu = k_B T\left(J_+ - J_-\right)\ln\frac{J_+}{J_-} \geq 0.$$

This equation represents the amount of work done by the system to maintain a non-equilibrium steady state or similarly, the heat dissipation. It is a restatement of the second law of thermodynamics. Qian and Beard (2005), Kauffman (2000), Schneider and Sagan (2005), among others point out the significance of this. The work comes about by exploiting the molecular (and ionic) flux.

This simple abstraction does not consider the fact that many molecular species in biochemical systems are large nano-scale machine parts that do not readily participate in chemical equilibrium relations. So the reactions are only forward or in the model the reverse reaction rate constant would be far smaller than the forward rate constant. Those concepts can be addressed with electrical network analogies. Before we address this problem, we examine a small chemical reaction loop shown in Figure 2.

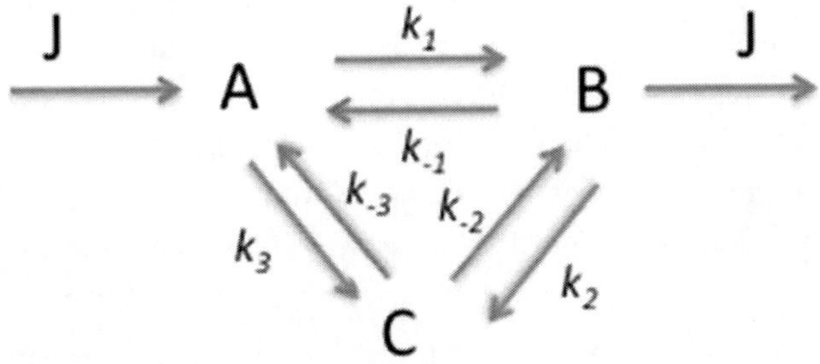

Figure 2. Loop reaction network.

The concentrations of A, B and C are given by

$$c_A = \frac{C_T}{1 + \frac{k_1}{k_{-1}} + \frac{k_2}{k_{-2}}} + J\frac{k_2 + k_{-2} + k_3}{D}$$

$$c_B = \frac{C_T \frac{k_1}{k_{-1}}}{1 + \frac{k_1}{k_{-1}} + \frac{k_1 k_2}{k_{-1} k_{-2}}} + J\frac{k_{-2} + k_3 + k_{-3}}{D}$$

$$c_C = \frac{C_T \frac{k_1 k_2}{k_{-1} k_{-2}}}{1 + \frac{k_1}{k_{-1}} + \frac{k_1 k_2}{k_{-2} k_{-2}}} + J\frac{k_2 - k_{-3}}{D}$$

$$D = k_1 k_3 + k_3 k_{-1} + k_{-1} k_{-3} + k_2 (k_1 + k_3 + k_{-3}) + k_{-2} (k_1 + k_{-1} + k_{-3})$$

From which we can write the fluxes

$$J_{AB} = J\frac{k_1 (k_2 + k_{-2} + k_3) + k_{-1} (k_{-2} + k_3 + k_{-3})}{D}$$

$$J_{AC} = J_{CB} = J\frac{k_2 k_3 + k_2 k_{-3} + k_{-2} k_{-3}}{D}$$

which has the ratio

$$\frac{J_{AB}}{J_{AC}} = \frac{k_1 k_2 + k_{-1} k_{-3}}{k_2 k_{-3}}$$

This is the same as Ohm's law in electrical circuits, and allows us to use electrical circuit analogies. Electrical engineers have developed an extensive toolset to deal with complex networks. Many of these same tools can be used to explore chemical networks. We will show that the chemical networks can be represented as an n-dimensional space and the rate of free energy utilization is a slope on a hyper-surface in this n-dimensional space. First we develop the concept of chemical flux in more detail focusing on energy balance.

Now imagine we have a larger molecular network consisting of $N+N'$ molecular species where N' are fixed or clamped (Qian et al., 2003; Qian and Beard, 2005). In this $N+N'$ set we have $i=1,2,...,N$ components with dynamic concentrations and we have $i=N+1,N+2,...,N+N'$ components with fixed, or clamped, concentrations. These total components participate in $M+M'$ reactions. Where $j=1,2,...,M$ are internal reactions and $j=M+1,M+2,...,M+M'$ external reactions with a fixed boundary flux. For any given reaction j^{th} we can write the kinetic equation

$$\kappa_1^j X_1 + \kappa_2^j X_2 + \cdots + \kappa_{N+N'}^j X_{N+N'} \underset{k_-^j}{\overset{k_+^j}{\rightleftarrows}} \nu_1^j X_1 + \nu_2^j X_2 + \cdots + \nu_{N+N'}^j X_{N+N'}.$$

Of course many of the integer coefficients may be zero. The stoichiometry of this set of reactions can be represented by a $(N+N')X(M+M')$ matrix, ς.

$$\varsigma = \begin{bmatrix} \ddots & \vdots & \cdot^{\cdot^{\cdot}} & \vdots & \cdot^{\cdot^{\cdot}} \\ \cdots & \tilde{\varsigma}_{NM} & \cdots & \bar{\varsigma}_{NM'} & \cdots \\ \cdots & \vdots & \cdots & \vdots & \cdots \\ \cdots & \varsigma_{N'M} & \cdots & 0 & \ddots \\ \cdot^{\cdot^{\cdot}} & \vdots & \ddots & \vdots & \ddots \end{bmatrix}$$

The upper-left block is coefficients for the dynamic molecular species and internal dynamic reactions; the upper-right block is the molecular species that participate in external exchange but with "clamped" reactions; the lower-left block represents clamped species in the internal dynamic reactions; and the lower-right block represents no input or flux to/from clamped species. The matrix is the starting point for flux balance analysis described extensively by Palsson (2006) and also plays an important role in metabolic control analysis (Fell and Sauro, 1985; Kholodenko and Westerhoff, 1995, Heuett et al., 2008).

Assume we have M internal reactions in the system and M' fixed boundary fluxes that extract mass into and out of the system. If $M'=N'=0$ the system is closed and steady state equilibrium. If the system has N dynamic biochemical species we can represent the concentration of each these molecular species as c_i. Then we have

$$\mu_i = \mu_i^0 + k_B T \ln \frac{c_i}{c_t}$$

and the total concentration is

$$c_t = \sum_{l=1}^{N+N'} c_l .$$

Using the conventional expressions we can now write the total enthalpy, entropy and free energy in terms of concentration as:

$$H = \sum_{i=1}^{N+N'} c_i h_i^0$$

$$G = \sum_{i=1}^{N+N'} c_i \mu_i$$

$$S = \sum_{i=1}^{N+N'} c_i \left(s_i^0 - k_B \ln \frac{c_i}{c_t} \right)$$

where h_i^0 is the enthalpy of species *i*. The last equation can be written as $S = (H - G)/T$ to remind us that temperature gradients are needed to produce an entropy change. The chemical potential difference can be written as

$$\Delta\mu_j = \sum_{i=1}^{N+N'} \varsigma_{ij} \mu_i \qquad \text{Eq. [2]}$$

where ς_{ij} is an element in the stoichiometric matrix. This represents the flow or chemical flux. As described, we can now write Kirchhoff's laws, the electrical circuit analogs, for these molecular networks. For nonequilibrium steady state the product of the metabolic fluxes and the stoichiometric matrix satisfy *Kirchhoff's current law*, $\varsigma J = 0$, where $\boldsymbol{J}$ is a vector consisting of internal reaction loops. For each loop in the molecular circuit we can write the Kirchhoff's loop equation,

$$\sum_{j=1}^{M} \left(\Delta\mu_j - \sum_{i=N+1}^{N+N'} \mu_i \varsigma_{ij} \right) = 0 \qquad \text{Eq. [3]}$$

where $0 \le j \le M$. The second term in the difference is the chemical potential or the equivalent of a voltage source in a electrical circuit.

Network Thermodynamics and Topology

With the introduction of flux and flows, we can draw on electrical network analogies in more detail, because chemical circuits and electrical circuits are strictly isomorphic. But we temporarily diverge from that to allow us to link this molecular systems biology to a larger relational biology. We start by reexamining the larger-scale system relations given by Demirel and Sandler (2002).

The flow of mass and the rate of chemical reactions, J_i are caused by forces, X_i. These forces are the electrochemical potential or potential gradients for ions or molecular concentration potentials. We write the forces as: $X_i = grad(\mu_i)$ or conventionally in operator notation $\nabla \mu_i$. Give this, the overall chemical affinity we represent with A, which is related to the stoichiometric matrix and the chemical potential:

$$A_j = -\sum_{i=1}^{n} \varsigma_{ij} \mu_i$$

where, of course ς_{ij} is the relevant element from the stoichiometric matrix and n is the number of types of molecular species being considered in the reaction system. Any force can produce a flow $J_i(X_i)$, and flows and forces are nonlinear relations and functions of each other. These flows and forces can be written as a Taylor series at equilibrium:

$$J_i = J_{i,eq}(X_i = 0) + \sum_{j=1}^{n} \left(\frac{\partial J_i}{\partial X_j} \right)_{eq} X_j + \frac{1}{2!} \sum_{j=1}^{n} \left(\frac{\partial^2 J_i}{\partial X_j^2} \right)_{eq} X_j^2 + \cdots$$

$$X_i = X_{i,eq}(J_k = 0) + \sum_{k=1}^{n} \left(\frac{\partial X_i}{\partial J_k} \right)_{eq} J_k + \frac{1}{2!} \sum_{k=1}^{n} \left(\frac{\partial^2 X_i}{\partial J_k^2} \right)_{eq} J_k^2 + \cdots$$

If we neglect the higher order terms we can simplify to

$$J_i = \sum_{i=1}^{n} L_{ik} X_i \quad (i, k = 1, 2, \cdots, n)$$

$$X_i = \sum_{k=1}^{n} R_{ik} J_k$$

where the coefficients are given by:

$$L_{ik} = \left(\frac{\partial J_i}{\partial X_k} \right)_{X_j} = \left(\frac{J_i}{X_k} \right)_{X_j = 0} \qquad (i \neq k)$$

$$R_{ik} = \left(\frac{\partial J_i}{\partial J_k} \right)_{J_i} = \left(\frac{X_i}{J_k} \right)_{J_i = 0}$$

and are called the phenomenological coefficients and represent the *conductance* and *resistance* respectively. This provides us a link into electrical engineering circuit theory and, potentially, all it's associated developments. These two relations form the basis of linear nonequilibrium thermodynamics by closing the system for energy and mass.

The above relations for conductance and resistance are known as Onsager's equations (Onsager, 1931), and are important because we can now devise simple *components* of networks that allow us to write relations for unstable systems (i.e., oscillators, e.g., citric acid cycle), stable energy storing systems (i.e., molecular systems, e.g., ATP), dissipative reciprocal systems, reversible systems, etc.

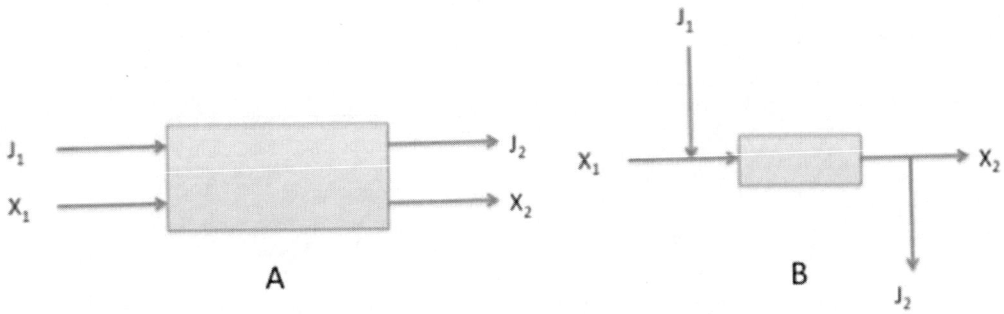

Figure 3. Simple two-port example of two flow, two force system.

Consider the two-port system shown in Figure 3. As described by Peusner (1983), we can write the resistive and the conductance equations for this system as:

$$X_1 = R_{11}J_1 + R_{12}J_2$$
$$X_2 = R_{21}J_1 + R_{22}J_2$$
$$J_1 = L_{11}X_1 + L_{12}X_2$$
$$J_2 = L_{21}X_1 + L_{22}X_2$$

Peusner (1989) also write two sets hybrid equations showing the mixing of forces and flows:

$$X_1 = H_{11}J_1 + H_{12}J_2$$
$$J_2 = H_{21}J_1 + H_{22}X_2$$
$$J_1 = P_{11}X_1 + P_{12}J_2$$
$$X_2 = P_{21}X_1 + P_{22}J_2$$

This notation allows one to map into the mathematics Tellegen's Theorem which relates network theory to general systems theory – of course from there we are at a good launching point into relational biology (Mikulecky, 2001; Rosen, 1970; Rosen, 1991, Rietman, Coltz and Tuszynski, 2011). As seen in the above relations, at junctions for components we simply add the flows and forces. In the case of electronic circuits the forces are voltage and the flows are currents or really electron flow.

We can write the thermodynamic relations for individual "components" in our circuit and then add them up for bulk molecular circuits (or electronic circuits) by using *Kirchhoff's* laws. Changes in thermodynamic variables and forces, essentially thermodynamic changes of state, can be written as vector displacements (Margenau and Murphy, 1956). This allows thermodynamics to be mapped to an Euclidean R^n space or a configuration space C (Figure 4).

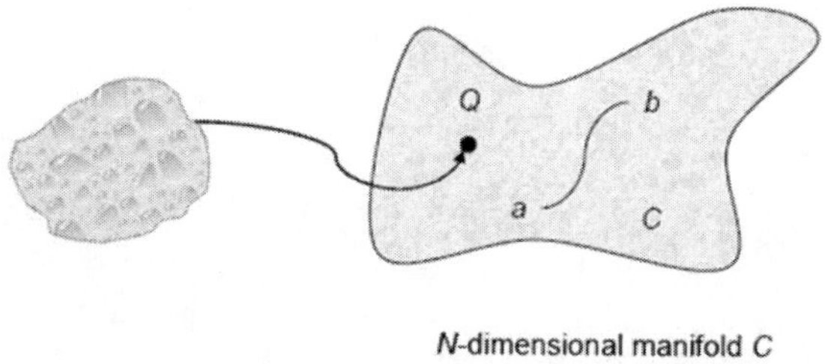

Figure 4. Thermodynamic configuration space on a N-dimensional manifold.

The advantage of using the mathematics of manifolds and differential geometry for thermodynamics has been described at length by Weinhold (1975, 1976). Gibbs wrote the energy state U of a system as a function of entropy S, volume V, and mass quantities expressed as mole numbers $(N_1,....,N_c)$

$$U = U(S, V, N_1,...., N_c) \qquad \text{Eq. [4]}$$

Temperature is the partial derivative of U with respect to S and pressure is the partial of U with respect to V. The partials of U with respect to N_i are chemical potentials. These various partial derivatives are tangents on the energy surface. Using the language of mechanics the relationship between T is said to be conjugate to S and P is conjugate to V. So in general a conjugate relationship for the variable R_i (see above system of equations) is given as:

$$X_i = \left(\frac{\partial U}{\partial m_i}\right)_{m_j} \qquad U = U(m_1, m_2, \cdots) \qquad \text{Eq. [5]}$$

Though this model is valuable, distances, angles, and areas, for example have no intrinsic meaning on this surface. Furthermore, thermodynamic inner products are state-dependent so a Riemmanian metric is required. We now follow Peusner (1985) and "flatten" our metric space into a network. Technically this is known as a simplicial mapping (Hocking and Young, 1961; Penrose, 2004). More specifically, we describe the n-dimensional surface as a network and preserve important *extensive variables* (S, V, N_i) and *intensive variables* (i.e., the partial derivatives resulting in T or P) in their correct association with each other and with the Hessian matrix.

Equation [5] above can be written briefly as

$$X_i = \left(\frac{U_i - U_{i0}}{\partial m_i} \right)_{m_j}$$

and interpreted as finite displacement. On our network, overlaid on the topological surface, this represents the force X_i between two nodes held at potentials U_i and U_{io} in the direction of the change in the extensive variable. This single path change is shown in Figure 5 along with more complicated 2-, 3-, and 4-path changes. As usual we just use Kirchhoff's laws and simply add the variables at the nodes. The sum of energy changes inside the network must match the sum of changes outside the network and is known as Tellegen's Theorem.

$$\sum_{ports} X_i dm_i = - \sum_{branches} X_b dm_b$$

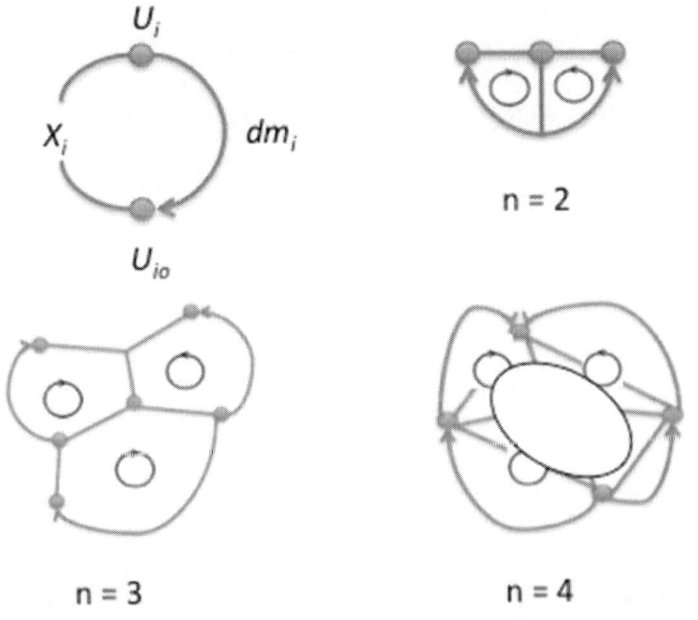

Figure 5. Drawn after Peusner (1985). Schematics of small topological networks of the first law of thermodynamics. Different nodes interact through potentials. n represents the number of ports in the network.

Now from a different approach we show that the geometrical representation on the n-dimensional surface and the network representation are equivalent. In the geometrical space the squared displacement of a vector is given by the dot product $\|\xi\| = \langle \sum d_i, \sum d_i \rangle$ where ξ are the Euclidean components of the n-dimensional displacement vector.

In the network view, this norm can be expressed as the dot product of the node intensive variable, meaning the derivatives of *P* or *T*, and the derivatives of the conjugate extensive variables, S, V or N_i, $\|\xi\|^2 = \langle dm, dX \rangle$. Noting of course that the dot product is explicitly written as

$$\langle dm, dX \rangle = \sum_i dm_i dX_i$$
$$= \sum g_{ik} dX_k dX_i = g(u^i, u^k) dX_k dX_i$$
$$= \sum r_{ik} dX_k dX_i = r(u_i, u_k) dX_k dX_i$$

where g and r are Hessian matrices whose elements are given by

$$g_{ik} = \frac{\partial^2 G}{\partial X_i \partial X_k}, \quad r_{ik} = \frac{\partial^2 U}{\partial m_i \partial m_k}$$

Typically these elements are found by taking the dot product of the contravariant and covariant Euclidean vectors u^i, u_i (see Penrose, 2004). G is the same a Gibbs free energy. In the case of networks these represent conductance and resistance respectively

We find that we can represent the thermodynamics of local planar networks with tangent vectors $\xi_i = \frac{\partial \xi}{\partial x^i}$. As an example, entropy would increase at greater slopes, and this corresponds to studies by Martyushev and Seleznev (2006), Schneider and Sagan (2005) and Chaisson (2001, 2005), that entropy production is *maximum* for biological systems. Later we will see how this maxima translates to cancer as being a way of producing entropy faster than healthy cells.

Maximum Entropy Production in Biological Systems

In the first section of this paper we saw that a molecular replicator with the highest kinetics would win in a molecular competition with other replicators. Then we added a feature to the analysis showing cooperation among the replicators. The same is actually true with cancer cells. The cancer cells with the highest metabolic rate will become dominant, and there could be many different genomes contributing to this population of cells. Thermodynamics of course is the first principle to consider. Together, thermodynamics and kinetics (energy-rate) describes much of the biological world (Pross, 2008; Dobovisek et al., 2011). But the biological world consists of open-thermodynamic systems. So we need to include energy-density. In this section we include further discussion on entropy, tie some of the earlier concepts together, and conclude with comments on cancer cells having a thermodynamic and kinetic advantage over healthy cells.

Consider a system of molecular replicators (or an ecosystem of early stage cancer cells). This is an open system with external energy. To keep the system (molecular or cellular) functioning there must be a transfer of energy from a high-energy source, like a battery, to lower energy – typically heat dissipation. The difference between the high-energy source and the lower energy of course is a slope on the *n*-dimensional energy manifold described earlier. Between the high-energy density at the source and much lower energy density at the sink is the biological system or the replicator system. The kinetics of this system effectively sculpts the landscape to enable fast transfer of energy down the manifold slope. This sculpting exploits geodesics on the manifold.

Entropy is discussed and reviewed by many investigators, but we will review Shamra and Annila (2007) and Annila (2012) because of the clarity of exposition and because the emphasis on the meaning of Boltzmann's law. Entropy change due to reaction velocities, ν_i and chemical affinities (or chemical potentials), $A_j = \sum_i \mu_i$ is given by:

$$\frac{dS}{dt} = \frac{1}{T}\sum_j \nu_j A_j.$$

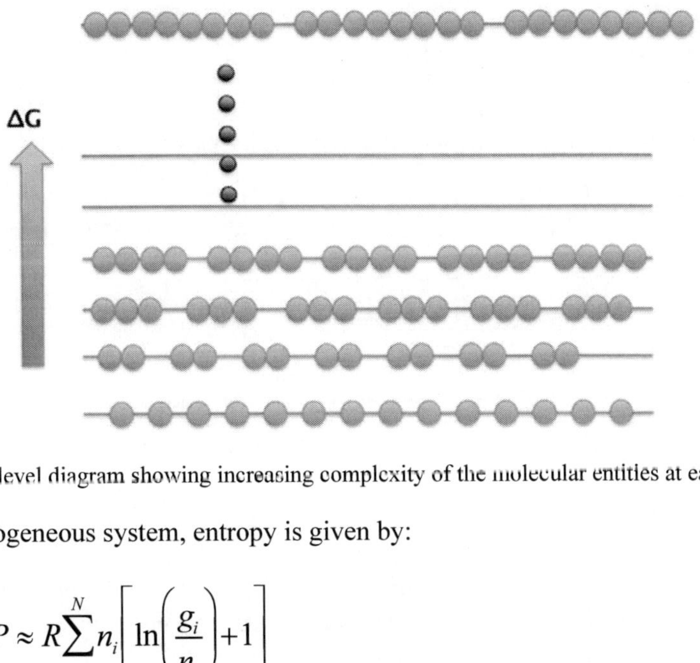

Figure 6. Energy level diagram showing increasing complexity of the molecular entities at each level.

For a homogeneous system, entropy is given by:

$$S = R\ln P \approx R\sum_{i=1}^{N} n_i\left[\ln\left(\frac{g_i}{n_i}\right)+1\right]$$

where indistinguishable particles n_i, are said to be non-interacting. The total number of these particles, $N = \sum_i n$ are distributed among kinetic energy levels, E_i relative to thermal background *RT*. The occupancy number at each level, g_i is said to be constant. In reality, of course multiple reactions are operating at the same time and products become substrates for another reaction and this changes the occupancy numbers. Further, the total number of types of molecular species may also change via mutation or errors in chemical synthesis. If there can be many types of chemical compounds on the overall reaction surface (e.g., energy manifold), we can distribute those compounds according to the probability distribution

$$P_{jb} = \frac{(N - n_1 - \cdots - n_{j-1})!}{(n_j / j)!(N - n_1 - n_2 - \cdots - n_j)!}$$

The term n_j / j allows for compound stoichiometry (i.e., assembly rules). The factor $N!$ will cancel out identical compounds. The subscript *jb* signifies energy level *j* and *b* is a measure of the size of the compounds at that level. If we have N_k available substrates, N_j reaction products, an external energy source ΔQ_{jk} that couples to our system, and Gibbs free energy difference between the substrates and products is given by $\Delta G = G_j - \sum_k G_k$, then, as shown by Shamra and Annila (2007), we can write the probability as

$$P_j = \left[\prod_k N_k \exp\left(\frac{\Delta Q_{jk} - \Delta G_{jk}}{RT} \right) \right]^{N_j} \equiv \Gamma_j^{N_j}.$$

Considering all levels we can write the overall probability as

$$P = \prod_{j=1} \frac{\Gamma^{N_j}}{N_j!}.$$

Recalling that $S = R \ln P$ and after some simplification, we can write the entropy as:

$$S \approx \frac{1}{T} \sum_j N_j \left(A_j + RT \right)$$

where

$$A_j = \sum_k \left(\mu_k + \Delta Q_{jk} - \mu_j \right).$$

We can use Liouville's theorem to understand the dynamics on the manifold (Penrose, 2004). To describe the flow of molecular species associated with entropy changes we differentiate the above equation to get $\partial(\ln P)/\partial t = (1/P)(\partial P/\partial t)$ or

$$\frac{dN_j}{dt} = \nu_j \frac{A_j}{RT} = -\sum_k \frac{dN_k}{dt}$$

where ν_j is the rates of the chemical reactions. The Liouvillian my be written as:

$$L \approx \sum_j \frac{dN_j}{dt} \frac{A_j}{RT}$$

This is similar to the Onsager reciprocal relations we described earlier in discussion with electronic and molecular circuits.

The distribution shown in Figure 6 is important to understand entropy. When the *number* of types of complex units (e.g., molecular, cellular) N_k participating in the energy dynamics is large, then there will be a larger number of levels in the distribution. The change in Gibbs free energy will then be high and there will be a higher flow of substance (e.g., molecular, electronic, photonic) from the high-energy source, or high-energy density, to a lower energy sink. This flow takes place along geodesics on an n-dimensional manifold. The steeper the slope, the higher the rates of energy flow. Energy-density-rate becomes the key metric to understand living systems, and the number of types of complex units in the energy distribution is the way to think about diversity of living systems in an ecosystem, or the diversity of cancer cells in a tumor.

Comments on Energetics of Cancer

The easiest way to understand cancer is from the molecular/genomic perspective. One unifying feature of all cancers is genomic instability and aneuploidy (Hanahan and Weinberg, 2011), typically observed as copy number variation in the genome. Above we have shown that as the number of types of molecular complexes increase so does the entropy. Also as the numbers of complexes on a particular energy level increase, this may also increase the entropy, indirectly, by increasing the chemical potential or the Gibbs free energy. In the cases of cancer cells the increase in copy number for genes gives an obvious increase in the numbers of molecular complexes, in the form of additional protein load, and of course this sculpts the energy landscape appropriately.

Imagine we have a duplication event that creates an extra copy of a chromosome. The genes in this chromosome will be expressed at the appropriate time according to the now different energy landscape of the cell and therefore different dynamics. These extra proteins and RNA molecules will, if nothing else, effect the molecular reactions they participate in by mass balance or Le Chatelier's principle. Typically, in a healthy cell this increase in numbers of proteins would result in proteotoxic stress (Torres et al., 2010). The cell will attempt to compensate for the extra proteins using its quality control systems such as chaperone-mediated sequestration, or ubiquitination. The cell will undergo a phase transition at cell cycle and produce a mutator phenotype, or result in apoptosis.

Assuming we now have a mutator phenotype, we can describe the molecular network using the mathematics developed earlier for molecular competition. Following Sole and Deisboeck (2004), assume we have a master sequence (e.g., a healthy genome) and a population of genomes in another "compartment." Mutations take place in the master sequence x_1 and the "average" sequence of the other genomes is x_2. We can now write a set of equation very much similar to the previous set:

$$\frac{dx_1}{dt} = f_1 x_1 A - x_1 \Phi(x_1, x_2)$$
$$\frac{dx_2}{dt} = f_1 x_1 (1-A) + f_2 x_2 - x_2 \Phi(x_1, x_2)$$

where A represents a mutation matrix, Φ represents the average fitness. The fitness of the two populations are represented by f_1 and f_2. Assume $f_1 > f_2$ then the master sequence replicates faster than the mutant sequences. The fixed points in this space on are the line $x_1^* = 1 - x_2^*$ and

$$x_2^* = \frac{f_1(1-A)}{f_1 - f_2}.$$

Realizing that the partial derivatives

$$\partial\Phi(x_1, x_2)/\partial x_1 = x_2 f_1; \quad \partial\Phi(x_1, x_2)/\partial x_2 = x_1 f_2$$

then the Jacobi matrix for this system is

$$L_\mu(x^*) = \begin{pmatrix} f_1 A - \Phi(x) - x_1 \dfrac{\partial\Phi}{\partial x_1} & -x_1 f_2 \\ f_0(1-A) - x_2 f_1 & f_2 - \Phi(x) - x_2 \dfrac{\partial\Phi}{\partial x_2} \end{pmatrix}$$

and describes the equilibrium for the system. When the eigenvalues of this matrix are negative we have $f_1 A < f_2$ there will be an error threshold in the master sequence and we have a catastrophe leading to cancer. Stephens et al., (2011) have described a biological model for massive genomic rearrangement acquired in a single catastrophe event.

This catastrophe event is known as a dynamical phase transition. Unfortunately, this analytical model does not help in developing an intuition to what may be taking place at the cellular level. For that we draw on Rietman et al., (2013) who describe an integrated synthesis model connecting several areas of biophysics and biochemistry. Cancer cells are known to exhibit what is known as the Warburg effect (Warburg, 1956). Even though glycolysis produces only two ATP molecules per glucose vs. 26 glucose molecules produced in the mitochondria by oxidative phosphorylation, these cells prefer to process glucose in the cytoplasm rather than in the mitochondria. One scenario to induce cells to prefer glycolysis to phosphorylation is similar to Raleigh-Benard (RB) convection, a phenomenon at the interface of fluid dynamics and thermodynamics.

In the case of RB convection there is a thermal gradient in a fluid layer between a thermal source and sink. As a result of this, there will be symmetry break and small fluid oscillators appear that are able to transfer energy from the hot side to the cold side more quickly – a steeper slope on the energy manifold. A similar phenomenon can occur in a cell. If there is a persistent high glucose concentration just outside a cell (e.g., over eating), in order to take advantage of this chemical potential the cell will begin to manufacture additional resources to take advantage of this. The cell will produce the appropriate protein machinery. Glycolysis is a cyclic process. As shown by Gatenby and Frieden (2010) these protein oscillators accumulate at the nuclear boundary – just as oscillators accumulate in RB convection. This increase in glycolysis will perturb the pH balance in the cell because of excess lactic acid. Persistent pH imbalance as small as a hundredth of a unit can result in depolarization of the mitochondria and breakdown of the mitochondria membrane. GTP is manufactured during the citric acid cycle in mitochondria and is vitally important for end caps on microtubules to keep them from depolymerizing too rapidly. Microtubules, of course are the components of the cytoskeleton.

The next step on the road to aneuploidy starts with spindle pole formation with microtubules that are either too short and/or in sufficient numbers of them by only a small amount. When the cell now enters mitosis a mitotic catastrophe can occur. For lucky cells this will result in apoptosis. For unlucky cell, because of the microtubule insufficiency, there can be an event where chromosomes are effectively distributed incorrectly. We now have a potential mutator phenotype ready to begin the series of operations described above to produce further mutated progeny with imbalanced numbers of chromosomes.

This describes a scenario by which a single cell can become cancerous. But the immune system may easily be able to address this problem (Wilkie and Hahnfeldt, 2013a, 2013b). But once we have a population of self-replicating aneuploidy cells (e.g., micro-tumor) that have escaped immune system suppression, the cellular environment has clear passed through a phase transition – meaning the surrounding cells look different and have a clear phenotypic difference. At this point, as described by Davies et al., (2011) this phase transition results in a thermodynamically more stable population of cells. In fact using Figure 6 above we can easily see that tumors will have a thermodynamic advantage not only from an ability to rapidly process glucose but because they have a *distribution* of genomes. The number of ways to dissipate an energy potential is greater for tumor cells than for healthy cells. This literally makes them analogous to an invasive species (Davies and Lineweaver, 2011), a thermodynamically more stable cell type than healthy cells (Rietman, et al., 2012).

Acknowledgments

We acknowledge grant number U54CA149233 (to LH) from the National Cancer Institute. The content is solely the responsibility of the authors and does not necessarily represent the official views of the National Cancer Institute.

References

Annila, A., (2012), "Space, Time and Machines", Int. J. of Theoretical and Mathematical Physics, 2 (3), 16-32.

Avery, J., (2003), *Information Theory and the Evolution of Life*, World Scientific, Singapore

Beard, D. A., Babson, E., Curtis, E. and Qian, H., (2004), "Thermodynamic Constraints for Biochemical Networks," *J. Theor. Biol*. 228, 327-333.

Bonner, J. T., (1988), *The Evolution of Complexity by Means of Natural Selection*, Princeton University Press.

Brooks, D. R., and Wiley, E. O., (1988) *Evolution as Entropy: Toward A Unified Theory of Biology*, Univ. Chicago Press, Chicago, IL.

Bharcucha-Reid, A. T., (1988), *Elements of the Theory of Markov Processes and their Applications*, Dover.

Casti, J., (1992), *Reality Rules: II, Picturing the World in Mathematics – The Frontier*, Wiley, New York, NY.

Chaisson, E. J., (2001), *Cosmic Evolution, The Rise of Complexity in Nature*, Harvard U. Press.

Chaisson, E. J., (2005), *Epic of Evolution, Seven Ages of the Cosmos*, Columbia University Press, New York, NY.

Coombes, S. and Bressloff, P. S., (2005), *Bursting, the Genesis of Rhythm in the Nervous System*, World Scientific, London, UK.

Davies, P. C. W., Demetrius, L. and Tuszynski, J. A., (2011), "Cancer as a dynamical phase transition", *Theoretical Biology and Medical Modeling*, 8, 30.

Davies, P. C. W. and Lineweaver, C. H., (2011), "Cancer tumors as Metazoa 1.0: tapping genes of ancient ansestors", *Physical Biology*, 8, 015001.

Davies, P. and Gregersen, N. H., (editors) (2010), *Information and the Nature of Reality, From Physics to Metaphysics*, Cambridge University Press, Cambridge, UK.

Demetrius, L., (1974), "Demographic Parameters and Natural Selection", *Proc. Natl. Acad. Sci. USA*, 71, 4645-4647.

Demetrius, L., (1997), "Directionality Principle in Thermodynamics and Evolution", *Proc. Natl. Acad. Sci.* USA, 94, 3491-3498.

Demetrius, L., (2000), "Thermodynamics and Evolution", *J. Theor. Biol*. 206, 1-16.

Demetrius, L., (2013), "Boltzmann, Darwin and Directionality Theory", *Physics Reports*, 530 (1), 1-85

Demetrius, L., Schuster, P., Sigmund, K., (1985), "Polynucleotide Evolution and Branching Processes", *Bulletin of Mathematical Biology*, 47 (2), 239-262.

Demirel, Y. and Sandler, S. I., (2002), "Thermodynamics and bioenergetics", Biophysical Chemistry, 97, 87-111.

Dobovisek, A., Zupanovic, P., Brumen, M., Bonacic-Losic, Z., Kuic, D., Juretic, D., (2011), "Enzyme kinetics and the maximum entropy production principle", *Biophysical Chemistry*, 154, 49-55.

Doya, K., Ishii, S., Pouget, A. and Rao, R. P. N., (2007), *Bayesian Brain, Probabilistic Approaches to Neural Coding,* MIT Press, Cambridge, MA.

Eigen, M., (1971), "Self-Organization of Matter and the Evolution of Biological Macromolecules", *Naturwissenchaften* 56, 465-523.

Eigen, M., (1992), *Steps Toward Life*, Oxford University Press, New York, NY.

Fell, D. A. and Sauro, H. M., (1985) "Metabolic control and its analysis. Eur. J. Biochem., 148, 555-561.

Fuchs, C. A., (2011) *Coming of Age with Quantum Information, Notes on a Paulian Idea*, Cambridge U. Press, Cambridge, UK.

Gatenby, R. A. and Frieden, B. R., (2010), "Coulomb Interactions between Cytoplasmic Electric Fields and Phosphorylated Messenger Proteins Optimize Information Flow in Cells", PLoS One, vol. 5, e12084.

Gillespie, J. H., (1991), *The Causes of Molecular Evolution*, Oxford University Press, New York, NY.

Gregory, T. R., (editor), (2005), *The Evolution of the Genome*, Elsevier, Boston.

Gurel, D. and Gurel, O., (1983), *Oscillations in Chemical Reactions*, Springer-Verlag, New York, NY.

Heuett, W. J., Beard, D. A. and Qian, H., (2008), "Linear analysis near a steady-state of biochemical networks: Control analysis, correlation metrics and circuit theory", *BMC Systems Biology*, 2, 44.

Hilborn, R. C., (1994), *Chaos and Nonlinear Dynamics, An Introduction for Scientists and Engineers*, Oxford Univ. Press, New York, NY.

Hocking, J. G. and Young, G. S., (1961), *Topology*, Dover Publications, New York, NY.

Jantsch, E., (1980), *The Self-Organizing Universe*, Pergamon Press, New York, NY.

Kauffman, S. A., (1993), *The Origins of Order: Self-Organization and Selection in Evolution*, Oxford University Press, New York, NY.

Kauffman, S. A., (2000), *Investigations*, Oxford University Press, New York, NY.

Kuppers, B.-O., (1983), *Molecular Theory of Evolution*, Springer-Verlag, Berlin.

Kuppers, B.-O., (1990), *Information and the Origin of Life*, MIT Press, Cambridge, MA

Leff, H. S. and Rex, A. F., (1990), *Maxwell's Demon, Entropy, Information and Computing*, Princeton Series in Physics, Princeton, NJ.

Liberles, A., (1968), *Introduction to Theoretical Organic Chemistry*, Macmillan Company, New York, NY.

Martyushev, L. M., and Seleznev, V. D., (2006), "Maximum Entropy Production Principle in Physics, Chemistry, and Biology" *Physics Reports*, 426, 1-45.

Muller, S. J., (2007), *Asymmetry: The Foundation of Information*, Springer, New York, NY

Nicolis, G. and Prigogine, I., (1989), *Exploring Complexity*, W. H. Freeman and Company, New York.

Nielsen, N. A. and Chuang, I. L., (2000), Quantum *Computing and Quantum Information*, Cambridge University Press, Cambridge, UK.

Onsager, L., (1931), "Reciprocal Relations in Irreversible Processes. I.", *Physical Review*, 37, 405-426.

Palsson, B. O., (2006), *Systems Biology: Properties of Reconstructed Networks*, Cambridge U. Press, Cambridge, UK.

Pastor-Satorras, R. and Sole, R. V., (2001), "Field theory for a reaction-diffusion model of quasispecies dynamics", *Physical Rev. E*. 64, 051909.

Peusner, L., (1983), "Hierarchies of Irreversible Energy Conversion Systems: A Network Thermodynamics Approach. I. Linear Steady State Without Storage", *J. Theor. Biol.*, 102, 7-39.

Peusner, L., (1985), "Network Thermostatics", J. Chem. Phys, 83 (3), 1276-1291.

Penrose, R., (2004) *The Road Reality: A Complete Guide to the Laws of the Universe*, Knopf, London.

Pross, A., (2008), "How can a chemical system act purposefully? Bridging between life and non-life", *J. of Physical Organic Chemistry*, 21, 724-730.

Qian, H. and Beard, D. A., (2005), "Thermodynamics of stoichiometric biochemical networks in living systems far from equilibrium" *Biophysical Chemistry*, 114, 213-220.

Qian, H., Beard, D. A. and Liang, S., (2003), "Stoichiometric network theory for nonequilibrium biochemical systems" *Eur. J. Biochem*, 207, 415-421.

Rietman, E., (1993) *Creating Artificial Life: Self-Organization*, Windcrest/McGraw-Hill, New York, NY.

Rietman, E. A., Karp, R. L. and Tuszynski, J. A., (2011), Review and Applications of Group theory to Molecular Systems Biology, *Theoretical Biology and Medical Modelling*, 8, 21.

Rietman, E. A., Colt, J. Z. and Tuszynski, J. A., (2011), "Interactomes, manufacturomes and relational biology: analogies between systems biology and manufacturing systems", *Theoretical Biology and Medical Modelling*, 8, 19.

Rietman, E. A., McDonald, J. T., Hlatky, L., (2013), Organism mutation stability and cancer. In: Shoja MM, Agutter PS, Tubbs RS, Ghanei M, Ghabili K, Harris A, Loukas M (eds). *Hypotheses in Clinical Medicine*. Nova Science Publishers, Inc.; 2013. 594 p.

Rietman, E. A., Friesen, D. E., Hahnfeldt, P., Gatenby, R., Hlatky, L. and Tuszynski, J. A., (2013), "An Integrated Multidiciplinary Model Describing Initiation of cancer and the Warburg Hypothesis", Theroetical Biology and Medical Modelling, 10, 39

Rosen, R., (1970), *Dynamical System Theory in Biology, Volume 1, Stability Theory and Its Applications*, Wiley Interscience, New York, NY.

Rosen, R., (1991), *Life Itself, A Comprehensive Inquiry into the Nature, Origin and Fabrication of Life*, Columbia University Press, New York, NY.

Schneider, E. D. and Sagan, D., (2005) *Into the Cool: Energy Flow, Thermodynamics and Life*, University of Chicago, Press, Chicago, IL.

Schuster, P., (1986) "Dynamics of Molecular Evolution" *Physica* 22D, 100-119.

Seife, C., (2006), *Decoding the Universe, How the New Science of Information, is Explaining Everything in the Cosmos from our Brains to Black Holes*, Viking Press, New York, NY.

Sharma, V.and Annila, A., (2007), *Biophysical Chemistry*, 127, 123-128.

Sole, R. V., (2003), "Phase transitions in unstable cancer cell populations" *Eur. Phys. Journal*, B, 35, 117-123.

Sole, R. V. and Deisboeck, T. S., (2004), "An error catastrophe in cancer?" *J. Theoretical Biology*, 228, 47-54.

Stephens, P. J., et al., (2011) "Massive Genomic Rearrangement Acquired in a Single Catastrophic Event during Cancer Development" *Cell*, 144, 27-40.

Torres, E. M., Williams, B. R., Tang, Y-C and Amon, A., (2010), "Thoughts on Aneuploidy" *Cold Spring Harb. Symp. Quant. Biol.* 75, 445-451

von Baeyer, H. C., (2004), *Information, the New Language of Science*, Harvard University Press, Cambridge, MA.

Warburg, O., (1956), "On the Origin of Cancer Cells" Science, 123, (3191), 309-314.

Weinhold, F., "Metric Geometry of equilibrium thermodynamics I, II, III, IV, V" *J. of Chem. Phys*. 63, 2479, 2484, 2488, 2496 (1975); 65, 558 (1976).

Wilkie, K. P. and Hahnfeldt, P., Mathematical Models of Immune-Induced Cancer Dormancy and the Emergence of Immune Evasion, *Interface Focus*, 3:4, 20130010, doi:10.1098/rsfs.2013.0010, 2013.

Wilkie, K. P. and Hahnfeldt, P., Tumor-Immune Dynamics Regulated in the Microenvironment Inform the Transient Nature of Immune-Induced Tumor Dormancy, *Cancer Res*. 73:12, 3534–3544, 2013.

Zurek, W. H., (editor) (1990), *Complexity, Entropy, and the Physics of Information*, Volume VIII Santa Fe Institute Studies in the Sciences of Complexity, Addison-Wesley Publishing, Redwood City, CA.

In: Recent Advances in Systems Biology Research
Editors: A. X. C. N. Valente, A. Sarkar and Y. Gao
ISBN: 978-1-62948-736-6

Chapter 5

Analysis of Protein-Protein Interaction Networks for Identification of Functional Modules

Tolga Can
Department of Computer Engineering, Middle East Technical University, Ankara, Turkey

Abstract

Genome-scale protein-protein interaction (PPI) networks for an increasing number of organisms are being constructed by researchers worldwide. Screens such as yeast two-hybrid (Y2H) and tandem affinity purification (TAP) allow for identification of interactions in a high-throughput manner and interactions detected by small scale experiments such as co-immunoprecipitation can be collected from literature by text-mining methods. As a result, databases devoted for collecting such data have emerged and provide researchers with a wealth of PPI data.

This chapter focuses on analysis of PPI networks for discovering novel functional modules within these networks for the ultimate goal of understanding the modular organization of the cell. The discovered modules provide models for systems biologists to further verify and understand the dynamics of signaling, metabolic, and regulatory mechanisms.

Methods to identify functional modules can be categorized with respect to whether they employ knowledge from known modules, i.e., model based versus non-model based approaches, and also with respect to the types of modules they identify, i.e., complex detection methods versus signaling pathway detection methods. In this chapter, we describe state of the art methods from each category and discuss further research directions.

1. Protein-Protein Interaction Databases

Proteins physically interact with each other to carry out biological functions. There are various types of interactions within a cell. Interactions that govern signaling pathways such as phosphorylation are categorized as transient interactions; whereas, interactions that form protein complexes are categorized as stable interactions. In addition to physical protein interactions, many databases and computational tools contain some indirect interactions such as membership in the same pathway, or complex.

There are several protein-protein interaction (PPI) databases which are studied and compared by many researchers (De las Rivas and Fontanillo, 2010; Mathivanan et al., 2006). De las Rivas and Fontanillo (2010) provide a comprehensive review on protein-protein interaction databases starting with the definition of *interaction* and then providing an analysis of the overlap of various human PPI databases. An interesting finding is that, of the total eighty thousand interactions available in these databases, only three protein interactions appear in all of the databases and databases such as HPRD (Prasad et al., 2009) and MINT (Chatr-aryamontri et al., 2007) contain considerable amount of interactions that are unique to these databases. For example, 50.7% of the interactions in HPRD is only found in that database. The main reason behind this discrepancy is the different techniques that are used to collect protein-protein interactions. For example, the Database of Interacting Proteins (DIP) (Xenarios et al., 2000) contains protein-protein interactions mostly from high-throughput interaction assays such as yeast two-hybrid. On the other hand, databases such as MINT and IntAct (Kerrien et al., 2012) mostly contain interactions extracted from literature hence contain higher quality interactions from low throughput studies. These observations suggest that an integrative approach that combines data collected in multiple databases would provide a better coverage of the interactome. PINA (Wu et al., 2009) is one such network analysis framework which builds a large network using data collected from the IntAct, MINT, BioGRID (Stark et al., 2010), DIP, HPRD, and MIPS/MPact (Guldener et al., 2006) databases. There are also databases such as STRING (Szklarczyk et al., 2011) which contain computationally predicted interactions. Based on the prediction methodology, these interactions may include indirect non-physical interactions and therefore they are much denser compared to physical interaction networks. On the other hand, predicted networks provide better coverage of the true interactome, since predictions are carried out genome-wide. It is important to note that the quality of the results of the analysis methods described in this chapter depend very much on the quality and coverage of the protein-protein interaction dataset that is analyzed. Some of these methods assume that the analyzed dataset is complete and accurate and delegate the database quality problem to database curators, while others take specific measures to deal with noisy protein-protein interaction data.

2. Ab initio Methods for Identification of Modules in PPI Networks

As genome-scale protein-protein interaction networks became available, methods for automated analysis of these networks were needed. The graphical abstraction of PPI networks allowed researchers employ existing methods in graph theory. For example, graph clustering is a common problem in network analysis. Graph related problems constitute a considerable part of theoretical computer science, especially in complexity analysis. While some problems on graphs such as Hamiltonian paths are inherently difficult, other problems such as single source shortest paths can be solved efficiently. Analysis of protein-protein interaction networks emerge as a suitable area where existing techniques from graph theory can be applied. Therefore, in recent years significant amount of research has been conducted on genome-scale protein-protein interaction networks as well as other types of biological networks such as metabolic networks and gene regulatory networks. While there are several biological problems that are related to analysis of protein-protein interaction networks, such as automated function prediction and identification of frequent interaction patterns, in this section, we focus on the problems of identification of molecular complexes and identification of signaling pathways from PPI networks.

Given a large-scale PPI network with thousands of nodes and tens of thousands of edges, the first thing to do is to divide the graph into groups of nodes in a reasonable way. This is called graph partitioning or graph clustering problem. In graph partitioning, the graph is divided into non-overlapping regions. Partitioning a graph into two near equal subgraphs such that the edges that cross different partitions are minimum is a computationally difficult problem (Symoys, 1995). In graph clustering, overlapping clusters may be generated and the goal is to identify subgraphs that have high density of edges. A number of methods that solve this problem are given in Section 2.1. PPI networks contain transient interactions as well as permanent interactions that are observed as part of protein complexes. Transient interactions are usually activation or inhibition of a protein by another protein in a signal transduction pathway. Such mixed nature of PPI networks necessitates a careful dissection of large scale networks into functional components. In addition to subgraphs that are dense, signaling pathways exhibit themselves as paths or as flow networks with complex topologies in the PPI networks. Therefore, just searching for dense regions is not sufficient to identify these functional modules in PPI networks. Techniques that find paths from a source node to a target node in a signal transduction pathway are therefore developed in a different manner than graph clustering approaches and are discussed in detail in Section 2.2.

Ab initio techniques mainly apply graph theoretic approaches on PPI networks without specific data from known complexes or known pathways. In the objective function, which directs the search, simple heuristics such as graph density or probability of a path are used. Model based approaches, on the other hand, make extensive use of the knowledge of existing complexes/pathways by building statistical models and they either filter out modules that are different from the models or conduct a more directed search on PPI networks using these models. Nevertheless, ab initio methods that are not model-based are useful for discovering modules that show novel topological structures.

2.1. Identification of Molecular Complexes

There are dozens of molecular complex identification methods published in literature. There are reviews and comparative studies that evaluate the performance of these methods in different settings (Moschopoulos et al., 2011). Most of existing molecular identification methods are non-model based unsupervised techniques that use *graph clustering* methods to identify dense subnetworks with certain topological properties. Several clustering algorithms such as MCL (Markov CLustering) (van Dongen, 2000), SPC (Super Paramagnetic Clustering) (Blatt et al., 1996), and MCODE (Molecular COmplex DEtection) (Bader and Hogue, 2003) are developed to find clusters in PPI networks. MCL algorithm simulates a flow on the input PPI network. MCL calculates successive powers of the adjacency matrix to reveal clusters in the graph. It has only one parameter which influences the number of clusters generated. SPC is a hierarchical algorithm inspired from an analogy with the physical properties of a ferromagnetic model subject to fluctuation at nonzero temperature. MCODE gives weights to each node in the PPI network by its local neighborhood density. MCODE uses a modified version of the clustering coefficient measure by introducing the idea of k-cores. A k-core is a subgraph in which all nodes have at least k edges in the induced subgraph. MCODE starts from the top weighted node and include neighborhood nodes with similar weights in the cluster. MCODE has a post-processing step to remove or add new nodes. Iteratively, MCODE continues with the next highest weight node in the network. MCODE is able to provide overlapping clusters.

There are also complex identification techniques which use additional information such as Gene Ontology annotations (Cho et al., 2007; Xu et al., 2011) and gene expression profiles (Li et al., 2012) to increase the biological specificity of identified complexes.

2.2. Identification of Signaling Pathways

Identification of signaling pathways from protein-protein interaction networks usually involves the user to specify a source and a target gene in the protein-protein interaction network. Hence, unlike complex identification methods, instead of analyzing the whole PPI network for pathways, a local region is analyzed between two nodes and either a simple path, or a complex network of parallel paths are identified as the signaling pathway between the source and the target gene. This restriction to a local region seems to make the problem of finding the pathways easier; however, the number of possible paths between two nodes grows exponentially with the number of nodes that connect these two nodes and based on the optimization function, finding the best path between the two nodes may become intractable quickly as the length of the path reaches 10-15 nodes. One of the first methods to identify signaling pathways from PPI networks is by Scott et al., (2006) in which they find the largest weighted path of a specific length between two nodes in a weighted PPI network. Since the problem is intractable even for small paths, they propose a dynamic programming solution based on color coding which does not guarantee to find the best path. Although the dynamic programming solution has an exponential time complexity, it is more efficient compared to searching all the possible paths. In addition, with the dynamic programming formulation, the search for the best pathway, which starts at one of the nodes of a given a set of starting nodes, I, is conducted in the whole PPI network. The user also specifies a fixed path length k. The method then tries to find the best path of length k that starts from one of the nodes in set I and ends at a vertex v. The vertex v is not given by the user; hence, all possible nodes in the network are tried as the end vertex, v. Below, we give the details of the color coding method which finds the best simple path of a specific length k starting from a set of nodes, I. The dynamic programming recurrence relation described below can be extended to find rooted trees and parallel paths as well (Scott et al., 2006)

A simple path between two nodes in a graph is a path in which each vertex in the path is visited only once and which contains no cycles. The probability of a simple path is defined as the product of edge weights along the path. For an easier formulation of the problem similar to shortest paths, the logarithm of the edge weights can be used in the formulation and the product of edge weights can be turned into summation of the logarithms of edge weights. If one converts the probability of an edge w, to a new edge weight as -log(w), then as the probabilities are between 0 and 1, the new weights will be positive values between 0 and Infinity. In this formulation, smaller probabilities will have larger weights and higher probabilities will have smaller weights, resulting in the best path to be the shortest path. The problem is, then, formally defined as follows. Given an undirected weighted graph $G = (V, E)$ with $|V| = n$ and $|E| = m$, find, for each node $v \in V$, a minimum-weight simple path of length k that starts from a node in I and ends at v. For the dynamic programming formulation, we define $W(v, S)$ as the minimum weight of a simple path of length $|S|$, which starts at a node in I, visits each node in S, and ends at v. Starting with an initial set with one node, S can be iteratively grown until $|S| = k$ by using the following recurrence.

$$W(v, S) = \min_{u \in S - \{v\}} W(u, S - \{v\}) + w(u, v), |S| > 1$$

$$W(v, \{v\}) = 0 \text{ if } v \in I \text{ and } \infty \text{ otherwise}$$

The recurrence is computed to fill in a table of $W(v, S)$iteratively for all v and S. The time complexity of this dynamic programming formulation is $O(kn^k)$. The number of possible simple paths of length k constructed with different nodes of the graph G is exactly:

$$O(n^k) = n(n-1)(n-2)\ldots(n-k+1)$$

Scott et al., propose the color coding technique to reduce this complexity to $O(2^k)$. Instead of treating each node of the graph as a unique entity, Scott et al., randomly assign k different colors to the nodes of the graph. The problem is then to search for a path that is composed of k different colors. A simple path with all the k different colored nodes is called a *colorful* path. The dynamic programming formulation can be modified to find the best colorful path starting from a set of nodes I, as follows:

$$W(v, S) = \min_{u:\, c(u) \in (S - \{c(v)\})} W(u, S - \{c(v)\}) + w(u, v), |S| > 1$$

$$W(v, \{c(v)\}) = 0 \text{ if } v \in I \text{ and } \infty \text{ otherwise}$$

In the formulation above, instead of the node ids the colors of the nodes are maintained in the set S. Therefore, the number of all the possible subsets is 2^k. The running time complexity is $O(2^k km)$ and the space complexity is $O(2^k km)$, both of which are much smaller functions compared to $O(n^k)$. Applied on a yeast network, the algorithm was able to find pathways of length up to 10. The authors report that it took about two hours to find paths of length 10. The color coding technique improves the running time 3-fold for 8 node pats and 14-fold for 9 node paths compared to the original dynamic programming formulation. However, note that, color coding may not always find the best path due to the random coloring that is applied. Therefore, a number of different runs of the algorithm with different color assignments can be used to improve its results.

There are other studies which analyze PPI networks for discovering signal transduction pathways. Steffen et al., (2002) proposed a method called NetSearch to find the signaling pathways by integrating PPI data with gene expression data in a non model-based banner. They did not employ any pathway based knowledge and tried to extract subnetworks from PPI networks in which the subnetwork members exhibit correlated expression profiles. Wang et al., (2011) proposed another method named CASCADE_SCAN based on the steepest descent method to identify signaling pathways from PPI networks. In CASCADE_SCAN, instead working on known start and end nodes, Wang et al., identify seed proteins in a given PPI network by computing densities around nodes. Seed proteins are selected as the nodes with the highest densities in the network. They extend the seed proteins to construct signaling pathways. Lan et al., (2011) integrate protein-protein interaction networks with genetic screens and expression profiling assays to construct signaling pathways. They formulate the pathway construction problem as a minimum flow optimization problem and solve it efficiently using existing linear programming solvers. Sing (2011) proposed a method named InfluenceFlow which integrates protein-protein interaction networks with RNA interference experiments for constructing signaling pathways. The constructed pathways have tree topologies. The amount of effect of RNA interference screens on the proteins is used to impose an order on the proteins of the signaling pathway. Influence flow assumes that the signal flows from proteins which show little response to RNA interference screens to those which show significant response.

3. Model-Based Approaches for Identification of Modules in PPI Networks

Model-based approaches utilize known complexes and pathways from well-studied model organisms to identify novel complexes or pathways. Statistical models capture the characteristics of existing modules using graph theoretic or biological information contained within the modules. In this section, we describe state of the art model based approaches to identify molecular complexes and signaling pathways in protein-protein interaction networks.

3.1. Identification of Molecular Complexes

There are relatively few studies that use model based approaches for complex identification in PPI networks. One of the most popular of these model-based methods is the RNSC (Restricted Neighborhood Search Clustering) algorithm (King et al., 2004). King et al., (2004) use learned graph theoretic and functional properties on known MIPS complexes as filters to remove falsely identified molecular complexes. Molecular complexes are identified by a regular clustering method which tries to find dense subgraphs within the PPI network.

Such clustering methods are common in non-model based identification of molecular complexes. However, King et al., use a model of known complexes in terms of graph theoretic and functional measures to increase the biological specificity of the identified clusters. The three criteria they use to filter the identified clusters are cluster size, cluster density, and functional homogeneity criteria. Cluster size indicates the number of subunits in the molecular complex. Graph theoretically, it is the number of nodes in the subgraph. Cluster density is a measure of how close the identified subgraph is to a clique. A clique is a graph in which all the pairs of nodes in the graph are connected by an edge. Cluster density is measured by the ratio of the number of edges in the subgraph to the number edges in a clique of the same number of nodes. Functional homogeneity is found by using the functional group information provided by von Mering et al., (2002) as the probability that a given set of proteins exhibits a certain functional group significantly. By comparing the identified molecular complexes with a benchmark dataset of known protein complexes, King et al., show that using model based filters increase the biological specificity of the identified clusters significantly.

Qi et al., (2008) model known complexes by a probabilistic Bayesian Network (BN) model. They train the BN model to represent topological and functional properties of known complexes. In addition to the cluster size and cluster density used by King et al., (2004), Qi et al. use additional topological features such as degree, edge, clustering coefficient, degree correlation statistics. As biological information, they include the protein/size statistics of a complex in their model. The structure of the BN model is decided manually. After building a single Bayesian Network model on a training set of known protein complexes, they use this model to assess the log-likelihood of an identified cluster. This approach is similar to the filtering idea used by King et al. However, instead of employing a simple threshold based filter, they use a more elaborate statistical approach to assess the likelihood of an identified cluster to be a real molecular complex. By integrating additional topological measures into the BN model, they are able to identify not only clique like complexes but molecular complexes with arbitrary topologies. Their results show that using a model based approach improves both the precision and recall of discovered complexes significantly compared to a non-model-based approach, MCODE (Bader and Hogue, 2003). One shortcoming of Qi et al. method is the use of a single Bayesian Network model to represent the whole repertoire of molecular complexes. It is quite possible that different molecular complexes exhibit quite different structural topologies; and therefore, a single BN model will not be able to accommodate these different topological properties. As a future research direction, it may be possible to partition the space of known molecular complexes based on their topological similarities and train a different BN model for each partition.

Shi et al., (2011) try to solve the problem of noise and overlapping clusters in PPI networks. Similar to Qi et al. work they use topological and biological features of known complexes to construct a model to increase the quality of the identified clusters. However, different from Qi et al. work, they use a "semi-supervised" approach and recursively train artificial neural network models. As additional biological features, they use polarity of amino acids information and Gene Ontology annotations. The Gene Ontology annotations are used to compute a semantic similarity measure between proteins that share an edge in the PPI network and this measure is used as the edge weight. Another important aspect in Shi et al., work is the "semi-supervised" approach in which the neural network is *retrained* on the fly as the complexes are discovered. Their results show that they can achieve a higher precision and accuracy compared to the Bayesian Network model proposed by Qi et al., (2008).

3.2. Identification of Signaling Pathways

Similar to the scarcity of model-based methods for complex identification, there are very few model based approached for identification of signaling pathways from protein-protein interaction networks. PathFinder (Bebek and Yang, 2007) is one such method which uses a set of trained association rules on known signaling pathways. In order to cope with the inherent noise in PPI networks, Bebek and Yang first eliminate non-specific interactions by using microarray expression profiles, subcellular localization and sequence information. For example, interactions between two proteins with very different expression profiles or two proteins at different subcellular locations are removed. PathFinder also deals with missing interactions and recovers them by employing protein family information. PathFinder gets the start and end proteins of the signaling pathway as input and finds a path between the nodes that represent these proteins in the PPI network. The knowledge in existing pathways are captured by mapping functional annotations of known proteins on the PPI network and learn functional interaction patterns in known pathways by using a well-known data mining technique called association rule mining. Bebek and Yang validated their method on a yeast PPI network and successfully identified well-known MAP kinase pathways.

A recent study by Supper et al., (2009) models signaling pathways in a bow-tie topology in contrast to a single source-single target topology. By analyzing existing pathways in the KEGG database (Kanehisa et al., 2006), they observe that in many signaling pathways, multiple signals are detected by multiple sensory proteins. These external signals are transduced into the cell and are integrated in a few core proteins. These few core proteins activate many transcription factors due to this signaling process. The overall process can be visualized as a bow-tie topology. By using this model, Supper et al. successfully construct the innate immune response pathway in human.

Summary

Protein-protein interaction networks are invaluable resources to uncover the modular organization of cellular processes. The complexity of genome-scale protein-protein interaction networks necessitates computational techniques which can automatically identify modular structures such as protein complexes and signal transduction pathways. There are various graph theoretical methods to discover strongly connected components and paths with high probability.

However, due to the noisy nature of high-throughput protein-protein interaction data, in order to uncover biologically specific modules, it is important to integrate additional biological data such as gene expression profiles and functional annotations. In addition, the structural and topological diversity of known pathways makes it difficult to discover the whole repertoire of modules by methods with simple heuristics. Therefore, methods that learn the structural and topological landscape of existing modules are promising for identification of molecular complexes and signaling pathways from protein-protein interaction networks.

References

Bader G, Hogue C. An automated method for finding molecular complexes in large protein interaction networks. *BMC Bioinformatics* (2003), 4:2.

Bebek G, Yang J. PathFinder: mining signal transduction pathway segments from protein-protein interaction networks. *BMC Bioinformatics* (2007), 8:335.

Blatt M, Wiseman S, Domany, E. Superparamagnetic clustering of data. *Physical Review Letters*. (1996) 76:3251–3254.

Chatr-aryamontri A, Ceol A, Palazzi LM, Nardelli G, Schneider MV, Castagnoli L, Cesareni G. MINT: the Molecular INTeraction database, *Nucl. Acids Res.* (2007) 35(suppl 1): D572-D574.

Cho YR, Hwang W, Ramanathan M, Zhang A. Semantic integration to identify overlapping functional modules in protein interaction networks. *BMC Bioinformatics*, (2007), 8:265.

De Las Rivas J, Fontanillo C.Protein–Protein Interactions Essentials: Key Concepts to Building and Analyzing Interactome Networks. *PLoS Comput Biol* (2010) 6(6): e1000807. doi:10.1371/journal.pcbi.1000807.

Güldener U, Münsterkötter M, Oesterheld M, Pagel P, Ruepp A, Mewes HW, Stümpflen V. MPact: the MIPS protein interaction resource on yeast. *Nucl. Acids Res.* (2006) 34: D436-D441.

Kanehisa M, Goto S, Hattori M, Aoki-Kinoshita KF, Itoh M, Kawashima S, Katayama T, Araki M, Hirakawa M. From genomics to chemical genomics: new developments in KEGG. *Nucleic Acids Res* (2006), 34 Database:D354-D357.

Kerrien S, Aranda B, Breuza L, Bridge A, Broackes-Carter F, Chen C, Duesbury M, Dumousseau M, Feuermann M, Hinz U, Jandrasits C, Jimenez RC, Khadake J, Mahadevan U, Masson P, Pedruzzi I, Pfeiffenberger E, Porras P, Raghunath A, Roechert B, Orchard S, Hermjakob H. The IntAct molecular interaction database in 2012 *Nucl. Acids Res.* (2012) 40(D1): D841-D846.

King AD, Pržulj N, Jurisica I. Protein complex prediction via cost-based clustering. Bioinformatics (2004) 20(17): 3013-3020.

Lan A, Smoly IY, Rapaport G, Lindquist S, Fraenkel E, Yeger-Lotem E. ResponseNet: revealing signaling and regulatory networks linking genetic and transcriptomic screening data. *Nucl. Acids Res.* (2011) 39 (suppl 2): W424-W429.

Li M, Wu X, Wang J, Pan Y. Towards the identification of protein complexes and functional modules by integrating PPI network and gene expression data. *BMC Bioinformatics*, (2012), 13:109.

Mathivanan S, Periaswamy B, Gandhi TK, Kandasamy K, Suresh S, Mohmood R, Ramachandra YL, Pandey A. An evaluation of human protein-protein interaction data in the public domain. *BMC Bioinformatics* (2006), 7(Suppl 5):S19.

Moschopoulos CN, Pavlopoulos GA, Iacucci E, Aerts J, Likothanassis S, Schneider R, Kossida S. Which clustering algorithm is better for predicting protein complexes? *BMC Research Notes*, (2011) 4:549.

Prasad, TSK et al., Human Protein Reference Database - 2009 Update. *Nuc. Acids Res.* (2009), 37, D767-72.

Qi Y, Balem F, Faloutsos C, Klein-Seetharaman J, Bar-Joseph Z. Protein complex identification by supervised graph local clustering. *Bioinformatics* (2008), 24:i250-i268.

Scott J, Ideker T, Karp RM, Sharan R. Efficient algorithms for detecting signaling pathways in protein interaction networks. *J Comput Biol.* (2006) 13(2):133-44.

Shi L, Lei X, Zhang A. Protein complex detection with semi-supervised learning in protein interaction networks. Proteome Science (2011), 9(Suppl 1):S5.

Shmoys, DS. Cut problems and their application to divide and conquer. *Approximation Algorithms for NP-Hard Problems*, (1995) Hochbaum D., *Ed.* PWS Publishing.

Stark C, Breitkreutz BJ, Chatr-Aryamontri A, Boucher L, Oughtred R, Livstone MS, Nixon J, Van Auken K, Wang X, Shi X, Reguly T, Rust JM, Winter A, Dolinski K, Tyers M. The BioGRID Interaction Database: 2011 update. *Nucl. Acids Res.* (2010) Nov 11.

Steffen M, Petti A, Aach J, D'haeseleer P, Church G. Automated modelling of signal transduction networks. *BMC Bioinformatics* (2002), 3:34.

Supper J, Spangenberg L, Planatscher H, Dräger A, Schröder A, Zell A. BowTieBuilder: modeling signal transduction pathways. *BMC Systems Biology* (2009), 3:67.

Szklarczyk D, Franceschini A, Kuhn M, Simonovic M, Roth A, Minguez P, Doerks T, Stark M, Muller J, Bork P, Jensen LJ, von Mering C. The STRING database in 2011: functional interaction networks of proteins, globally integrated and scored. *Nucl. Acids Res.* (2011) Jan;39 (Database issue):D561-8.

van Dongen, S. Graph Clustering by Flow Simulation. PhD thesis, University of Utrecht, May 2000.

von Mering C, Krause R, Snel B, Cornell M, Oliver SG, Fields S, Bork P. Comparative assessment of large scale data sets of protein–protein interactions. *Nature* (2002), 417, 399–403.

Wang K, Hu F, Xu K, Cheng H, Jiang M, Feng R, Li J, Wen T. CASCADE_SCAN: mining signal transduction network from high-throughput data based on steepest descent method. *BMC Bioinformatics* (2011), 12:164.

Wu J, Vallenius T, Ovaska K, Westermarck J, Makela TP, Hautaniemi, S. Integrated network analysis platform for protein-protein interactions, *Nature methods* (2009), 6, 75-77.

Xenarios I, Rice DW, Salwinski L, Baron MK, Marcotte EM, Eisenberg D. DIP: the database of interacting proteins. *Nucl. Acids Res.* (2000), 28:289-291.

Xu B, Lin H, Yang Z. Ontology integration to identify protein complex in protein interaction networks. *Proteome Science*, (2011), 9(Suppl 1):S7.

In: Recent Advances in Systems Biology Research
Editors: A. X. C. N. Valente, A. Sarkar and Y. Gao

ISBN: 978-1-62948-736-6

Chapter 6

Mass Spectrometry in Systems Biology

***Lei Cheng*[1], *Robert A. Fenton*[1] *and Trairak Pisitkun*[1, 2*]**

[1]Department of Biomedicine and Center for Interactions of Proteins in Epithelial Transport, Aarhus University, Denmark

[2]Faculty of Medicine, Chulalongkorn University, Bangkok, Thailand

Abstract

Protein mass spectrometry (MS) is one of the essential technologies for systems biology research. The recent advances in MS techniques have made MS-based proteomics a powerful technique to identify, quantify, and analyze a large number of proteins in a functional context. In this chapter, we discuss the recent MS techniques for protein identification and quantification and the application of these techniques to study biological processes within the cell i.e., post-translational modifications and protein-protein interactions.

Keywords: Proteomics, phosphorylation, protein-protein interactions

Introduction

In the current era, emerging technologies have opened up a new frontier in biological research namely 'systems biology'. Rather than focusing on an individual element of a biological process, systems biology strives to explore all elements of a biological process and understand the interconnection between them. Here we describe a state-of-the-art method in systems biology viz. protein mass spectrometry (MS). MS is generally employed to solve the following systems biology tasks:

* Correspondence to Trairak Pisitkun, Faculty of Medicine, Chulalongkorn University, Bangkok, Thailand, E-mail: trairak@gmail.com.

1) Identify the protein components of the system being analyzed
2) Determine how these components interact i.e., by quantification of changes in protein abundance, protein post-translational modifications (PTMs), protein translocation, and protein-protein interactions (PPIs)

The main focus of this chapter is in the recent advances in the technical aspects of protein identification and quantification by MS and the application of these techniques to PTMs and PPIs studies.

Protein Identification by Mass Spectrometry

Liquid chromatography-MS/MS (LC-MS/MS) or high performance liquid chromatography-MS/MS (HPLC-MS/MS) is an analytical technique in which the separation ability of LC/HPLC is combined with tandem mass spectrometry (MS/MS). LC-MS/MS allows highly sensitive and continuous analysis of complex samples on MS instruments. Thus, LC-MS/MS is a typical setup of modern mass spectrometers configured for protein identification tasks in a complex mixture of proteins.

Figure 1 illustrates a summarized view of a typical "shotgun" proteomic approach, which is the most common strategy used in large-scale protein identifications from complex samples. "Shotgun" proteomics, also referred to as the bottom-up technique, starts out with the proteolytic digestion of a complex mixture of proteins, such as cell lysate or tissue homogenate. Trypsin is the favorable proteolytic enzyme (protease) of choice and hydrolyzes peptide bonds following lysine (K) or arginine (R) moieties. Due to the high degree of protein complexity and vast dynamic range of proteins expressed in complex biological mixtures, fractionation (or separation) is usually performed before and/or after digestion in order to reduce the complexity of proteins and/or peptides, respectively. Enrichment for a group of proteins or peptides is usually performed either before or after fractionation. Described later in this chapter are examples of enrichment for protein complexes in protein-protein interaction studies and for phosphopeptides in phosphoproteomic studies. The subsequently yielded peptide mixture is separated by HPLC and then analyzed by a mass spectrometer. Reversed phase (RP) materials, such as C18 materials, are often used to isolate and fractionate peptide samples. In RP chromatography, peptides are separated according to their total hydrophobicity. The more hydrophobic the peptide is, the longer it will be retained on an RP column. The more hydrophilic peptides are usually eluted earlier during the elution gradient. The time at which a specific peptide is eluted (retention time) under the same HPLC setup can also be considered as an identifying character for that peptide. Through HPLC, peptides are often directly introduced into the tandem mass spectrometer via the electrospray ionization (ESI) technique. This technique produces peptide ions from the continuous peptide flow along an elution gradient and allows the simultaneous MS analysis of eluted peptides. The nano-flow HPLC separates very low quantities of complex peptide mixtures under high pressure and enables highly sensitive MS analysis.

Tandem MS, also known as MS/MS (MS^2) or higher order MS (MS^n), involves multiple steps of mass selection from survey mass analyses (MS^1 scans), with fragmentation of the selected analyte occurring in between. MS^1 scans yield mass-to-charge (m/z) spectra of parent peptide ions.

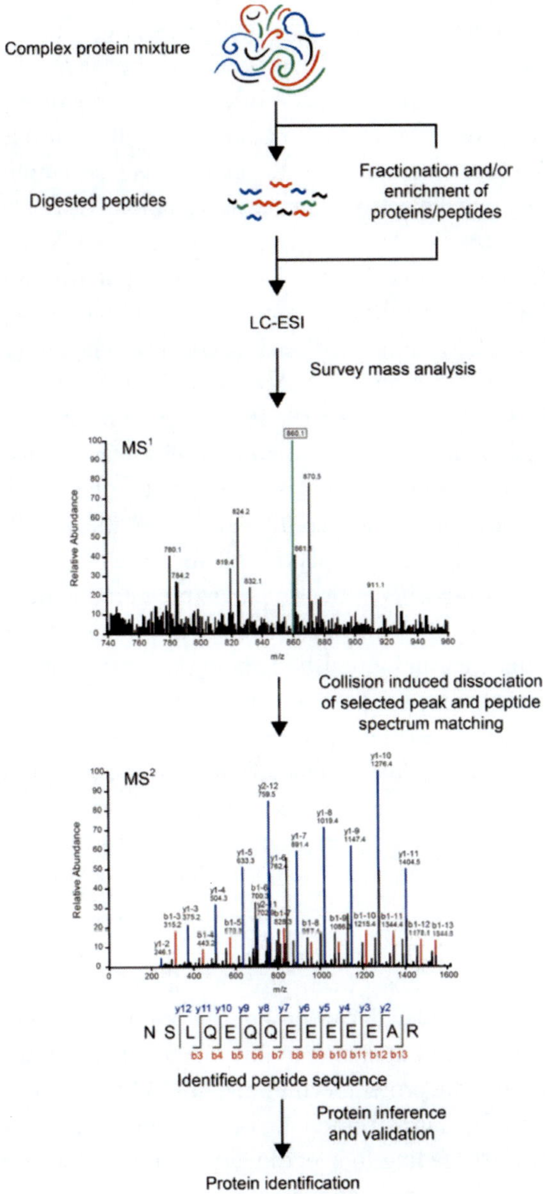

Figure 1. A "shotgun" proteomic workflow. The workflow begins with the proteolytic digestion of a complex mixture of proteins using proteases such as trypsin. Various fractionation and/or enrichment techniques are usually applied before and/or after the proteolytic digestion in order to reduce the complexity of samples and/or enrich for a group of proteins or peptides. Peptides are often introduced through nano-flow HPLC into the tandem mass spectrometer via the electrospray ionization (ESI) technique. Tandem MS starts out with survey mass analyses (MS^1 scans) that yield mass-to-charge (m/z) spectra of parent peptide ions. Parent m/z peaks from the MS^1 scan are selected (e.g., the green peak in this spectrum) for collision-induced dissociation (CID), which creates a series of b- and y-type ion fragments in the second stage mass analysis (MS^2). Peptide identification is typically performed by peptide spectrum matching via computer-assisted search algorithms. Subsequently, protein identification is inferred from the identity of the identified peptides, a process that often requires statistical validation.

Parent m/z peaks from the MS^1 scan are selected one at a time for collision-induced dissociation (CID), which fragments the parent peptide backbone. This creates complementary b- and y-type ions as a series of fragments in the second stage mass analysis (MS^2). The b-ions contain the N-terminus through the cleavage site; the y-ions contain the C-terminus through the cleavage site. The difference in m/z between adjacent b- or y- ion series is exactly the residue mass of the amino acid present in one fragment, but absent in the other, thus allowing unique identification of that amino acid. Although CID is the most common fragmentation method, other fragmentation techniques can be employed including electron capture dissociation (ECD) and electron transfer dissociation (ETD), both producing c- and z-type ion series. Information on sequence and even PTMs of a particular peptide can be obtained, either by de novo sequencing [1] or via computer-assisted comparison against theoretical spectra of all possible tryptic peptides (based on sequence data from genome sequencing projects) [2]. Many automated peptide spectrum matching algorithms for peptide identification have been developed. MASCOT (http://www.matrixscience.com/) and SEQUEST (http://fields.scripps.edu/sequest/) are among the most commonly used algorithms. If an identified peptide is unique to a given protein then the identity of the proteins can be inferred from the identity of these peptides. The greater the number of unique peptides identified for a protein, the higher the confidence is for the protein identification. Under circumstances where only one unique peptide is identified for a particular protein, it is necessary to minimize false-positive identification by introducing computational methods such as target-decoy analysis [3]. Another challenge in protein inference consists of dealing with peptide spectrum matches ambiguously mapping to several protein entries in the protein database. Statistical validation of protein identifications can be obtained through an estimation of false discovery rate (FDR) [4].

With the current technology, data obtained from a large-scale proteomic project can yield the identification of thousands of proteins from a single LC-MS/MS analysis. With extremely complex samples, it is possible to add pre-fractionation steps on a subcellular level (e.g., by differential centrifugation), protein level (e.g., by 1D or 2D SDS-PAGE), or peptide level (e.g., by strong cation exchange) [5]. SDS-PAGE separation is most commonly used as it has the advantage of effectively removing SDS and other detergents that interfere with MS analysis. At the same time, subsequent protein identification via LC-MS/MS can be confirmed by reassessing the molecular weight of the original protein sample according to its location on the protein gel. A disadvantage to this method is that extracting the proteins from the gel, or directly digesting the proteins within the gel (in-gel digestion), requires cutting the gel into multiple small pieces. Furthermore, the accessibility of the proteolytic enzymes to the proteins inside the gel varies leading to irreproducible and low recovery of peptides extracted from the gel. Chromatographic pre-fractionation steps such as strong cation exchange (SCX) or hydrophilic interaction liquid chromatography (HILIC) are also currently widely used to fractionate extremely complex samples. This dimension of separation can be automated online with a secondary RP separation enabling 'MudPIT' or 2D LC-MS analysis [6-8] that allows the identification of proteins with a much lower abundance in each sample. Even though more initial starting material is often required, together with faster and more sensitive MS instruments and better data analysis software, development and utilization of pre-fractionation methods has taken researchers one step closer to a complete proteome coverage.

Apart from the 'bottom-up' proteomic approach described above, the 'top-down' technique has also gained significant progress. Unlike 'bottom-up', the 'top-down' approach usually does not involve any enzymatic cleavage and instead the intact protein is subjected to MS analysis. The advantage of this method is that no primary amino acid sequence or PTM information is lost during the analysis. However, the analysis of large proteins, multiple charged large fragment ions, and complicated MS/MS spectra presents a great challenge to this approach [9, 10].

Quantitative Mass Spectrometry

The ability to identify and characterize proteins of interest is important and MS-based proteomic methods have excelled in these areas over the past years. However, it is becoming increasingly clear that the ability to quantify the changes in the abundance of these proteins and their functional PTMs in a biological context (such as the changes of protein phosphorylation upon the activation of certain pathways), will provide further insights into relevant biological/physiological questions. Traditionally, quantitative proteomics was usually achieved by two-dimensional gel electrophoresis-based methods. However, these methods are time-consuming and only highly abundant proteins can be visualized and quantified on the gel. Furthermore, the correct identification of the protein bands usually requires downstream MS characterization. Currently, a number of MS-based methods have been developed for quantitative proteomic analyses that have allowed more sensitive, simultaneous and large-scale quantification of proteins and protein PTMs from complex biological samples. Current MS-based quantitative methods include label-based quantification, label-free quantification as well as the more recent targeted quantification. Regardless of the quantitative method employed, MS-based quantification is most often a relative quantification, with direct comparisons between two or more sample sets. Absolute quantification can also be achieved if an internal standard of known quantity is 'spiked' into the starting sample and analyzed at the same time.

Label-Based Quantification

Label-based quantification is widely employed for MS-based proteomic quantification and is based on MS measurement of peptides that are differentially labeled, in most cases isotopically (e.g., 2H vs.1H, 13C vs.12C, or 18O vs.16O). Isotopic labeling of peptides does not affect HPLC retention time of the peptides, so differentially labeled versions of identical peptides will be eluted simultaneously during HPLC separation prior to MS analysis. A number of label-based methods have been developed and successfully applied in many different projects. Based on the chemical reaction and design of the quantification protocols, peptide relative quantification data can be obtained from either the precursor scan (MS^1 level) or from the MS/MS scan (MS^2 level). Stable isotope labeling with amino acids in cell culture (SILAC), isobaric tags for relative and absolute quantification (iTRAQ), and stable isotope dimethyl labeling are currently the most popular approaches to label-based quantification. Simplified diagrams of these three stable isotopic labeling approaches are schematically shown for easy comparison in Figure 2.

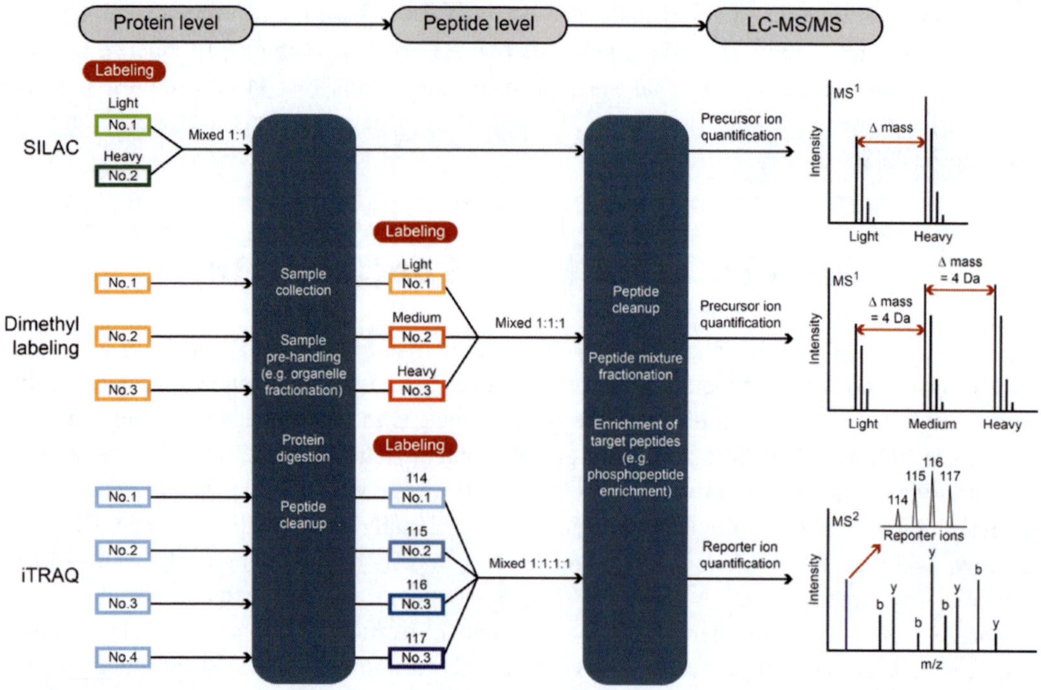

Figure 2. Schematic representation of three stable isotopic labeling approaches: stable isotope labeling with amino acids in cell culture (SILAC), stable isotope dimethyl labeling, and isobaric tags for relative and absolute quantification (iTRAQ). In this example, SILAC, dimethyl labeling, and iTRAQ approaches are applied on experiments with 2, 3, and 4 biological treatments/conditions, respectively. In SILAC method, two cell populations are grown in culture media containing either a 'light' or a 'heavy' form of a particular amino acid (e.g., 12C or 13C labeled L-lysine, respectively). After metabolic labeling and experimental manipulation, the cultured cells are harvested, equally mixed, and prepared for and analyzed by LC-MS/MS. Peptide abundances in the two samples are quantified at the precursor ion (MS^1) level. The mass difference (Δ mass) between the 'light' and 'heavy' peaks of a certain peptide is determined by the composition of isotopic-labeled amino acids used, e.g., 13C6-lysine and 13C615N 4-arginine have mass differences of 6 and 10 Da, respectively, compared to their light-isotope counterparts. In dimethyl labeling and iTRAQ methods, different samples are labeled at the peptide level before being mixed and further processed for and analyzed by LC-MS/MS. The quantification for dimethyl labeling method is performed at the MS^1 level. The example MS^1 spectrum of dimethyl labeling method shows the three most commonly used light, medium, and heavy labeling conditions i.e., labeling masses of 28 Da, 32 Da, and 36 Da, with sequentially mass differences of 4 Da. In contrast with SILAC and dimethyl labeling methods, iTRAQ quantification is performed at the MS^2 level after the reporter ions are liberated from their precursor ion during the peptide fragmentation. The example MS^2 spectrum of iTRAQ method shows a group of 4-plex reporter ions (114–117 Da) at the low mass area of the spectrum.

SILAC: The SILAC approach is employed when samples generated from cultured cells are utilized. Isotopic-labeled amino acids are added to cell culture medium during cell growth, with the proteins being subsequently labeled through metabolic incorporation of the amino acids. Complete incorporation of labeled amino acids (i.e., all proteins present in the cell are metabolically labeled) is usually achieved after several cell-doublings [11]. A light form and a heavy form of isotopic-labeled amino acids are typically utilized (Figure 2), however a medium form can be added for the quantification of up to three sets of cells. The differentially labeled cells can then be mixed equally, digested, and followed by LC-MS/MS analysis. Relative quantitative information can be obtained by comparing peak intensities or peak areas generated from these differentially labeled peptides at the MS^1-scan (precursor ion scan) level. The advantage of this method arises from the labeling step, which being the very first stage of sample preparation, significantly reduces sample preparation biases. However, SILAC is time-consuming, can alter cell proliferation, is a relatively expensive approach and cannot be applied to human clinical samples or animal tissues. SILAC labeling can complicate mass spectra at the MS^1 level, as the same peptide can be represented as multiple peaks, which might lead to difficulties in data analysis and interpretation. The SILAC approach has also been successfully applied to animal models, e.g., SILAC mouse [12], which has proved to be the most "real" *in vivo* labeling for MS-based quantification. However, the isotopic incorporation of such a SILAC mouse is difficult to optimize, time-consuming, and extremely expensive.

iTRAQ: In contrast to SILAC, the iTRAQ approach labels different samples at the peptide level. A similar approach to iTRAQ is the labeling of peptides with tandem mass tag (TMT) reagents. Both methods are based on the covalent labeling of primary amines at the N-termini and lysine side-chains of peptides with isobaric tags, followed by MS/MS analysis [13, 14]. The iTRAQ isobaric tag consists of a reporter group (based on N-methylpiperazine), a mass balance group (carbonyl), and a peptide-reactive group (NHS ester). iTRAQ labeling does not interfere with peptide identification as the total mass of the different isobaric tags is kept constant at 145 Da (for 4-plex version of iTRAQ) using differential isotopes 13C, 15N, and 18O. Thus, differentially labeled peptides still have the same m/z and are selected together for MS/MS analysis. The reporter groups yield a series of reporter ions at the low mass area of the MS/MS spectrum (e.g., 114, 115, 116, or 117 Da for 4-plex iTRAQ, Figure 2) and their peak intensities correspond to the relative quantities of the peptide in different samples. Due to their low masses, the spectra of these reporter ions do not normally interfere with the peptide identification. This type of quantification method is very flexible and can be adapted to various biological samples. The labeling procedure is simple and usually up to 8 different biological conditions can be investigated simultaneously. This is advantageous for research projects involving multiple data points such as a time course design, e.g., monitoring protein expression at a series of different time points after stimulation. iTRAQ labeling does not lead to complicated MS^1 spectra, since the differentially labeled peptides are of the same mass and they are co-eluted from HPLC prior to MS analysis. However, compared to the SILAC approach, iTRAQ labeling is incorporated at a later stage of sample preparation which will increase the quantification biases. Once the labeled peptides are pooled, further experimental biases is minimized.

Dimethyl labeling: This is another chemical labeling method for quantitative proteomics. It is a fast and simple approach, with samples labeled with different isotopes at the peptide level. It does not require expensive reagents (like iTRAQ reagents). Dimethyl labeling can also be used to label and quantify a wide range of samples, including the ones that are not feasible for metabolic labeling (such as using SILAC). However, only up to 3-4 different biological states can be quantified at the same time, and the quantification is based on the MS^1-scan level similar to SILAC. Dimethyl labeling also leads to complicated precursor ion spectra from different isotopes making subsequent extraction of peaks for quantification, especially for complex samples, more difficult. The basic chemical reaction involved in stable isotope dimethyl labeling is the reaction of formaldehyde with primary amines at the peptide N-terminus and side-chain of a lysine residue. This reaction forms a Schiff base that is reduced by sodium cyanoborohydride forming a secondary amine [15]. When different combinations of formaldehyde and cyanoborohydride isotopic isomers are used, labeled peptides gain different additional masses - up to five different possibilities. The three most commonly used labeling conditions produce a mass increase of 28 Da, 32 Da, and 36 Da corresponding to the light, medium, and heavy labeling, sequentially different by 4 Da (Figure 2). Stable isotope dimethyl labeling has also been reported to assist peptide identification and sequencing [16, 17].

Label-Free Quantification

Various MS-based label-free quantitative methods exist. With the label-free approach, quantification is based on comparison of separate LC-MS/MS analyses of different samples. There is no chemical labeling step, which helps reduce sample handling and sample loss. At the same time, it is cost effective. However, it is highly dependent on the reproducibility of sample preparation and LC-MS/MS analysis. One significant advantage of this approach is that there is no limitation on the number of samples or sample conditions used in an experiment. In label-free quantification, peptides can be quantified according to integrated peak intensities/areas, integrated ion counts, number of peptide identifications, or spectral counting. Figure 3 illustrates a simplified diagram of label-free quantification of two samples using integrated peak intensities/areas (comparing area under curves of MS^1 extracted ion chromatograms). Quantification based on integrated ion counts or peak intensities/areas is relatively straightforward, and has been adapted to a number of studies where linear responses and reproducibility have been shown [18-20]. When it comes to complex samples, quantification relies on efficient peptide separation and sufficient resolution of mass to distinguish the peptides with very small mass differences eluted at the same time (co-eluted peaks). Since this type of quantification is usually based on the MS^1 scan, it requires high sensitivity and mass accuracy during the MS^1 analysis. On the other hand, the quantification that is not based on the MS^1 scan, such as spectral counting or peptide identification counting, is better at coping with co-eluted MS^1 peaks but with less quantitative accuracy; especially for low abundant peptides. These methods are based on counting the MS/MS spectral numbers or peptide identification numbers, either with or without normalizing the counted numbers to the theoretical MS-observable total peptide numbers for each protein (allowing a quantitative comparison between different proteins without biases from protein size and protease used) [21-23].

MS^E is another notable technique that can be used for label-free quantification. MS^E analysis is usually conducted on Q-TOF instruments. It refers to data-independent acquisition (DIA) MS analysis where all ions from one survey scan are fragmented and analyzed at the same time (unlike data-dependent acquisition (DDA) as shown in Figure 1). This is carried out by alternative low and high collision energy during the analysis. Information on both peptide quantification and identification can be obtained in a single experiment. Both relative and absolute quantification can be obtained using MS^E depending on whether an internal standard is introduced [24, 25]. Data analysis can be difficult due to the more complex spectra obtained especially for complex samples, however, there are reports showing that this method yields high sequence coverage and reliable quantification [26, 27].

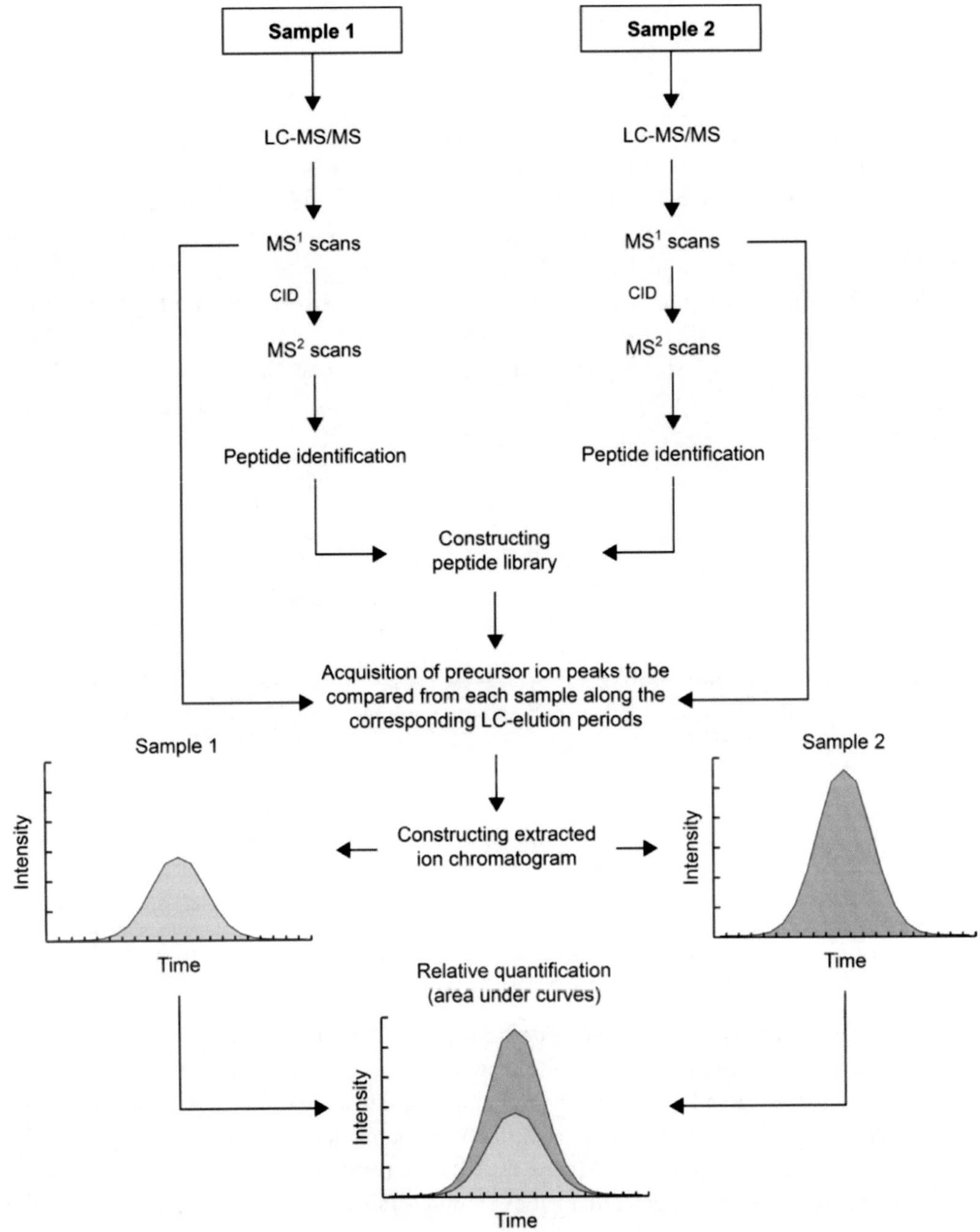

Figure 3. A simplified diagram of extracted ion chromatogram-based label-free quantification. In this example, two different samples are separately analyzed by LC-MS/MS using a typical "shotgun" proteomic workflow. A library of all peptides identified from both samples is constructed in order to organize the acquired data. The relative quantification of peptides of interest is performed by plotting the intensity of the precursor (MS^1) signal of a chosen peptide ion as a function of retention time so-called "extracted ion chromatogram" and then comparing the area under the two chromatographic curves.

Targeted Quantification

Targeted quantification by LC-MS/MS aims to quantify specific peptides based on their abundance across different samples. It can be applied with or without stable isotope labeling. Unlike large-scale quantification, targeted quantification is mainly used for the verification of data obtained from large-scale discovery studies or for biomarker studies [28-30]. Selected reaction monitoring (SRM), previously also referred to as multiple reaction monitoring (MRM), is the most commonly used targeted quantitative approach. SRM is predominantly conducted in triple quadrupole instruments. Precursor ions are selected with a narrow window and transmitted from the first quadrupole (Q1) to the second quadrupole (Q2), where they are fragmented. Subsequently, several daughter ions are selected and individually detected in the third quadrupole (Q3), see Figure 4. Each precursor ion/daughter ion pair is termed a transition.

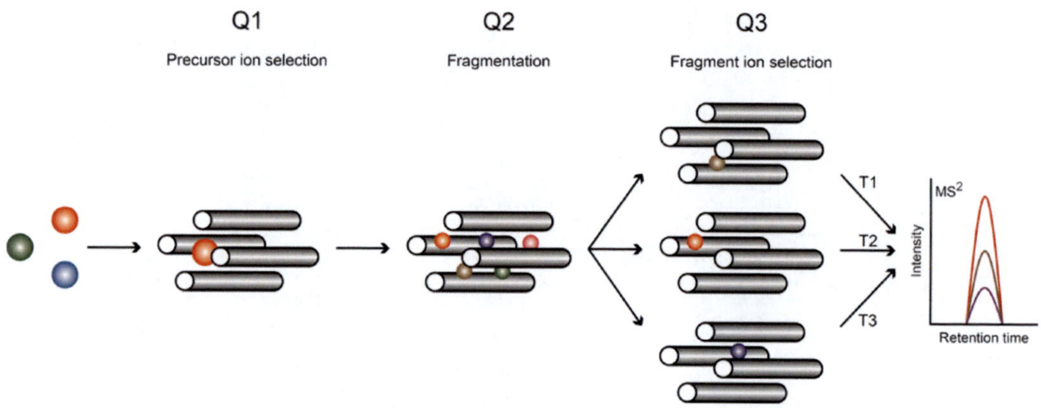

Figure 4. A simplified diagram of selected reaction monitoring (SRM) approach by triple quadrupole mass analysis. The first quadrupole (Q1) is used to select a pre-specified precursor ion for fragmentation in the second quadrupole (Q2). The third quadrupole (Q3) then selects several daughter ions for the quantification. In this example, three daughter ions from a precursor ion (3 transitions, T1-T3) are quantified.

The selection of precursor and daughter ions is based on m/z as well as LC retention time of the targeted peptide transitions. One transition is usually not enough to verify successful peptide selection, especially in a very complex biological sample. For a peptide with several transitions set up, the resulting spectra will be a series of transitions with varying intensities (e.g., transitions T1, T2, and T3 in Figure 4). Relative peptide quantification can be obtained by comparing the sum of the peak intensities from all transitions. SRM detection relies on the detection of both the precursor ion and its daughter ions, so it is a highly specific detection method that is also very sensitive and reproducible. However, the SRM method has a limitation in that it depends on the previous or simultaneous DDA type of peptide identification in order to obtain the transition information for the peptides of interest.

Another novel quantitative technique called SWATH MS, which complements the classical SRM method, was recently introduced. This technique is based on DIA analysis, which allows the acquisition of all fragment ions from all precursor ions in consecutive precursor ion isolation windows across the mass range of interest. This method is optimally carried out on the TripleTOF instrument, utilizing its fast scan time and high mass accuracy [31]. The recorded data from one sample injection contain information on all detectable precursor and fragment ions. Thus, this approach can benefit targeted quantification since no upfront development of transition list is required on specific targets and targets are 'mined' post-acquisition. By utilizing the fragment ion spectral libraries combined with the targeted data extraction method, SWATH MS has proved to possess the same high sensitivity and reproducibility as SRM, but for larger-scale peptide quantification [31]. Moreover, SWATH MS acquired data sets can be re-mined for further targeted quantification of newly identified peptides.

Study of Protein Post-Translational Modifications (PTMs)

Protein PTMs such as phosphorylation, acetylation, methylation, ubiquitination, sumoylation, and glycosylation can play important roles in biological processes within the cell. Changes in PTMs can affect protein conformation, protein localization, protein interaction with other proteins, and protein degradation. Protein PTMs often occur through the covalent attachment of small molecules to certain amino acid residues in the protein causing an increment of protein mass that is detectable by MS. MS analysis of protein PTMs is often a challenging task. For example, protein glycosylation usually leads to the attachment of many different glycan structures to the protein and protein ubiquitination may result in the addition of single or multiple ubiquitin molecules on a modified lysine residue. Due to the large number of different possibilities, it is difficult to predict the modifying mass for certain PTMs, thus adding difficulty to their characterization. Protein phosphorylation is among one of the relatively easy PTMs that can be addressed using MS-based methods. It usually occurs on serine, threonine, or tyrosine residue resulting in a mass increase of 80 Da on the modified site. Protein phosphorylation has thus been widely studied by MS-based methods and has evolved as a prominent branch in the proteomic field; so-called "phosphoproteomics". Reversible protein phosphorylation is a key PTM responsible for various cellular regulatory mechanisms and is typically involved in complex cellular signaling pathways. Thus, the study of phosphorylation has always been a major focus in systems biology. Here we discuss in greater details the techniques involved in phosphoproteomic research and aspects of data analysis that can bring a more confident and fruitful interpretation of the results.

Phosphoproteomics

Improved MS techniques, the development of various enrichment strategies for phosphoproteins/peptides, and the ability to monitor phosphorylation events on a global scale have all made MS-based phosphoproteomics a popular research area. With the additional power of quantitative proteomics, not only can MS-based phosphoproteomics broaden our catalog of phosphoproteins in the system, but it also allows the analysis of phosphorylation-based signaling networks in a global and quantitative manner. A typical MS-based phosphoproteomic workflow consists of three steps: sample preparation, phosphoprotein /peptide enrichment, and MS analysis (Figure 5).

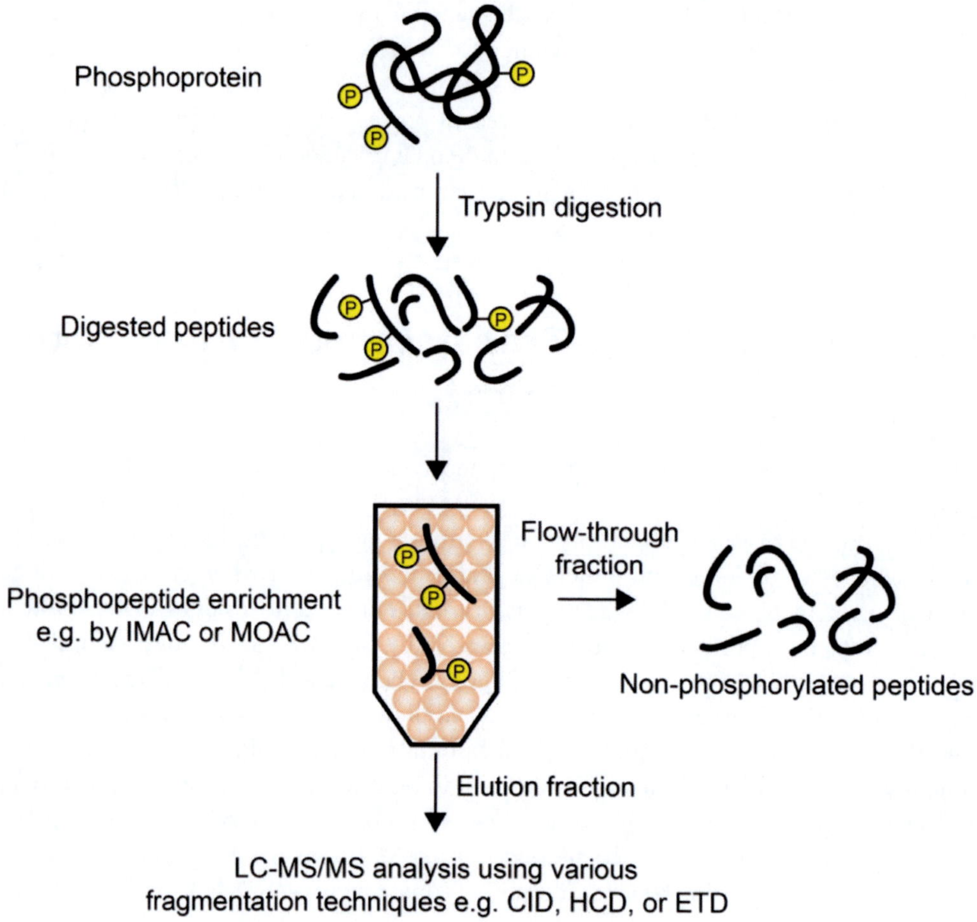

Figure 5. A typical MS-based phosphoproteomic workflow. A particular phosphoprotein in a complex mixture of proteins is digested with trypsin and the resulting peptide mixture is enriched for phosphopeptides using immobilized metal affinity chromatography (IMAC) and/or metal oxide affinity chromatography (MOAC). The phosphopeptides are eluted and analyzed by LC-MS/MS.

Sample preparation and phospho-enrichment: The most fundamental aspect to consider during phosphoproteomic sample preparation is to keep protein phosphorylation in its original state. For instance, protein phosphatase inhibitors should be added throughout the sample preparation procedure to avoid the removal of transient phosphorylation by naturally occurring protein phosphatases [32].

The study of protein phosphorylation is challenging since phosphorylated proteins are often of very low abundance and protein phosphorylation is substoichiometric. Phosphorylated peptides from a complex mixture often exhibit low ionization efficiency during MS analysis. Thus, careful sample preparation and efficient phosphopeptide enrichment are required prior to MS analysis. Phospho-enrichment can be conducted on both the protein level and the peptide level, and each approach has advantages and disadvantages. Phosphoproteins are normally enriched using antibody-based methods. The enrichment of phosphoproteins is suitable for targeted protein analysis, e.g., the analysis of possible phosphorylation sites in one or a few target phosphoproteins. But the specificity of these antibody-based methods is not always high, and the characterization for certain phosphorylation sites is still difficult using this level of enrichment. Phosphopeptide enrichment methods are more widely adapted to the 'bottom-up' proteomic approach. They can also be performed after phosphoprotein enrichment for further purification on the peptide level. A number of phosphopeptide enrichment methods have been shown to be efficient and sensitive towards phosphopeptides from complex mixtures. Different phosphopeptide enrichment methods can be categorized into two major groups: chemical derivatization methods and affinity enrichment methods. Chemical derivatization methods such as β-elimination combined with Michael addition reaction and phosphoramidate chemistry (PAC) have been successfully applied for phosphopeptide enrichment. The β-elimination-based method excludes tyrosine-phosphorylated peptides, while PAC targets all phosphorylated residues. However, the disadvantages of chemical derivatization methods include complex experimental procedures, sample loss, and unexpected side reactions [33, 34]. Calcium phosphate precipitation (CPP) is another enrichment method that precipitates phosphopeptides from tryptic digests. This method is easy to apply but the specificity is not ideal. It is usually employed as a pre-enrichment step in combination with other enrichment methods [35]. Immobilized metal affinity chromatography (IMAC) is the most widely used method for affinity enrichment of phosphopeptides. It is based on the affinity of the negatively charged phosphate groups for the positively charged metal ions such as Ga^{3+} and Fe^{3+} [36]. However, non-specific binding from acidic peptides, due to competing affinity of the side chain carboxyl groups to the positively charged metal ions, remains a major problem. Recent studies have demonstrated that metal oxide affinity chromatography (MOAC), e.g., using titanium dioxide (TiO_2), is a stable and robust alternative to IMAC for phosphopeptide enrichment [37, 38]. Compared to IMAC, it is simpler and easier to prepare and apply, and TiO_2 is more tolerant towards contaminating reagents including salts and detergents [38]. The carboxyl groups of acidic amino acids (aspartic acid and glutamic acid) also exhibit affinity for TiO_2 materials, but it has been demonstrated that non-specific binding can be greatly reduced by adding DHB or hydroxy acid to the sample loading solution [32, 37, 39]. A well-established protocol using self-packed micro-columns of TiO_2 powder is also available [40]. As tyrosine phosphorylation is relatively rare, an additional enrichment step using phosphotyrosine-specific antibodies has been successfully adapted for the large-scale analysis of protein tyrosine phosphorylation. Compared to non-enrichment approaches, this approach has led to the identification of significantly larger numbers of tyrosine phosphorylation sites [41, 42]. In addition to the methods mentioned above, chromatographic prefractionation techniques such as SCX or HILIC can also be applied as charge-based enrichment methods for phosphopeptides. These different enrichment techniques are complementary [43]. Therefore, comprehensive identification of phosphopeptides in a given sample may require a combination of different enrichment techniques either in series or in parallel.

MS analysis: During mass spectrometry, traditional CID usually results in the poor fragmentation of phosphopeptides in the MS^2 spectrum because of a predominant loss of phosphoric acid, H_3PO_4 (−98 Da), from the parent ion (neutral loss). The recent development and application of MS techniques including precursor ion scanning, neutral loss scanning, and multi-stage activation (MSA) have made the identification of phosphorylation sites in complex biological samples a routine procedure [44-46]. CID and MSA combined with a high mass accuracy analyzer [47, 48] as well as more recent fragmentation methods including higher-energy collisional dissociation (HCD) and ETD have been applied to phosphoproteomic analysis [49-51]. These techniques have allowed better fragmentation of phosphopeptides, improved assignment of phosphorylation sites, and increased sensitivity of MS analysis.

One of the remaining challenges in phosphoproteomic data analysis is how to confidently assign correct phosphorylation sites. A number of different approaches have been established for this purpose including Ascore [52], highly similar PTM score [53], and Mascot Delta Score (MD-score) [54]. Currently, there is no gold standard for judging the correctness of phosphorylation site assignment, but a new matrix called false localization rate (FLR) is promising and underway [55].

The development of quantitative MS methods including stable isotope labeling, label-free approaches, and the more recent SRM type of targeted analysis have made quantitative phosphoproteomics possible and helped generate more biologically relevant protein phosphorylation data. In addition, the development of bioinformatics tools (e.g., motif and network analysis tools [56, 57]) and databases (e.g., http://www.phosphosite.org/) have allowed better interpretation and utilization of phosphoproteomic results. The development, maintenance, and utilization of these resources are still moving forward.

Study of Protein-Protein Interactions (PPIs)

Proteins rarely function alone and often need to bind or interact with other proteins in order to carry out their biological function. PPIs are crucial for a variety of biological processes, such as signal transduction and metabolic or genetic regulation. Disruption of these PPIs can lead to disease. Complete understanding of when, where, how, and under what biological/physiological circumstances proteins interact in dynamic "*interactome*" networks is essential to comprehend cell function and facilitate therapeutic strategies for restoration of normal PPIs within diseased cells.

The advances in MS-based proteomic methods have allowed sensitive characterization of proteins and also quantitative analysis of relative protein amount between different samples making these methods highly suitable for protein-protein interaction studies. Affinity purification-mass spectrometry (AP-MS) is the most widely adapted workflow for MS-based protein-protein interaction studies. It enables the identification of interacting proteins involved in different protein complexes, and provides easy and fast means to study protein-protein interactions. The basic idea for AP-MS is to use suitable ligand-target binding systems to purify target protein together with its interacting partners, and subsequently identify the interacting complex using MS analysis. Antibody-based affinity purification is most commonly used. Without subsequent MS analysis, this method can only be used to validate hypothetical binding/interacting proteins using a western blotting approach, which is low-throughput and time consuming. When combined with MS analysis, all potential binding/interacting proteins can be identified together in a single experiment. Figure 6 displays a typical antibody-based AP-MS workflow.

In this case, an antibody recognizing a specific epitope tag of the target protein is used to pull down the interacting complex. MS analysis of the pull-down products leads to the identification of target protein, all the binding/interacting partners of target proteins as well as non-specific proteins binding to the antibody or being detected by the antibody. On the other hand, the affinity purification of control sample (containing the target protein without the epitope tag) usually pulls down only non-specific proteins. By comparing the results from MS analysis of both affinity purification experiments, a list of potential binding/interacting proteins can be obtained for further validation. Beside the antibody-based procedure, other affinity tags such as His-tag (using nickel or cobalt to pull down) and Strep-tag (using streptavidin to pull down) are also commonly used for affinity purification. One disadvantage of these approaches is that they are dependent on highly specific antibodies against the native protein, or require genetic modification of proteins, which in itself can alter protein conformation and subsequent interacting proteins.

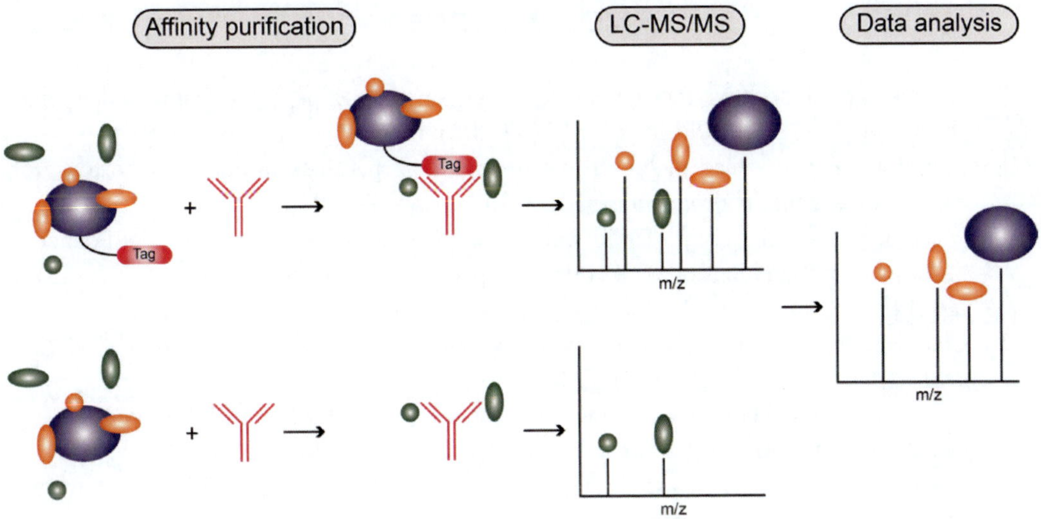

Figure 6. A typical antibody-based affinity purification-mass spectrometry (AP-MS) workflow. Affinity purification is performed on two different samples containing a target protein with or without a specific epitope tag. An antibody recognizing the specific epitope tag is used to pull down the interacting complex. The antibody only captures non-specific proteins (green objects) in the sample containing the target protein without the epitope tag. LC-MS/MS analysis of both affinity purification experiments helps identify the potential binding/interacting proteins.

AP-MS analysis can also be carried out in a quantitative manner, which helps generate more reliable MS data for protein-protein interaction studies and provide further insights on the dynamics of these interaction events. Both label-free and label-based methods such as SILAC and iTRAQ can be adapted [58]. A recent method using SILAC in combination with RNA interference and AP-MS allows the detection of endogenous PPIs without the need for an affinity tag [59]. The available bioinformatic tools and databases, such as Cytoscape [60] and STRING [61], can also provide great aid in interpreting the acquired AP-MS data and carrying out functional associations for further biological validation.

A challenge in this field is to investigate the weaker and more transient PPIs. Although these PPIs are often of great biological relevance, the interactions are often weak and can be easily disrupted during the sample preparation and affinity purification steps prior to MS analysis. Several cross-linking strategies have been developed that can be employed to stabilize the weak or transient interaction partners before AP-MS analysis [58].

Acknowledgments

This work was funded by the Lundbeck Foundation, Danish Medical Research Foundation, Novo Nordisk Foundation, Carlsberg Foundation, Aarhus University Research Foundation, and the Ratchadapiseksomphot Endowment Fund of Chulalongkorn University (RES560530124-HR).

References

[1] Allmer, J., Algorithms for the de novo sequencing of peptides from tandem mass spectra. *Expert review of proteomics*, 2011. 8(5): p. 645-57.

[2] Eng, J.K., et al. A face in the crowd: recognizing peptides through database search. *Molecular & cellular proteomics* : MCP, 2011. 10(11): p. R111 009522.

[3] Elias, J.E. and S.P. Gygi, Target-decoy search strategy for increased confidence in large-scale protein identifications by mass spectrometry. *Nature methods*, 2007. 4(3): p. 207-14.

[4] Claassen, M., Inference and validation of protein identifications. *Molecular & cellular proteomics* : MCP, 2012. 11(11): p. 1097-104.

[5] Wu, L. and D.K. Han, Overcoming the dynamic range problem in mass spectrometry-based shotgun proteomics. *Expert review of proteomics*, 2006. 3(6): p. 611-9.

[6] Schirmer, E.C., J.R. Yates, 3rd, and L. Gerace, MudPIT: A powerful proteomics tool for discovery. *Discovery medicine*, 2003. 3(18): p. 38-9.

[7] Nagele, E., et al. 2D-LC/MS techniques for the identification of proteins in highly complex mixtures. *Expert review of proteomics*, 2004. 1(1): p. 37-46.

[8] Washburn, M.P., D. Wolters, and J.R. Yates, 3rd, Large-scale analysis of the yeast proteome by multidimensional protein identification technology. *Nature biotechnology*, 2001. 19(3): p. 242-7.

[9] Sze, S.K., et al. Top-down mass spectrometry of a 29-kDa protein for characterization of any posttranslational modification to within one residue. *Proceedings of the National Academy of Sciences of the United States of America*, 2002. 99(4): p. 1774-9.

[10] Drabik, A., A. Bodzon-Kulakowska, and P. Suder, Application of the ETD/PTR reactions in top-down proteomics as a faster alternative to bottom-up nanoLC-MS/MS protein identification. *Journal of mass spectrometry : JMS*, 2012. 47(10): p. 1347-52.

[11] Ong, S.E., et al. Stable isotope labeling by amino acids in cell culture, SILAC, as a simple and accurate approach to expression proteomics. *Molecular & cellular proteomics : MCP*, 2002. 1(5): p. 376-86.

[12] Kruger, M., et al. SILAC mouse for quantitative proteomics uncovers kindlin-3 as an essential factor for red blood cell function. *Cell*, 2008. 134(2): p. 353-64.

[13] Thompson, A., et al. Tandem mass tags: a novel quantification strategy for comparative analysis of complex protein mixtures by MS/MS. *Analytical chemistry*, 2003. 75(8): p. 1895-904.

[14] Ross, P.L., et al. Multiplexed protein quantitation in Saccharomyces cerevisiae using amine-reactive isobaric tagging reagents. Molecular & cellular proteomics : *MCP*, 2004. 3(12): p. 1154-69.

[15] Hsu, J.L., et al. Stable-isotope dimethyl labeling for quantitative proteomics. *Analytical chemistry*, 2003. 75(24): p. 6843-52.

[16] Hsu, J.L., et al. Beyond quantitative proteomics: signal enhancement of the a1 ion as a mass tag for peptide sequencing using dimethyl labeling. *Journal of proteome research*, 2005. 4(1): p. 101-8.

[17] Hennrich, M.L., et al. Dimethyl isotope labeling assisted de novo peptide sequencing. *Journal of the American Society for Mass Spectrometry*, 2010. 21(12): p. 1957-65.

[18] Bondarenko, P.V., D. Chelius, and T.A. Shaler, Identification and relative quantitation of protein mixtures by enzymatic digestion followed by capillary reversed-phase liquid chromatography-tandem mass spectrometry. *Analytical chemistry*, 2002. 74(18): p. 4741-9.

[19] Higgs, R.E., et al. Comprehensive label-free method for the relative quantification of proteins from biological samples. *Journal of proteome research,* 2005. 4(4): p. 1442-50.

[20] Wang, G., et al. Label-free protein quantification using LC-coupled ion trap or FT mass spectrometry: Reproducibility, linearity, and application with complex proteomes. *Journal of proteome research*, 2006. 5(5): p. 1214-23.

[21] Old, W.M., et al. Comparison of label-free methods for quantifying human proteins by shotgun proteomics. Molecular & cellular proteomics : MCP, 2005. 4(10): p. 1487-502.

[22] Liu, H., R.G. Sadygov, and J.R. Yates, 3rd, A model for random sampling and estimation of relative protein abundance in shotgun proteomics. *Analytical chemistry,* 2004. 76(14): p. 4193-201.

[23] Zhang, B., et al. Detecting differential and correlated protein expression in label-free shotgun proteomics. *Journal of proteome research*, 2006. 5(11): p. 2909-18.

[24] Silva, J.C., et al. Quantitative proteomic analysis by accurate mass retention time pairs. *Analytical chemistry*, 2005. 77(7): p. 2187-200.

[25] Silva, J.C., et al. Simultaneous qualitative and quantitative analysis of the Escherichia coli proteome: a sweet tale. Molecular & cellular proteomics : *MCP*, 2006. 5(4): p. 589-607.

[26] Levin, Y., E. Hradetzky, and S. Bahn, Quantification of proteins using data-independent analysis (MSE) in simple andcomplex samples: a systematic evaluation. *Proteomics*, 2011. 11(16): p. 3273-87.

[27] Geromanos, S.J., et al. The detection, correlation, and comparison of peptide precursor and product ions from data independent LC-MS with data dependant LC-MS/MS. *Proteomics*, 2009. 9(6): p. 1683-95.

[28] Yocum, A.K. and A.M. Chinnaiyan, Current affairs in quantitative targeted proteomics: multiple reaction monitoring-mass spectrometry. *Briefings in functional genomics & proteomics,* 2009. 8(2): p. 145-57.

[29] Barnidge, D.R., et al. Absolute quantification of the model biomarker prostate-specific antigen in serum by LC-Ms/MS using protein cleavage and isotope dilution mass spectrometry. *Journal of proteome research*, 2004. 3(3): p. 644-52.

[30] Prakash, A., et al. Expediting the development of targeted SRM assays: using data from shotgun proteomics to automate method development. *Journal of proteome research*, 2009. 8(6): p. 2733-9.

[31] Gillet, L.C., et al. Targeted data extraction of the MS/MS spectra generated by data-independent acquisition: a new concept for consistent and accurate proteome analysis. Molecular & cellular proteomics : *MCP*, 2012. 11(6): p. O111 016717.

[32] Thingholm, T.E., et al. TiO(2)-based phosphoproteomic analysis of the plasma membrane and the effects of phosphatase inhibitor treatment. *Journal of proteome research*, 2008. 7(8): p. 3304-13.

[33] McLachlin, D.T. and B.T. Chait, Improved beta-elimination-based affinity purification strategy for enrichment of phosphopeptides. *Analytical chemistry*, 2003. 75(24): p. 6826-36.

[34] Jaffe, H., Veeranna, and H.C. Pant, Characterization of serine and threonine phosphorylation sites in beta-elimination/ethanethiol addition-modified proteins by electrospray tandem mass spectrometry and database searching. *Biochemistry*, 1998. 37(46): p. 16211-24.

[35] Zhang, X., et al. Highly Efficient Phosphopeptide Enrichment by Calcium Phosphate Precipitation Combined with Subsequent IMAC Enrichment. Molecular & cellular proteomics : *MCP*, 2007. 6(11): p. 2032-42.

[36] Nuhse, T.S., et al. Large-scale analysis of in vivo phosphorylated membrane proteins by immobilized metal ion affinity chromatography and mass spectrometry. Molecular & cellular proteomics : *MCP*, 2003. 2(11): p. 1234-43.

[37] Larsen, M.R. and P. Roepstorff, Mass spectrometric identification of proteins and characterization of their post-translational modifications in proteome analysis. *Fresenius' journal of analytical chemistry*, 2000. 366(6-7): p. 677-90.

[38] Klemm, C., et al. Evaluation of the titanium dioxide approach for MS analysis of phosphopeptides. Journal of mass spectrometry : *JMS*, 2006. 41(12): p. 1623-32.

[39] Sugiyama, N., et al. Phosphopeptide enrichment by aliphatic hydroxy acid-modified metal oxide chromatography for nano-LC-MS/MS in proteomics applications. Molecular & cellular proteomics : *MCP*, 2007. 6(6): p. 1103-9.

[40] Thingholm, T.E., et al. Highly selective enrichment of phosphorylated peptides using titanium dioxide. *Nature protocols*, 2006. 1(4): p. 1929-35.

[41] Bergstrom Lind, S., et al. Immunoaffinity enrichments followed by mass spectrometric detection for studying global protein tyrosine phosphorylation. *Journal of proteome research,* 2008. 7(7): p. 2897-910.

[42] Zhao, B., et al. Large-scale phosphotyrosine proteomic profiling of rat renal collecting duct epithelium reveals predominance of proteins involved in cell polarity determination. American journal of physiology. *Cell physiology*, 2012. 302(1): p. C27-45.

[43] Bodenmiller, B., et al. Reproducible isolation of distinct, overlapping segments of the phosphoproteome. *Nature methods*, 2007. 4(3): p. 231-7.

[44] Tholey, A., J. Reed, and W.D. Lehmann, Electrospray tandem mass spectrometric studies of phosphopeptides and phosphopeptide analogues. J*ournal of mass spectrometry : JMS*, 1999. 34(2): p. 117-23.

[45] Schlosser, A., et al. Analysis of protein phosphorylation by a combination of elastase digestion and neutral loss tandem mass spectrometry. *Analytical chemistry*, 2001. 73(2): p. 170-6.

[46] Bateman, R.H., et al. A novel precursor ion discovery method on a hybrid quadrupole orthogonal acceleration time-of-flight (Q-TOF) mass spectrometer for studying protein

phosphorylation. *Journal of the American Society for Mass Spectrometry*, 2002. 13(7): p. 792-803.

[47] Ulintz, P.J., et al. Comparison of MS(2)-only, MSA, and MS(2)/MS(3) methodologies for phosphopeptide identification. *Journal of proteome research*, 2009. 8(2): p. 887-99.

[48] Haas, W., et al. Optimization and use of peptide mass measurement accuracy in shotgun proteomics. Molecular & cellular proteomics : MCP, 2006. 5(7): p. 1326-37.

[49] Stensballe, A., et al. Electron capture dissociation of singly and multiply phosphorylated peptides. Rapid communications in mass spectrometry : *RCM*, 2000. 14(19): p. 1793-800.

[50] Molina, H., et al. Global proteomic profiling of phosphopeptides using electron transfer dissociation tandem mass spectrometry. *Proceedings of the National Academy of Sciences of the United States of America*, 2007. 104(7): p. 2199-204.

[51] Jedrychowski, M.P., et al. Evaluation of HCD- and CID-type fragmentation within their respective detection platforms for murine phosphoproteomics. Molecular & cellular proteomics : *MCP*, 2011. 10(12): p. M111 009910.

[52] Beausoleil, S.A., et al. A probability-based approach for high-throughput protein phosphorylation analysis and site localization. *Nature biotechnology*, 2006. 24(10): p. 1285-92.

[53] Olsen, J.V., et al. Global, in vivo, and site-specific phosphorylation dynamics in signaling networks. *Cell*, 2006. 127(3): p. 635-48.

[54] Savitski, M.M., et al. Confident phosphorylation site localization using the Mascot Delta Score. Molecular & cellular proteomics : *MCP*, 2011. 10(2): p. M110 003830.

[55] Chalkley, R.J. and K.R. Clauser, Modification site localization scoring: strategies and performance. Molecular & cellular proteomics : *MCP*, 2012. 11(5): p. 3-14.

[56] Schwartz, D. and S.P. Gygi, An iterative statistical approach to the identification of protein phosphorylation motifs from large-scale data sets. *Nature biotechnology*, 2005. 23(11): p. 1391-8.

[57] Linding, R., et al. Systematic discovery of in vivo phosphorylation networks. *Cell*, 2007. 129(7): p. 1415-26.

[58] Gingras, A.C., et al. Analysis of protein complexes using mass spectrometry. *Nat. Rev. Mol. Cell Biol.*, 2007. 8(8): p. 645-54.

[59] Selbach, M. and M. Mann, Protein interaction screening by quantitative immunoprecipitation combined with knockdown (QUICK). *Nat. Methods*, 2006. 3(12): p. 981-3.

[60] Shannon, P., et al. Cytoscape: a software environment for integrated models of biomolecular interaction networks. *Genome Res.*, 2003. 13(11): p. 2498-504.

[61] Szklarczyk, D., et al. The STRING database in 2011: functional interaction networks of proteins, globally integrated and scored. *Nucleic. Acids Res.*, 2011. 39(Database issue): p. D561-8.

In: Recent Advances in Systems Biology Research
Editors: A. X. C. N. Valente, A. Sarkar and Y. Gao

ISBN: 978-1-62948-736-6

Chapter 7

Theory of Reliability in Systems Biology: Aging versus Reliability

***Vitaly K. Koltover**[*]

Bio-Reliability Group,
Institute of Problems of Chemical Physics, Russian Academy of Sciences,
Chernogolovka, Moscow Region, Russia

Abstract

Biological systems perform their functions in the presence of a great number of random factors which disturb all functional strata. Therefore, similarly to technical devices, they are not perfectly reliable in operation. The field of systems biology and biophysics, in dealing with the problem of reliability ("robustness"), incorporates the theoretical and experimental investigations of quantitative characteristics and mechanisms of failures and renewal processes. It also includes elaboration of methods for testing reliability and predicting failures in biological systems. The reliability problems are closely related with aging and resistance of biosystems to deleterious environmental factors including ionizing radiation. In this report, in order to illustrate the ideas of the reliability trend, I present several examples of analysis of the reliability of biosystems at different functional levels, starting from the level of "biomolecular nanoreactors". Besides, the results of application of the systems reliability approach to aging are presented. This approach is based on the simple general principles that (i) all biomolecular constructions are designed in keeping with the genetic programs in order to perform the programmed, preset functions; (ii) all of them operate with the limited reliability; (iii) the timely replacement or prophylaxis of unreliable functional elements, i.e., the metabolic turnover, is the main line of assuring the high systems reliability; and (iv) there is a finite number of critical elements which perform the supervisory functions over the preventive maintenance but these "supervisors" also operate with the limited, preset, reliability. On this basis, the universal features of aging, such as the exponential growth of mortality rate with time and the correlation of longevity with the species-specific resting metabolism are naturally explained. The stochastic malfunctions of the

[*] E-mail address: koltover@icp.ac.ru

mitochondrial electron transport nanoreactors, that produce the superoxide radical ($O_2^{\bullet-}$), seem to be of first importance. Basing on the reliability-theory approach, one can estimate that the longevity of human brain could reach 250 years should the antioxidant defense against the free-radical failures be perfect. Thus, aging is a stochastic consequence of the limited, programmed reliability of biomolecular constructions. Furthermore, the systems reliability approach serves as heuristic methodology for development of new preventive medicine, including novel radiation protectors based on the stable magnetic isotopes. Much as a growth in noise predicts the wear-out failures in electronics, one can expect that the failures in living systems are also preceded by increasing in noise of the relevant physiological parameters. Along this line, values of the variances and the variation coefficients may serve as the simple indexes in testing reliability of biological constructions.

Keywords: systems biology, reliability, robustness, aging, longevity, failures, program, stochastic, free radicals, prophylaxis, preventive maintenance

Introduction

Studies of the structural and functional mechanisms of biosystems at different levels of complexity have become the traditional problems of biophysics. Biological systems perform their functions in the presence of a great number of random factors which disturb all functional strata starting at the molecular organization level. Therefore, due to such randomness, failures may be induced at different levels of any biosystems. The new field of systems biology and biophysics, in dealing with the problem of the reliability of biosystems, incorporates the investigation of: systematization and classification of failures in biosystems; quantitative characteristics of failures; physical and chemical mechanisms of failures; possible ways to evaluate molecular failures in functional breaks; mechanisms of renewal processes; application of mathematical reliability theory; new methods for testing the reliability of biosystems. This problem of reliability of biosystems obviously has a direct bond to such problems as aging and resistance of biosystems to deleterious environmental factors including ionizing radiation, and so on [1-7]. The regular conferences, which were initiated by the special Committee on Reliability of Biological Systems of Academy of Sciences of the former USSR, to deal with the problem of reliability of biological systems, have given a strong impetus to research in this direction, starting from 1975 [8-10]. It has also spurred the similar studies on reliability ("robustness") on the other side of the former "iron curtain" [11-18]. In this report, in order to illustrate the ideas of the reliability trend, I would like to present several examples of analysis of the reliability of biosystems at different functional levels. And I will present the results of a reliability-theory approach to the overall mechanisms of aging.

1. Basic Terms and Ideas

In essence, each and all biological systems are devices constructed to perform the preset functions according to the information planes, i.e., – the genetic programs. Similarly to technical devices, biological constructions are not perfectly reliable in operation: malfunctions happen alternating with the normal operation functions. In engineering, reliability is defined as the ability of a device to perform its function for a given time under given, specific conditions (see, for example, books [19, 20]). The same intuitive definition of reliability fits biological systems [1-3].

Hence, to study reliability of a device, we need, first of all, to specify the function, i.e., we need to know what this device has been designed for. However, in many instances a biological object performs not a single function but several functions and, moreover, some of them may be mutually exclusive. For example, in any population, an individual should perform two functions: he should survive in the given conditions and he should advantage to survivorship of his population or even the species as the whole. Thus, even this very first step of the bioreliability analysis can meet difficulties. Nevertheless, one of the operation functions of the object may be allocated, to a first approximation, as the main or "quasi-main" function.

The next thing to be done is to define what a normal operation of the biological device under study is and what its failure is. The simplest way of the definition is to introduce (i) a time-dependent random vector $\boldsymbol{Y(t)}$, each k-element of which ($k = 1, 2,\ldots, m$) corresponds to the relevant functional parameter of the device and (ii) the relevant m-dimensional admissible limits of the parameters. If values of the elements Y_k are in the limits, then the state of the device is defined as normal operation during the given time. If any of them occurs beyond the limits, then we say that the device is in the state of failure. Inasmuch as the object performs its functions in the presence of random factors, all Y_k are random functions of time. Moreover, the admissible limits of the functional parameters may also be time-dependent and stochastic.

Accordingly, the quantitative characteristic of reliability is probability R of the non-failure operation in the given interval of time $(0, t)$:

$$R(t) = P(\tau > t),$$

where τ is a random value of the failure-free operation. In reliability tests, this probability is experimentally estimated as the ratio $n(t)/n_0$, where $n(t)$ is a number of the objects which are alive (normally operate) in the moment t and n_0 is the initial size of the sample. The graph of the function R with time is named the reliability function. The probability of failure in the same time-interval is correspondingly defined as the probability of the opposite event, $G(t) = P(\tau \leq t) = 1 - R(t)$, with the probability density function $g = dG/dt$.

Another important characteristic of reliability is the so-called hazard or failure-rate function:

$$h(t) = -R'/R = -d(\ln R)/dt.$$

This parameter has the meaning of the conditional probability of failure per unit time provided the object operated failure-free up to the given moment. The sequence of the random moments of time in which the failures take place is named a flux of failures [19, 20]. Of special importance in the theory of reliability are the so-called elementary or Poisson fluxes of failures. In the case of a Poisson-type flux, the probability of failure-free operation during the given time t follows an exponential distribution with the time-independent parameter $\lambda > 0$:

$$R(t) = P(\tau > t) = \exp(-\lambda t).$$

In this case, the failure rate h is constant (λ) and reliability of the device does not depend on the operation time. Hence, a device, reliability of which is described by the exponential distribution, does not age during its operation. Moreover, if there are a great number of fluxes of random events, each being of low intensity and mutually independent, then the sum of them asymptotically converges on a Poisson flux (the so-called Khinchin-Palm's in-limit theorem). It works for complex systems with a large number of components none of which contributes very heavily to the total failure probability even if the distributions for the individual components are not exponential. Among other processes, it applies to the distribution of the time between failures for a complex system when the failed component is replaced immediately by another component of the same type. Therefore the exponential distribution plays a central role in engineering [19, 20] and in biology too [1-3].

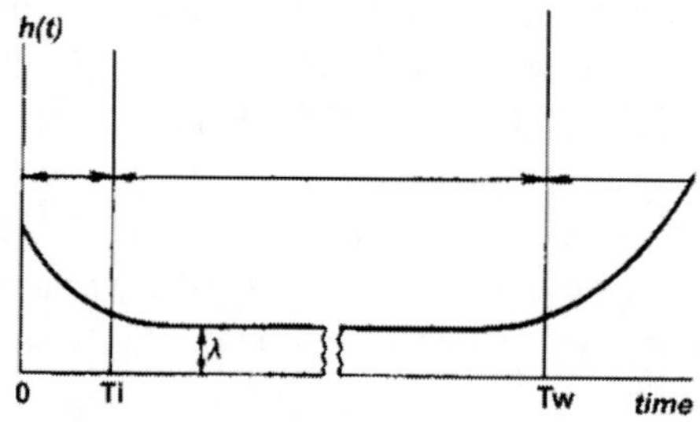

From [19].

Figure 1. A graph of the failure-rate function $h(t)$ versus time for a typical technical device. The intervals ($t < T_i$), (T_i, T_w), and ($t > T_w$) correspond to the phases of initial failures ("breaking-in period"), normal operation and wear-out, respectively.

Figure 1 shows a graph of a hazard function for a typical technical device. Inasmuch as some functional elements of the device can be defective, the failure rate in the first period of life can be rather high. It is the time of the so-called initial failures ("breaking-in period"). As for the "breaking-in" of biological systems, it has obviously accomplished at the earliest evolution phase or even before the beginning of evolution. At least it should be true for the simplest biological constructions ("molecular machines", "biomolecular nanoreactors") like enzymes, ribosomes, mitochondria, and cells. For modern technical devices such as large-scale integrated circuits, the initial failure rates are negligible as well. As the defective elements are replaced with the good ones, the failure rate goes down and stabilizes at a rather low constant level. It is a period of normal operation when rare failures come unexpectedly, due to accidental overloading. Correspondingly, it is a Poisson flux of failures with the constant failure-rate parameter. With time, when wear-out of functional elements comes into play, the failure rate of the device begins to grow up. It is a period of the wear-out failures.

An analogy between failures of technical devices and deaths of living systems is appealing. The reliability function $R(t)$ for probability of failure-free operation during a given time is quite analogous to the survival function (the probability of being alive at a moment t), the probability density function $g(t)$ is analogous to the so-called death-rate function, and the failure-rate function $h(t)$ is analogous to the mortality rate function. Therefore, the same mathematical theory of reliability is essentially applicable to the mathematics of mortality.

The breaking-in period for technical devices is formally similar to the period of infant mortality for people, marked by the high mortality rates. There is also a formal analogy between chance failures of technical devices during normal operation period and deaths of wild animals. For technical devices the exponential distributions of failures take place with the constant failure-rate parameters. In a similar manner, survival curves of wild animals fall exponentially with the constant mortality-rate parameters. It means that the wild animals die mostly not because of senescence but owing to random accidents. Moreover, the mortality of people killed in accidents also follows the exponential distribution with the time-independent constant λ that reflects random environmental influences on the mortality. Furthermore, the failure rate of a device begins to grow up with time when wear-out of functional elements comes into play and, similarly, aging of people and of animals results in growth of the mortality rate.

Here, it is pertinent to call attention to the survival functions of genera of the extinct animals from the paper of Van Valen [21]. Figure 2 shows these curves for three groups of specialized Mesozoic animals. It is quite clear that the logarithm of the percentage of surviving for all three genera linearly falls with time. Hence it appears that the kinetics of the survivorship follows the exponential distribution. From the reliability-theory point of view it means that the species succumbed due to "chance failures". This finding corresponds to the well-known idea of A. Weissmann, who, in 1882, stated that the germ cells of species can be considered as potentially immortal compared to an aging soma (a review of Weismann's ideas can be found in [22]).

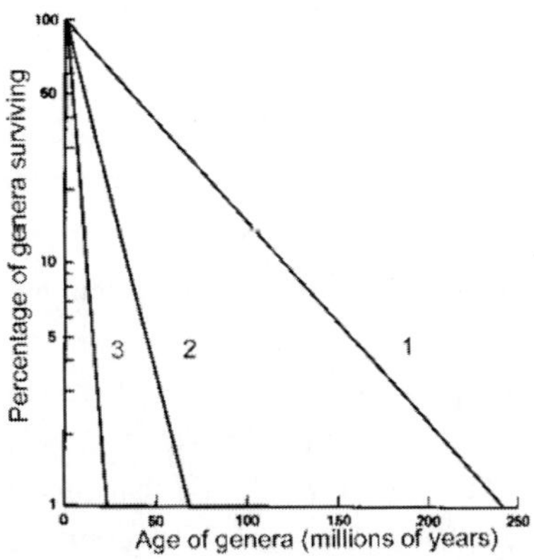

From [21].

Figure 2. Survival functions for extinct genera of the clams (1), Mesozoic rudists (2), and mammals (3) plotted using a logarithmic ordinate.

In biology, as well as in engineering, there are several lines of creating the reliable biological devices from the unreliable functional components. One of them is redundancy, when redundant components of the same type are introduced in the system to fulfill one and the same function. There are different kinds of redundancy, among them – structural redundancy (insertion of superfluous amount of functional elements), functional redundancy (exploitation of elements capable to carry out additional functions besides their basic ones), informational redundancy (surplus information supplement), temporal redundancy (superfluous time of functioning) [19, 20]. Indeed, all essential biomolecular constructions are present in cells in superfluous amounts. The redundant amounts of enzymes, mitochondria and other organelles represent the examples of structural reservation. The elimination of hydrogen peroxide (H_2O_2) and organic peroxides from cells by two different enzyme systems, by catalase and glutathione peroxidase, is the example of functional reservation. There are many examples of the information type of reservation in biology. The nucleotide sequences of DNA coding certain histones in animal cells are repeated tens times per genome (for example, up to the hundred times in the Sea hedgehog's genome). The numbers of recurrent sequences of DNA coding mRNA in the brains of animals of different species were reported to be, on the average, 1.3 per genome for a mouse, 1.5 for a cow and 2.1 for a man (see the references in [3, 4]. Thus, in this example, the extent of information redundancy positively correlates with the maximum life span of the species (3, 30 and 120 years for mice, cows and people, respectively) [3, 4].

Another line of enhancement of reliability is to supply repair and replacement of components. Repair and renewal processes proceed in all complex biological systems starting from the cell level. The template principle of organization of living systems implies that information DNA structures are of the first operation importance in the cell hierarchy. Hence, of importance is the special enzyme system of reparation of DNA from damages induced by ionizing radiation and other deleterious environmental factors. The special antioxidant enzymes, the function of which is to catch active forms of oxygen in aerobic cells should be mentioned as another important case of the repair system.

The template synthesis provides cells with novel biomolecules in place of ones which have been damaged. In engineering, however, the failure rate of the system may become intolerably high if components are replaced only as they have been damaged or wearied-out. In order to provide the level of failures as low as possible, components are to be replaced for novel ones ahead the phase of their wear-out begins. In engineering, the preventive maintenance replacement of components, i.e., – a schedule of planned maintenance actions, may increase the reliability as great as 100-fold [19]. It is naturally to assume that the well known turnover of cell components enables the similar preventive maintenance. Moreover, quantities of proteins, lipids and other biomolecules in cells are too enormous to discriminate damaged molecules from undamaged ones promptly. Hence, the preventive replacement of functional elements which follow the pattern preset in the cell genome seems to be the main line of providing the high reliability of biological systems.

There are many simple models in the mathematical theory of reliability that describes a monotonic increase of the failure rate with time. Thereupon, it is not hard to write a formal "reliability theory of aging and longevity" (e.g., see reviews [3, 11]). *Objet de vertu* of the theory of reliability lies in the possibility to serve as an investigative approach for searching for realistic mechanisms of aging.

2. Reliability of Enzymes

First of all, let's deal with the reliability of enzymes. The function of an enzyme, as a molecular machine, consists in catalyzing a specific biochemical reaction. From the point of view of the reliability theory, an enzyme is actually a "robot" specialized in accelerated assembly and disassembly of specific molecules. By way of derivation, it is interesting to note that this common word was originally proposed by Czech writer Isaac Azimov. It comes from the Slavonic word for "work" ("*rabota*" in Russian), which is further derived from the Slavonic word for "slave" ("*rab*" in Russian). Any enzyme catalysis is performed as a process of step-by-step electron-conformation interactions which are induced by attachment of a substrate molecule to the enzyme molecule. Regarding the membrane electron-transfer enzymes, this approach was initially formulated in our work with L. A. Blumenfeld [23]. The sequences of electron-conformation interactions represent the displacements and reorientations in enormous quantities of atomic groups. One can suggest that fluctuations of temperature and other "environmental parameters" bring about the possibility of random accidents in the sequences of electron-conformation interactions (the possibility of conformational relaxation changes in an incorrect direction). Such accidents may develop in one or another functional violation of the enzyme (inactivation, violation of selectivity, etc.), i.e., in the failures. Thus, the conformational fluctuations may set a limit to the enzyme reliability [1-3].

The analogy between the inactivation of enzymes and failures of technical devices is self-obvious. As a rule, inactivation of enzymes at work (i.e., their operation stability) obeys the rules of exponential kinetics. Yet, the thermal inactivation of enzymes does not directly concern the problem of reliability of living systems. A hen obviously lays her eggs not for the later use them in an omelet. Of more importance seems to be another kind of enzyme failures, namely – the violation of selectivity when an enzyme catalyzes a reaction with an analogue of the enzyme substrate instead of the reaction with the "substrate-in-law".

From the general principles of the theory of reliability, one can expect that a definite conformation and optimal conformation lability must correspond to the given conditions of the enzyme functioning in a cell, so that to provide not only high efficiency but also high reliability of the enzyme. Conversely, any environmental factors, capable of tuning the enzyme conformation away from its physiological optimum, should increase the probability of malfunctions, thereby decreasing the reliability of that enzyme [2]. This "theorem" may serve as a guide-line for designing special experiments on the problems of reliability [3]. In this respect, it would be of interest to study malfunctions of enzymes which are known that play key roles in controlling the long-term stability of cells, such as the enzymes of the translation apparatus. For example, malfunction in selectivity of aminoacyl-tRNA synthase may produce errors in translation of genetic information from DNA into proteins. It is worth remembering also numerous potentialities regarding temperature "tuning" of enzymes which may be used for the adaptation of bacteria, plants and poikilotherms to temperature conditions of the environment. Such effects of "tuning" may be observable in the presence of other variable factors such as pH, concentration of salts, etc. Unfortunately, these questions have not been explored to date.

Up to now, the enzymes of mitochondrial electron transport chains (ETC) seem to manifest the most important failures since free radicals of oxygen, the so-called superoxide radicals ($O_2^{\bullet -}$) are produced under these failures. The energetic demands of every operation in a living system are met by molecules of adenosine triphosphate (ATP) synthesized during oxidative metabolism in the cellular mitochondria. Normal functioning of electron-transport chains of mitochondria lies in the transport of electrons from the oxidation substrates to cytochrome oxidase and then to oxygen, with the reduction of oxygen molecules to water and synthesis of ATP. The mitochondrial nanoreactors have very ancient evolutionary origin and, hence, seem to be ones of the most reliable molecular machines. But, after all, their reliability characteristics are not perfect. Normal elementary acts of electron transfers on ETC alternate with accidental malfunctions which result in formation of $O_2^{\bullet-}$. It is well known that the superoxide radical is toxic. So, from the reliability point of view, the fact that this radical appears is to be considered as the random malfunction ("recurrent failure" in engineering) of ETC. This interpretation of superoxide radicals as the "free-radical failures" was first proposed in our papers [24, 25].

In accordance with the above-mentioned theorem, an essential increase in the free-radical failure rate should be expected when the catalytic conditions for the mitochondrial nanoreactors move away from the physiological optimum. Indeed, an inadequate supply of oxygen under short-term anoxia/ischemia conditions causes an increase in reactivity of the mitochondrial ubisemiquinones to oxygen, along with the relevant increase in production of $O_2^{\bullet-}$. It was experimentally proved that heart mitochondria become intensive generators of $O_2^{\bullet-}$ after transient anoxia/ischaemia conditions [27, 28].

There is a special antioxidant enzyme in mitochondria, the so-called superoxide dismutase (mtSOD), which catalyzes the reaction of dismutation of $O_2^{\bullet-}$ into hydrogen peroxide (H_2O_2) and oxygen (O_2) thus protecting cellular structures from $O_2^{\bullet-}$ and its toxic chemical products. This enzyme works in cooperation with other antioxidant enzymes, catalase and glutathione peroxidase, which catalyze decomposition of H_2O_2. As a matter of fact, the antioxidant enzymes provide the preventive maintenance against the reactive oxygen forms.

The elementary act of the superoxide radical occurrence, as well as the elementary act of its disappearance in the dismutation reaction, are stochastic processes. Inasmuch as all enzymes work with the limited reliability, $O_2^{\bullet-}$ can slip through the defense system. We have analyzed the stochastic dynamics of this system by a mathematical "Birth and Death" model, often used in the reliability theory. The calculations, based on the experimental data from the available literature [2, 3, 25], show that the rate parameter for the free-radical malfunctions in normal mitochondria $\lambda \approx 0.25\ s^{-1}$, about 1 radical every 4 sec. The rate parameter for elimination of $O_2^{\bullet-}$ by mitochondrial SOD (mtSOD), i.e., the probability of elimination of one $O_2^{\bullet-}$ per unit time, is

$$\mu \approx -\Delta n(t)/n(t)\Delta t = k_e[E] \approx 1.3 \cdot 10^4\ s^{-1},$$

where $\Delta n(t)$ is the number of the radicals eliminated during the time interval Δt, k_e and $[E]$ are the reaction rate constant and concentration of mtSOD, correspondingly, $\approx(2\text{–}3)\cdot 10^9\ M^{-1}s^{-1}$ and $\approx(0.4\text{–}0.5)\cdot 10^{-5}$ M. Then the probability of slipping of $O_2^{\bullet-}$ through the mtSOD defense has been estimated to be:

$$z = (\lambda/\mu)/(1+\lambda/\mu) \approx 1.9 \cdot 10^{-5},$$

i.e., about 2 radicals from every hundred thousand may penetrate the defense system. Thus, with the intense radical fluxes in cells, the probability of free-radical damage in biomolecular nanoreactors of cells can be high enough. Specifically, the free radical theory of aging was put forward, and inhibitors of free radical chain oxidation (antioxidants) were proposed for prolongation of life of humans and animals by Denham Harman about 50 years ago [29].

3. Reliability and Mortality of Organism as the Whole

Despite the complexity and phenotypic variety of living organisms, aging is governed by some common quantitative laws or, more precisely, the "patterns of relationship". First, there is the so-called species-specific maximal life-span potential. Indeed, it is common knowledge that there are neither mice nor rats exceeding 3-4 years of age, and that a human life-span does not exceed ≈120 years provided we take reliable data into account, not sensational press reports or legends [30, 31]. The limited lifetime of diploid cell strains *in vitro* is also a well-known phenomenon. Normal human fibroblasts, after performing about 50 doublings, die or mutate into cancer cells. American biologist Hayflick discovered this effect in 1961 and Russian biologist Olovnikov explained the Hayflick's limit suggesting the mechanism of under-reparation of telomere ends of DNA (see the references in [31]). Moreover, for placental mammals, the life-span potential is inversely related to the rate of the basal oxygen consumption of the species:

$$T \cdot V_0 = 4.12 \cdot 10^{10} \cdot B^{1.37},$$

where T is the maximal life-span value of the species (in sec), V_0 is the basal oxygen consumption (resting metabolic rate, in ml/sec), B is the brain mass (in kilograms). This correlation was first discovered by Rubner in 1883 and has been confirmed since then for different mammalian and non-mammalian species as the universal scaling relation (see [32]).

Besides, the growth of mortality rate with age obeys to a universal kinetics. Namely, if $n(t)$ is the number of live persons of the age t, Δn is the number of those who died during the time interval Δt (usually taken to be 1 year), then the mortality rate:

$$h(t) = -\Delta n(t) / n(t)\Delta t = h_0 \exp(\gamma t),$$

where parameters h_o and γ are independent of time, and the relevant expression for the survival function (probability of being alive at a moment t):

$$R(t) = \text{Prob}\{\tau > t\} = \exp\{(h_0 / \gamma)[1 - \exp(\gamma t)]\}.$$

This is the so-called "Gompertz law" of mortality that has been confirmed for people (of age approximately from 35 to 90 years), other mammals, flies, mollusks [30, 31], even for prokaryotes *Acholeplasma laidlawii* [33].

The mathematical theory of reliability provides the means for tackling this phenomenon. The Gompertzian mortality function corresponds to the Type 1 asymptotic distribution for the minimum value, known from the so-called statistics of extremes. Namely, the hazard function for a complex physical system consisting of a large number of components increases exponentially whenever the components are connected in series and are all subjected to a "wearing-out" process in which the risk of failure increases progressively [34]. This limit theorem of the statistics of extremes makes the Gompertzian mortality law appear as if it is almost universally valid, much like the central-limit theorem makes the normal distribution appear as the very appropriate model in the theory of errors.

The systems reliability approach, that was developed in our papers, is based on the following general principles. The 1st one is the template principle of organization of living systems implying that information structures rank first in the cell hierarchy. An organism works like a system of biomolecular constructions designed in accordance with the genetic program *(information plan)* in order to perform the preset programmed functions (*purpose*). Second, we believe that all biomolecular constructions operate with limited reliability, namely, for each and every biological device or bionanoreactor, starting from enzymes, normal operation acts alternate with accidental malfunctions. The 3rd principle states that preventive maintenance replacement of functional elements in cells and tissues is the main line of assuring the high systems reliability, just as in engineering. Following the preset genome pattern, unreliable elements are timely replaced for novel ones ahead the phase of their wear-out begins, the so-called metabolic turnover. The 4th principle states that there is a finite number (N) of critical elements which perform the supervisory functions over the organism's repair and renewal processes, i.e., – over the turnover. Since these critical elements of the highest hierarchic level exert the control over the systems reliability they can be called "longevity-assurance structures" (LAS). Inasmuch as all reliability facilities, among them – preventive maintenance, repair, and redundancy of functional elements, are genetically limited, stochastic damages in LAS accumulate up to the preset threshold dysfunction levels. As a result, each organism has the limited life-span.

Contrastingly to the so-called "robustness", there are many good models in the mathematical theory of reliability [19, 20]. Basing on the systems reliability approach, the simple mathematical model of aging was suggested in our papers [25, 26, 35]. First, it was taken that LAS accumulate stochastic flaws resulting in disarray of their functions. Account was also taken of another widespread peculiarity of living systems, i.e., the existence of threshold values for the most important functional parameters. According to this general idea, there is to be an upper limit value, m_c, at which LAS fails. The set of initial flaws, m_j ($j = 1, 2, \ldots, N$), represents a random sample of the exponential, though truncated, distribution with the density function:

$$f(m) = \alpha \exp(-\alpha m)/[1-\exp(-\alpha m_c)],$$

where $\alpha > 0$ is a parameter of this distribution, and $0 \leq m \leq m_c$. To underlay this hypothetical density function, some simple arguments were brought into account. As the flaws to LAS occur as rare events, the probability of getting an extensive flaw should obviously be less than the probability of getting a smaller one. The exponent was used as the simplest asymmetric distribution known from the mathematical statistics. The similar truncated exponential distributions have been known for many years ago as the simplest asymmetrical distributions in physics of linear polymer chains [36].

The organism has been assumed to perish the moment that any of the LAS develops the threshold dysfunction, i.e., the expected life-span:

$$\tau = \min \tau_j$$

Here $\tau_j = b(m_c - m_j)$ where $b > 0$ is the reciprocal of the dysfunction growth rate in LAS with time. As a matter of fact, the life-span of the organism is determined by the weakest link's longevity. Then, the survival function is given by the smallest value of the random sample of size N:

$$R(t) = \{1-[exp(\gamma t)-1]/[exp(\gamma T)-1]\}^N.$$

Here N is a number of LAS, $T = bm_c$ and $\gamma = \alpha/b$. For not very high values of time, the following approximation can easily be derived:

$$R(t) \approx \exp\{(h_0/\gamma)[1-\exp(\gamma t)]\},$$

with the relevant expression for mortality rate being:

$$h(t) = h_0 \exp(\gamma t), \text{ where } h_0 = \gamma N/[\exp(\gamma T)-1].$$

Hence, this reliability model gives the exponential growth of the mortality rate with time, i.e., – the Gompertz law of mortality. The limit life-span T has appeared as the direct result of the existence of the limit dysfunction, m_c, for the LAS. This limit is the life-span of an "ideal" organism with no flaws at $t = 0$.

All the hypothetical LAS are differently fallible since they are initially flawed at statistically varying degrees. However, they were postulated to have the same values for the reliability characteristics (α, b, m_c). This was certainly assumed for the sake of mathematical simplicity. However, an evolutionary mechanism could be proposed to support this assumption. The arrangement of the appropriate level of reliability for these structures falls into the basic cell maintenance processes of defense, restore and renewal because each of them is vitally important. All processes of this kind are metabolically expensive. Meanwhile, the energy budget of a cell is limited. If the organism perishes the moment that any one of them fails, there is no use in natural selection making some of them more reliable than others. The acquisition of the greater maintenance for any one than is necessary for others should obviously require an excess cost.

If we take that the maximum life-span for human populations, on the average, is about 95 years, the magnitude of γ varies from 0.0612 to 0.119 years^{-1} and the magnitude of h_0 varies from $0.820 \cdot 10^{-3}$ to $0.022 \cdot 10^{-3}$ years^{-1} [30, 31], then, using the expression for h_0, we find that $N \approx 5$–15. The values of N for dogs, mice and mares, calculated by the relevant values of the parameters (T, h_0, γ) fall in the same order of magnitude too [3, 35].

An analytical transition from the abstract "longevity-assurance structures" to real biomolecular structures seems to be no easier than similar transitions from the "generalized co-ordinates" in theoretical physics. It is yet noteworthy that this estimation corresponds, by the order of magnitude, to the number of the so-called "longevity-assurance genes" which have been recently discovered in nematodes, yeasts, drosophilae, mice, and other organisms (see, for example, the review [37, 38] and references therein). It is also noteworthy to remind that the existence of longevity-assurance genes in chromosomes was first predicted by American physicist Szillard in 1959 [39].

In the case of genetic damages due to superoxide radicals, the following equation was derived for the maximum lifespan [26]:

$$T = bm_c \approx m_c/[(qV/E)u + D].$$

In this equation q is probability of the malfunction in mitochondrial nanoreactors leading to the superoxide occurrence, V is respiration rate, E is activity of mtSOD in LAS, u is probability for the free-radical failures to provoke functional violations. The parameter u takes into account that deleterious effects of $O_2^{\bullet-}$, which slipped through the antioxidant defense, are of no concern if the preventive replacement of the damaged "biological constructions" is properly maintained in cells and tissues. D is index to incorporate other damage factors that are not associated with oxygen free radicals. As a matter of fact, the latter equation explains another universal quantitative law of aging, the "Rubner scaling relation".

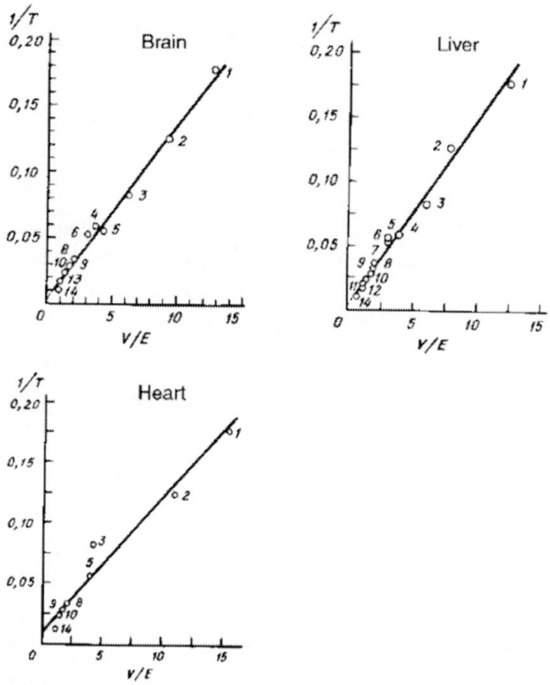

Figure 3. The correlations between the reciprocal maximum life-span (*1/T*) and the ratio of specific metabolic rate to SOD (*V/E*) for mammalian species: 1 - house mouse, 2 - deer mouse, 3 - common tree shrew, 4 - squirrel monkey, 5 - bush baby, 6 - moustache tamarin, 7 - lemur, 8 - African green monkey, 9 - *Rhesus monkey*, 10 - olive baboon, 11 - gorilla, 12 - chimpanzee, 13 - orangutan, 14 – man. The graphs were plotted using the experimental data from [40].

Brain ($r = 0.997$); $1/T = (0.0132 \pm 0.0002)(V/E) + (0.004 \pm 0.002)$

Liver ($r = 0.997$); $1/T = (0.0144 \pm 0.0003)(V/E) + (0.005 \pm 0.002)$

Heart ($r = 0.981$); $1/T = (0.0110 \pm 0.0009)(V/E) + (0.011 \pm 0.006)$

Using this equation, the experimental data of Cutler's group on the superoxide dismutase activity in the tissues of men and animals [40] were used to plot the reciprocal of maximum life-span ($1/T$) as a function of the ratio V/E. The straight lines were obtained, in agreement with the prediction of our theory (Figure 3). By using the free coefficient D, it was estimated that the longevity of human brain could reach 250 years, should the reliability of the antioxidant enzymes be absolutely perfect. By using the free coefficient D of the relevant equations for heart and liver, the limit longevity values were estimated to be 100 and 200 years, respectively [2, 3, 26]. Although these estimations are illustrative, it should be emphasized that "reactive oxygen species" do play a large role in pathogenesis of cardiovascular system [31].

There are two generally known concepts in gerontology [31]. Some men see things as the aging program. Others see that the life spans vary in the inverse proportion to the species-specific metabolic rates and say that aging proceeds stochastically. The systems reliability approach unites both these concepts. The living organisms are designed in accordance with the genetic programs, and the genetic program sets a limit to the lifespan. Yet, the stochastic free radical failures crucially modulate the programmed "melody" filed in the longevity-assurance genes. Thus, aging is the stochastic consequence of the programmed deficiency in reliability of biomolecular constructions.

The biochemical mechanisms of toxicity of $O_2^{\bullet-}$ are not clear (see, for example, a review [41] and references therein). It has long been known that free radical oxidation damages of DNA, proteins and lipid take place in cells and tissues. However, $O_2^{\bullet-}$ is not an oxidant but a reducing agent. Recent discoveries have identified the so-called sirtuins, NAD^+-dependent protein deacetylases, which prove to be the key regulators of many of most important cell processes including metabolism of glucose, secretion of insulin, and adaptation to oxidative stress [41]. The expression of the sirtuin genes depends on the redox state of the environment [42]. It can be assumed that $O_2^{\bullet-}$, as the powerful reducing agent, is capable to significantly affect the ratio of $NAD(P)H/NAD(P)^+$ and, by changing the activity of sirtuins, to slow down the renewal of biomolecular nanoreactors. The evident consequence of this will be accumulation of free-radical products and other "metabolic slag" in peripheral cells and tissues with the resulting impetus to autophagic or apoptotic cell death and, thereby, the age-associated clinical disorders.

4. Reliability and Mortality of Heterogeneous Populations

A well-known extension of the Gompertzian mortality law for people younger the age of ≈35 is the Makeham function which includes a constant term, in addition to the exponential one:

$$h(t) = \lambda + h_0 \exp(\gamma t) .$$

The constant λ is independent of time and reflects environmental influences on mortality as distinct from the senescent changes reflected by the exponential term [30]. Furthermore, there are reports on the deceleration of the mortality-rate function in cohorts of *Drosophila* flies at the advanced age [13, 43]. The similar findings for humans were taken up in the literature, notwithstanding the facts that the statistical data on mortality at geriatric ages are poor [13].

A qualitative attempt to highlight this limitation of the classical Gompertzian approach was undertaken by assuming a simple mixture of a few homogeneous populations [3]. In succeeding years, the question of how an inhomogeneity, i.e., – genetic or phenotypic variability of populations, may affect the behavior of the reliability model was examined quantitatively [44]. The parameters T and γ of the homogeneous reliability model were averaged over the ensemble assuming the normal distribution with the respective probability density functions, $g(T, T_0, \sigma_T)$ and $\varphi(\gamma, \gamma_0, \sigma_\gamma)$. The survival functions were calculated as follows:

$$R(t) = \iint \{1-[exp(\gamma t)-1]/[exp(\gamma T)-1]\}^N g(T)\varphi(\gamma)dTd\gamma.$$

At the advanced time values, the mortality rate function generated from this model may accelerate its run, slow it down, display a maximum or level it off depending upon the extent of heterogeneity of the parameters, thereby, behaving in quantitative agreement with the mortality rate curves of the real populations (Figure 4).

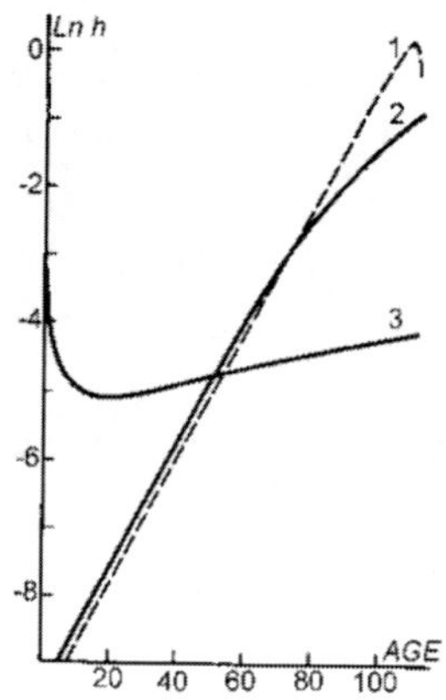

From [44].

Figure 4. Influence of the parameters of heterogeneity on the shape of mortality rate function for a hypothetical population. The graphs were calculated assuming the normal distribution for T and γ with $N = 20$, $T_0 = 110.0$ years, $\gamma_0 = 0.094$ year^{-1}, $\sigma_\gamma = 0.0014$ year^{-1}, $\sigma_T = 2.95$ (1), 7.07 (2) or 70.3 years.

It was shown that this reliability model allowed realistic mortality data for humans to be fitted with a rather high accuracy. In part, the life-tables of the 1969-1973 calendar periods for Swedish in the age range 35-105 years have been computed and the set of the fitting parameters $N = 46$, $T_0 \pm \sigma_T = 120.0$ (±0.3) years, $\gamma_0 \pm \sigma_\gamma = 0.095$ (± 0.001) years^{-1} gave an agreement between the reliability model and the overall mortality data (saving the infant mortality range) with an accuracy of not less than 13% [44]. Thus, the mortality rate functions, generated even from the simple reliability model, supply good quantitative fits to the experimental data, contrary to other models, which provide only qualitative descriptions.

This confirms the basic realism of the postulates underlying the reliability-theory approach proposed in [1-5].

The quantitative estimation of N obtained from the death statistics of the heterogeneous population ($N = 46$) differs from the previous estimation obtained for the case of homogenous population ($N \approx 10$). The origin of this diversity is quite obvious. Individual of a homogeneous population, being alike, die from a limited number of similar pathological reasons. However, it is not the case for heterogeneous populations. There are different individuals on the "tails" of the distribution function, from a centenarian who dies in the age of older than 100 years to a short-lived person who carries a mutant fatal gene in one of his 46 chromosomes. When estimated from the overall mortality data, LAS may be thought of as epigenetic nuclear complexes associated with nuclear chromosomes of diploid cells.

The simple reliability theory of mortality and aging has first been published in 1981 [25]. The modifications of the mathematical model which followed have mostly been attempted at ascertaining the nature of the critical elements that exert the systems reliability control. In our opinion, the "longevity assurance structures" are special genes or "gene networks", as it is now said. Their function is to supervise the prophylactic operation on the metabolic turnover including the control over synthesis of SOD and other antioxidant enzymes. Death occurs when reliability of operation of any of them drops below a tolerant threshold level. In humans and animals, such gene regulatory network are presumably located in the special group of hypothalamic neurons, maybe, of the suprachiasmatic nucleus, and supervise the sirtuin signaling pathway, NF-κB pathway in immune system, signaling by the reactive oxygen species from mitochondria, etc. The fatal loss of the systems reliability may be caused by critical reduction of telomere ends $(TTAGGG)_n$ of the relevant chromosomes [31]. It might be the most plausible hypothesis which pieces together the free-radical theory of aging and the theory of telomere marginotomy in the unified pattern.

Based on this reliability-theory approach, one can formulate a more complex mathematical model to take into account the preset pattern of the longevity-assurance genes, their functional interrelations (feedbacks) and other non-linear effects. It is to be hoped that a non-linear reliability-theory model will explain even so sophisticated cases as the demographic profiles of *Drosophila melanogaster* strains [43].

5. Reliability Testing in Biology

The extrapolation of experience with failures in engineering may be of use in the reliability testing of biological objects. For example, there are methods of the accelerated life testing when reliability of a technical device is quickly tested under increased loads [19, 20]. An exposure of a biological object to ionizing radiation followed by post-radiation recovery of cells and tissues serves an analogue to the accelerated reliability testing in engineering, when reliability of a device is tested under increased loads [1, 45]. Indeed, survival of cells and multi-cellular organisms depends on the functioning of the enzyme systems that protect sub-cellular structures and cells from spontaneous and induced injuries, especially on the enzyme systems of reparation of genetic structures. Evidently, any factor capable to influence on efficiency and reliability of cell nanoreactors shows up itself most vividly under the unfavorable drastic conditions of post-radiation recovery.

The recovery of yeast cells from radiation-induced damages was employed as the method of searching for novel antioxidants [45]. Following this line, we investigated the effects of magnetic and nonmagnetic isotopes of magnesium on post-radiation recovery of *S. cerevisiae* [46, 47]. There are three stable isotopes of magnesium, ^{24}Mg, ^{25}Mg and ^{26}Mg with natural abundance about 79, 10 and 11 %. Among them, only ^{25}Mg has the nuclear spin (I = 5/2) that produces the magnetic field. The question was if the nuclear spin of ^{25}Mg can impact on operation of the cell nanoreactors?

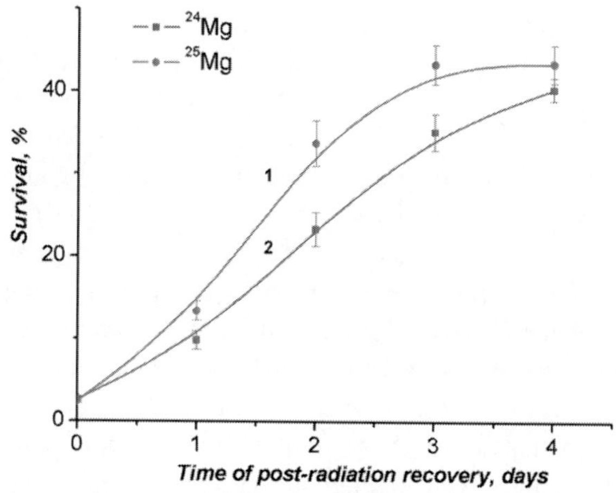

Figure 5. The difference in effects of magnetic and non magnetic isotopes of magnesium on post-radiation recovery of diploid yeast cells *S. cerevisiae*. The cells enriched with magnetic ^{25}Mg or nonmagnetic ^{24}Mg were irradiated by the short-wave UV light (λ = 240-260). Survival of cells was estimated as their ability to form colonies on the nutrition agar: 1 - recovery of the cells enriched with ^{25}Mg; 2 – recovery of the cells enriched with ^{24}Mg [46].

Figure 5 represents the kinetics of recovery of yeast cells from radiation injuries. The survival of the cells transferred to agar immediately after irradiation was no more than a few percent. In this case the injured genetic structures in most of the cells could not be repaired before the onset of mitosis and nonviable daughter cells were produced. Incubation in the nutrient-free water, in which the cells do not divide, provides them sufficient time for repair processes and leads to a corresponding increase in survival. From the kinetics curves represented on this figure, we notice that the cells enriched with the magnetic isotope of magnesium, ^{25}Mg, are recovered more effectively than the cell enriched with nonmagnetic ^{24}Mg. The value of the rate constant of the post-radiation recovery was twice higher for the cells enriched with ^{25}Mg than for the cells enriched with ^{24}Mg. This is the decisive evidence that the magnetic isotope essentially more effectively promotes the recovery of cells from radiation damages. In regard to the mechanisms of the effect, there are strong grounds for believing that magnetic ^{25}Mg more effectively performs the Mg^{2+} cofactor functions for the enzyme reactions of synthesis and hydrolysis of ATP, in comparison to nonmagnetic ^{24}Mg or ^{26}Mg. The "nuclear spin catalysis" makes possible the technological innovations based on the magnetic isotopy to exert control over efficiency and reliability of biomolecular nanoreactors [47].

Furthermore, there are many examples when a growth of noise predicts the wear-out failure of an electronic device. Indeed, by definition, a failure of a device happens once its functional parameters crosses the limits of the area of the admissible values. It means that the greater is probability of the crossing, the less is reliability of the device. Hence, the reliability decreases with growth of fluctuations of the functional parameters (Figure 6).

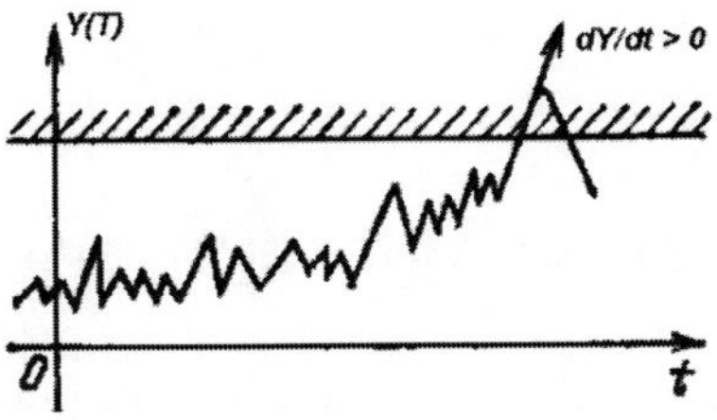

Figure 6. The growth of noise predicts the wear-out failure.

As applied to engineering, especially electronics, the theory of crossings is well elaborated field of applied mathematics [48]. There are the reasons to believe that the failures in living systems are also preceded by increasing in noise of the relevant physiological parameters [1-6]. Oscillations and chaotic fluctuations in species abundances along with gene expression noise are the examples [49-50]. In general, noise is characterized by its correlation functions and power spectra. However, even values of variances and of non-dimensional variation coefficients serve as simple indexes of the reliability for steady-state stochastic processes. Inasmuch as the phenomenon of growth in variability of transient processes is common for biological systems, the simple methods of testing the reliability at all functional levels can be set on this basis.

Conclusion

The systems reliability approach, that was developed in our papers, is based on the simple general principles. All biomolecular constructions are designed in keeping with the genetic programs in order to perform the programmed, preset functions. All of them operate with the limited reliability, namely, for each and every biological device or bionanoreactor, normal operation acts alternate with accidental malfunctions (failures). Just as in engineering, preventive maintenance, i.e., – the timely replacement or prophylaxis of unreliable functional elements, i.e., the metabolic turnover in cells and tissues, is the main line of assuring the high systems reliability. In keeping with the systems approach, we also postulate the existence of the critical elements which perform the supervisory functions over the preventive maintenance. On this systems reliability basis, the universal features of aging, such as the exponential growth of mortality rate with time and the correlation of longevity with the species-specific resting metabolism are naturally explained. Apart from the formal fitting to the mortality data, the systems theory of reliability serves as the investigative approach in searching for mechanisms of aging. Of first importance are the stochastic recurrent failures of the mitochondrial electron transport nanoreactors, since these malfunctions result in production of the superoxide free radicals which initiate the free-radical violations in cells and tissues. Thus, aging is a stochastic consequence of the limited reliability of biomolecular constructions. Besides, the systems reliability approach serves as heuristic methodology for development of new preventive medicine, including novel radiation protectors based on the stable magnetic isotopes. Much as a growth in noise predicts the wear-out failures in electronics, one can expect that the failures in living systems are also preceded by increasing in noise of the relevant physiological parameters. Along this line, values of the variances and the variation coefficients may serve as the simple indexes in testing reliability of biological constructions.

Acknowledgment

The motives that predetermined this work were my lectures on the international symposia "Systems Biology of Human Aging" (SBHA-2009), National Institute on Aging, 8-9 December 2009, Baltimore (USA), "Stochastic Models in Reliability Engineering, Life Sciences and Operations Management" (SMRLO'10), Shamoon College of Engineering, 8-11 February 2010, Beer Sheva (Israel), and "Systems Biology of Human Aging" (SBHA-2010), Drexel University, 6-7 December 2010, Philadelphia (USA). The support from the organizers of these symposia is greatly acknowledged as well as the financial support from Russian Foundation for Fundamental Investigations.

References

[1] Grodzinsky, D. M.; Vojtenko, V. P.; Kutlakhmedov, Y. A.; Koltover, V. K. *Reliability and Aging of Biological Systems,* Naukova Dumka, Kiev (fSU), 1987, 172 pp., (in Russian).

[2] Koltover, V.K. *Reliability of Electron-Transport Membranes and the Role of Oxygen Anion-Radicals in Aging (Doctoral Dissertation in Biology)*, Bogomolets Institute of Physiology, Akad. Nauk Ukr. SSR, Kiev (fSU), 1987, 351 p.

[3] (a) Koltover, V. K. *Theory of Reliability, Superoxide Radicals and Aging*, ICPH, USSR Acad. Sci., Chernogolovka, 1983, 42 p. (b) Koltover, V. K. Theory of reliability, superoxide radicals and aging. *Uspekhi Sovremennoj Biologii (Advances in Modern. Biology)*, 1983, 96 (4), 85-100 (in Russian).

[4] Koltover, V.K. Free radical theory of aging: View against the reliability theory. *Free Radicals and Aging* (eds. I. Emerit, B. Chance), Birkhauser, Basel, 1992, pp. 11-19.

[5] Koltover, V. K. Reliability concept as a trend in biophysics of aging. *J. Theor. Biol.*, 1997, 184 (2), 157-163.

[6] Kutlakhmedov, Y. A.; Korogodin, V. I.; Koltover, V.K. *Foundations of Radiation Ecology*, Vyshcha shkola, Kiev (Ukraine), 2003, 320 pp. (in Ukrainian).

[7] Koltover, V. K. Bioantioxidants: The systems reliability standpoint. *Toxicology and Industrial Health*, 2009, 25 (4-5), 295-299.

[8] Grodzinsky, D. M. (Ed.) *Systems of Reliability of Cell (Proceedings of the 1st all-union conference "Systems of Reliability of Cell", Kiev, 1975)*, Naukova Dumka, Kiev (fSU), 1977 (in Russian).

[9] Grodzinsky, D. M. (Ed.) *Reliability of Cells and Tissues (Proceedings of the 2nd all-union conference "Reliability of Cells and Tissues", Kiev, 1977)*, Naukova Dumka, Kiev (fSU), 1980 (in Russian).

[10] Grodzinsky, D. M.; Kutlakhmedov, Y. A.; Gudkov, I. N. (Eds.) *Reliability of Biological Systems (Proceedings of the 3rd all-union conference "Reliability of Biological Systems", Kiev, 1980)*, Naukova Dumka, Kiev (fSU), 1985 (in Russian).

[11] Doubal, S. Theory of reliability, biological systems and aging. *Mech. Ageing. Develop.*, 1982, 18 (4), 339-353.

[12] Witten, M. A return to time, cells, systems and aging: Rethinking the concept of senescence in mammalian organisms. *Mech. Ageing and Develop.*, 1983, 21, 69-81.

[13] Gavrilov, L. A.; Gavrilova, N.S. The reliability theory of aging and longevity. *J. Theor. Biol.*, 2001, 213, 527-545.

[14] Dimitrov, B. Ages in reliability and biosystems: interpretations, control, and applications. In: *Mathematical and Statistical Models and Methods in Reliability Statistics for Industry and Technology* (eds. V. V. Rykov, N. Balakrishnan, M. S. Nikulin), Springer, 2010, Part 3, 317-334.

[15] Kitano, H. Biological robustness. *Nat. Rev. Genet.*, 2004, 5, 826–837.

[16] Lehar, J.; Krueger, A.; Zimmermann, G.; Borisy, A. High-order combination effects and biological robustness. *Mol. Systems Biol.*, 2008, 4:215.

[17] Kaltenbach, H.-M.; Dimopoulos, S.; Stelling, J. Systems analysis of cellular networks under uncertainty. *FEBS Lett.*, 2009, 583, 3923-930.

[18] Larhlimi, A.; Blachon, S.; Selbig, J.; Nikoloski, Z. Robustness of metabolic networks: A review of existing definitions. *Biosystems*, 2011, 106 (1), 1-8.

[19] Bazovsky, I. *Reliability. Theory and Practice*, Prentice-Hall, London, 1961.

[20] Lloyd, D. K.; Lipov, M. *Reliability: Management, Methods and Mathematics*, Prentice Hall, New Jersey, 1977.

[21] Van Valen, L. A new evolutionary law. *Evol. Theory*, 1973, 1, 1-30.

[22] Kirkwood, T. B. L.; Rose, M. R. Evolution of senescence: late survival sacrificed for reproduction. *Phil. Trans. R. Soc. London B*, 1991, 332, 15-24.

[23] Blumenfeld, L. A.; Koltover, V. K. Energy transformation and conformational transitions in mitochondrial membranes as relaxation processes. *Mol. Biol.* (Moscow), 1972, 6 (1), 161-166 (in Russian).

[24] Koltover, V. K.; Kutlakhmedov, Y. A. Free radical mechanisms of failures and reliability of the cell defense systems. In: Grodzinsky, D. M. (Ed.) *Reliability of Cells and Tissues (Proceedings of the 2nd all-union conference "Reliability of Cells and Tissues", Kiev, 1977)*, Naukova Dumka, Kiev (fSU), 1980, pp. 41-51 (in Russian).

[25] Koltover, V. K. Reliability of enzymatic protection of a cell against superoxide radicals and the aging. *Doklady Biophysics (Doklady Akad Nauk SSSR)*, 1981, 256 (1), 3-5.

[26] Koltover, V. K. Reliability of enzyme systems and molecular mechanisms of ageing. *Biophysics (Moscow)*, 1982, 27 (4), 635-639.

[27] Nohl, H.; Koltover, V.; Stolze, K. Ischemia/reperfusion impairs mitochondrial energy conservation and triggers $O_2^{\bullet-}$ release as a byproduct of respiration. *Free Radical Res. Com.*, 1993, 18 (3), 127-137.

[28] Koltover, V. K. Reliability of mitochondrial electron-transport membranes and the role of superoxide radicals in aging. *Chemical Physics Reports (Khimicheskaya Fizika, Moscow)*, 1996, 15 (1), 109-115.

[29] Harman, D. Aging: A theory based on free radicals and radiation chemistry. *J. Gerontol.*, 1956, 11, 298-300.

[30] Sacher, G. A. Life table modification and life prolongation. In: *Handbook of the Biology of Aging* (eds. C.E. Finch, L. Hayflick), Van Nostrand Reinhold, New York, 1977, pp. 582-638.

[31] Anisimov, V. N. *Molecular and Physiological Mechanisms of Aging*, 2nd ed. Nauka, St. Petersburg, 2008, Vol. 1, p. 481, Vol. 2, p. 434.

[32] Schmidt-Nielsen, K. *Scaling. Why Is Animal Size So Important?* Cambridge Univ. Press., Cambridge, 1984.

[33] Kapitanov, A. B.; Aksenov, M. Y.; Tatishchev, O. S.; Koltover, V. K. Cell-culture of *Acholeplasma laidlawii* as an object for the investigation of the age-variation of biological membranes. *Doklady Biophysics (Doklady Akad Nauk SSSR)*, 1985, 281 (1), 186-189.

[34] Gumbel, E. J. *Statistics of Extremes*. Columbia University, New York, 1962.

[35] Koltover, V. K. Reliability of electron-transport membranes and the role of oxygen radical anions in aging: stochastic modulation of the genetic program. *Biophysics*, 2011, 56 (1), 125-128.

[36] Flory, P. J. *Principles of Polymer Chemistry*, Cornell Univ., Ithaka, 1953.

[37] Garinis, G. A.; van der Horst, G. T. J.; Vijg, J.; Hoeijmakers, J. H. J. DNA damage and ageing: new-age ideas for an age-old problem. *Nature Cell Biology*, 2008, 10, 1241-1247.

[38] Longo, V. D.; Shadel, G. S.; Kaeberlein, M.; Kennedy, B. Replicative and chronological aging in *Saccharomyces cerevisiae*. *Cell Metabolism*, 2012, 16, 18-31.

[39] Szillard, L. On the nature of aging process. *Proc. Nat. Acad. Sci. USA*, 1959, 45 (1), 30-45.

[40] Tolmasoff, J.; Ono, T.; Cutler, R. G. Superoxide dismutase: correlation with life span and specific metabolic rate in primate species. *Proc. Nat. Acad. Sci. USA*, 1980, 77 (5), 2777-2781.

[41] Koltover, V. K. Antioxidant biomedicine: from free radical chemistry to systems biology mechanisms. *Russian Chemical Bull.*, 2010, 59 (1), 37-42.

[42] Houtkooper, R. H.; Pirinen, E.; Auwerx, J. Sirtuins as regulators of metabolism and health span. *Nature Reviews*, 2012, 13, 225-238.

[43] Smith, E. D.; Tsuchiya, M.; Fox, L. A.; Dang, N.; Hu, D.; Kerr, E. O.; Johnston, E. D.; Tchao, Bie N.; Pak, D. N.; Welton, K. L.; Promislow, D. E. L.; Thomas, J. H.; Kaeberlein, M.; Kennedy, B. K. Quantitative evidence for conserved longevity pathways between divergent eukaryotic species. *Genome Research*, 2008, 18, 564-570.

[44] Koltover, V. K.; Andrianova, Z. S.; Ivanova, A. N. Simulation of survival and mortality curves of human populations based on the theory of reliability. *Russ. Biology Bulletin (Izv. Akad. Nauk Biol.)*, 1993, 20 (1), 95-101.

[45] Koltover, V. K.; Kutlakhmedov, Y. A.; Afanaseva, E. L. Recovery of cells from radiation-induced damages in the presence of antioxidants and the reliability of biological systems. *Doklady Biophysics (Doklady Akad Nauk SSSR)*, 1980, 254, 159-161.

[46] Koltover, V. K.; Korolev, V. G.; Kutlakhmedov, Y. A. Antioxidant prophylaxis of radiation stress. In: *Ionizing Radiation: Applications, Sources and Biological Effects* (eds. E. Belotserkovsky, Z. Ostaltsov), Nova Science Publishers, New York, 2012, pp. 117-128.

[47] (a) Koltover, V. K. Stable magnetic isotopes as a new trend in biomedicine. In: *Biomedicine* (ed. C. Lin), InTech-Europe (Rijeka, Croatia), 2012, pp. 105-122. (b) Koltover, V. K. Nuclear spin catalysis in nanoreactors of living cells. *Biophysics*, 2013, 58 (2), 187-192.

[48] Cramer, H.; Leadbetter, M. R. *Stationary and Related Stochastic Processes*. John Willey, New York et al., 1967.

[49] Li, B. L.; Charnov E. L. Diversity-stability relationships revisited: scaling rules for biological communities near equilibrium. *Ecol. Modelling*, 2001, 140 (3), 247-254.

[50] Munsky, B.; Neuert, G.; van Oudenaarden, A. Using gene expression noise to understand gene regulation. *Science*, 2012, 336 (6078), 183-187.

In: Recent Advances in Systems Biology Research ISBN: 978-1-62948-736-6
Editors: A. X. C. N. Valente, A. Sarkar and Y. Gao

Chapter 8

Computational Approaches in Synthetic and Systems Biology

Miha Moškon*, Jure Bordon, Miha Mraz, Nikolaj Zimic and Mattia Petroni
University of Ljubljana, Faculty of Computer and Information Science, Ljubljana, Slovenia

Abstract

Computational biology is an emerging scientific field which employs computational methods in the study of various biological systems. This chapter presents a review of methods that have been introduced to the fields of synthetic and systems biology in recent years. Approaches presented mainly rely on the establishment of computational models. These models allow us to observe the behaviour of a certain biological system in a given environment. Exact kinetic data that describe underlying dynamics are usually necessary to establish accurate computational models. Kinetic data are on the other hand hard or even impossible to obtain experimentally in some cases. Parameter estimation techniques that also rely on computational approaches can be used to accurately evaluate missing kinetic data. With the establishment of computational models, computational analyses can be conducted, such as performance, robustness, sensitivity or stability analysis. These techniques can be used further on to reduce the amount of experimental work and enable straightforward design of novel biological systems in the context of synthetic biology.

Keywords: Computational modelling, computer simulation, network inference, parameter estimation techniques, performance evaluation, sensitivity analysis, stability analysis, computational design

*E-mail address: miha.moskon@fri.uni-lj.si

1. Introduction

Use of engineering tools that rely on various computational approaches is more and more common in the field of systems biology [1]. Being a combination of chemistry, biology and engineering [2], these approaches gained even larger role in the field of synthetic biology. Computational approaches used in these two disciplines are mainly based on the establishment of various computational models. Given a predicted environment and initial conditions, these models are capable to mimic the behaviour of an arbitrary biological system. Regarding their design, two main approaches exist, i.e.,, *top-down* and *bottom-up* approach. Top-down design approach is a basis for the establishment of computational models using experimentally measured data from existent biological systems (usually obtained using DNA microarrays) [3, 4, 5]. This approach is also known as *network inference* and is based on reverse engineering of the computational model from experimentally measured data. The complementary, i.e., bottom-up approach, is on the other hand used to simplify the design of novel biological systems with predefined functionalities. These models are usually constructed with the association of biological modelling primitives, which can be obtained with the top-down approach [6].

The accuracy of computational models is tightly correlated with the accuracy of parameter values, which define the dynamic and behavioural properties of observed biological systems. These parameters are sometimes very hard or even impossible to determine accurately. In order to bypass this problem, several parameter estimation techniques can be used [7, 8] for bottom-up modelling. There are also several top-down modelling approaches that aim to reconstruct an existent biological system from experimental data without trying to estimate the parameter values (for example with the employment of fuzzy logic methods [9]).

Established models that reflect a satisfactory accuracy can be further used to get the specific insights into the behaviour of a certain biological system. When we are dealing with reverse engineered models, these systems already exists. The models are established to perform various analyses and deduct behavioural properties of the system which are hard or even impossible to obtain experimentally [10]. Moreover, modelling results can be used to optimize the behaviour of observed biological system with certain modifications [11]. On the other hand, bottom-up models can be applied to the computational design of biological systems, thus reducing the quantity of experimental work. Several computational approaches for automatic design already exist, such as heuristic optimization techniques which seek for optimal solutions regarding the given objective functions [12]. These functions can be established in different ways, e.g., with the evaluation of specific metrics that are able to quantify the behaviour of biological system [13].

In this chapter we review the computational approaches in the field of systems and synthetic biology. We especially focus on *gene regulatory networks* (GRNs), but most of the presented concepts can also be applied in the same way on *signal transduction* and *metabolic networks*. First we present state of the art modelling approaches in the field. We briefly review parameter estimation techniques that allow us to bridge the problem of parameter evaluation. Further on we present advanced model analysis techniques for performance, robustness and stability analysis of biological systems. In the end we present computational methods for computer-aided design, automatic design and optimization of biological systems.

2. Computational Modelling Approaches

The development in the field of computational modelling of biological systems goes in hand with the progress of systems and synthetic biology. One can select among various modelling approaches which were introduced recently, each with its own benefits and its own potential applications. The decision for an appropriate modelling strategy is vital and depends on several factors, i.e., the complexity of observed system, the availability of its kinetic data and the type of information the user would like to obtain. Several classifications exist with the aim to organize the modelling approaches in different groups and consequently ease the decision of choosing the appropriate ones. The first distinction can be made among *static* and *dynamic* models. Static models do not incorporate time component and only yield topologies or qualitative networks of gene interactions. Dynamic gene network models on the other hand describe the changes of gene expression in time [14]. Both of these two groups can be further classified in *continuous* and *discrete* models. Continuous models observe the abundances of chemical species as non-negative real values with certain upper limits. *Discrete models* however observe the system on the molecular level and therefore describe its current state with discrete quantity of molecular numbers. Further distinction among dynamic models classifies them in *qualitative models* (also referred to as *logical models* [15]), which are relatively simple and easy to infer even from imprecise data [16], but can only answer qualitative questions, and *quantitative models*, which are able to reflect higher accuracy, but on the account of their complexity and demand accurate values of kinetic data. These models are usually inappropriate for exact modelling of more complex biological systems, but are on the other hand more informative about the system's properties when used in accordance with their limitations. Quantitative models can be further divided among *deterministic models* which are only able to capture average response within a population of identical cells or average response in a single cell over a long time period and *stochastic models* which also consider the probabilistic nature of chemical reactions and can thus be used for the modelling of population heterogeneity and effects of intrinsic noise [17]. Both modelling types yield similar results when we are dealing with large systems (high numbers of molecules and large cell volumes) with fast promoter kinetics [18]. Considerable attention has been devoted to the concept of stochasticity in GRNs since the advent of synthetic biology [17, 19]. More details about stochastic approaches in synthetic biology can be found in [20]. Another distinction of modelling approaches can be made regarding the *level of details* (granularity) they consider, i.e., parts, topology, control logic [21] or coarse, average and fine grained models [22], where fine grained models can mostly be applied only to small systems and coarse grained models to large and complex systems.

Here we will present few examples of different modelling techniques belonging to different groups listed above (see Tab. 1). Since the field of computational modelling has expanded rapidly in last years, we will have to omit some of the modelling methods from this review.

Table 1. Different modelling techniques classified in groups described within the main text, where DG denotes Directed Graphs, (D)BYN (Dynamic) Bayesian Networks, (P)BN (Probabilistic) Boolean Networks, ODE Ordinary Differential Equations, TM Thermodynamic Models, CME Chemical Master Equations and PN Petri Nets. ST denotes static and DN dynamic models, CT continuous and DS discrete models, QL qualitative and QN quantitative models, DT deterministic and ST stochastic models, LG large and SM small networks. Tab. 1 is a modification of the table presenting the summary of properties of different modelling formalisms in [22]

	ST/DN	CT/DS,	QL/QN,	DT/ST,	LG/SM
DG	ST	DS	QL	DT	LG
BYN	ST	DS	QN	ST	LG
DBYN	DN	DS	QN	ST	LG
BN	DN	DS	QL	DT/ST	LG
PBN	DN	DS	QL	ST	LG
ODE	DN	CT	QN	DT/ST	SM
TM	ST	CT	QN	DT	SM
CME	DN	DS	QN	ST	SM
PN	ST/DN	CT/DS	QL/QN	DT/ST	SM

2.1. Directed Graphs

Directed graphs (DGs) can be regarded as the most basic presentation of biological regulatory networks. Simplicity of this approach makes it applicable to the modelling of large biological systems (i.e., for several magnitudes larger than with other modelling approaches), but on the account of its static nature. Various graph operations can be performed on these models in order to make relevant predictions about the structure and behaviour of observed regulatory system, e.g., search for paths among two genes or investigation of existence of cycles that may point at possible feedback relations [22]. A review of such approaches is presented in [23].

Mathematical description of DG is presented with two sets, i.e., set of vertices V and set of edges E. When dealing with gene regulatory networks, vertices usually refer to genes (or chemical species that are results of their expression) and edges to interactions among them. Each edge is defined as $\langle y,x,r \rangle$, where y is the head and x the tail of the edge. This can be interpreted as x regulates the expression of y. Regulation type is defined in r, i.e., y can be activated (+) or inhibited (–) by x (for graphical presentation see Fig. 1). Basic graph presentation can be extended with hypergraphs, in which each vertex is defined as $\langle y,X,R \rangle$. Here X corresponds to the vertices that cooperatively regulate y and R to the types of regulation for each one of them.

Deduction of DGs can be made on the basis of existent knowledge or with the application of different automated inference procedures. Various *clustering algorithms* can be used, which are motivated by the fact that two genes may regulate each other or may be coregulated by a third gene if they both reflect similar expression patterns [22]. Many novel alternative inference approaches were reported recently. We will briefly describe a solu-

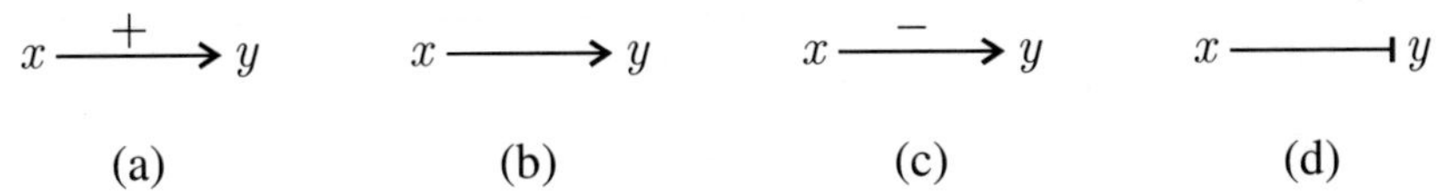

Figure 1. Symbols used in graphical presentations of directed graph models. Figs. (a) and (b) present a situation where tail vertex activates the expression of head vertex. Figs. (c) and (d) present a situation where tail vertex, inhibits the expression of head vertex. Fig. (a) is compatible with Fig. (c), and Fig. (b) is compatible with Fig. (d).

tion which combines *artificial neural-networks* with *genetic algorithms*, *particle swarm optimization* (PSO) and *fuzzy logic* into a *multi-layer evolutionary trained neuro-fuzzy recurrent network* (ENFRN) [24].

ENFRN is able to successfully extract the regulatory interactions from (noisy) data obtained with gene expression profiling. Evolutionary training of artificial neural network based on the PSO automatically generates an adaptive number of temporal fuzzy rules that describe the relationships between the input and the output genes. The training is performed on the basis of gene expression profiles in two phases: (a) the structure (topology) learning, in which fuzzy IF-THEN rules are generated and feedback configuration is established, and (b) parameter learning, in which free parameters which define the established fuzzy rules are tuned. Trained ENFRN structure is able to determine if specified input regulates the output, the kind of regulation and also provides a score, that specifies the confidence in retrieved relation.

2.2. Boolean Networks

Boolean networks (BNs) are the most basic dynamic extension of DG modelling approach in which a single gene is considered to be in one of two possible states, i.e., *on* or *off*. Direct regulatory interactions among genes are modelled as Boolean functions [22]. Let's presume, that the state of the system in time t is defined with the vector $\mathbf{x}(t) = [x_1(t), x_2(t),...,x_N(t)]^T$, where $x_i(t)$ is a boolean variable corresponding to activity (presence) or inactivity (absence) of gene i (or a chemical species that results in its expression) in time t. Model dynamics is determined with a set of Boolean functions $B = \{b_1,b_2,...,b_N\}$, where $x_i(t+1) = b_i(\mathbf{x}(t))$. BNs can be directly projected to DG models. Direct projection cannot be made in both directions, since BNs contain additional information to directed and signed network diagrams [16]. For example of such projection see Fig. 2. While the number of possible states of the system is limited, i.e., 2^N possible states for N Boolean variables, all initial states may eventually reach a *steady state* (point attractor) or a *state cycle* (dynamic attractor). All possible system trajectories from an initial configuration can be analysed with a *state transition graph*. However, these graphs can be constructed only for small systems or only for limited amount of initial configurations.

BNs do not consider intermediate levels of gene expression which results in possible information loss. In basic BNs, transitions between states, i.e., from $\mathbf{x}(t)$ to $\mathbf{x}(t+1)$, occur synchronously, meaning that all state changes happen simultaneously and in deterministic regime. Simplicity of this approach, its qualitative features and parameter-free

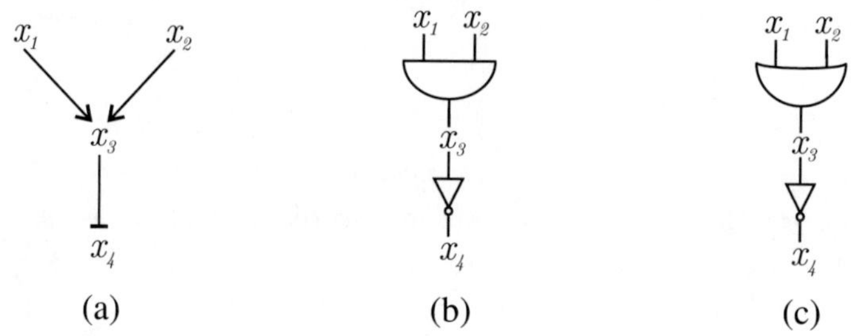

Figure 2. An example of directed graph with two possible projections to Boolean networks, i.e., $B = \{b_3(x_1,x_2) = x_1 \text{ OR } x_2, b_4(x_3) = \text{ NOT } x_3\}$ in Fig. (b) and $B = \{b_3(x_1,x_2) = x_1 \text{ AND } x_2, b_4(x_3) = \text{ NOT } x_3\}$ in Fig. (c).

nature makes it applicable for inference, dynamic modelling and efficient analysis of large scale regulatory networks [16]. Several simplifications can make this approach inappropriate for general modelling of biological systems [22]. However, BNs can be extended in some degree. *Probabilistic Boolean Networks* (PBNs) are able to incorporate the stochastic nature of chemical reactions [25]. PBN annotation extends a set of functions B into $B = \{B_1,B_2,...,B_N\}$, where B_i is a set of Boolean functions $\{f_1^{(i)}, f_2^{(i)},...,f_l^{(i)}\}$. It contains all possible functions that may define the state of Boolean variable x_i. Each of these functions can be chosen to update the state vector in each iteration according to its probability. Moreover, basic and probabilistic BNs can be extended with the introduction of non-synchronous transition schemes, in which the states of the nodes are updated in dependence on the time-scales of individual biological events regarding their underlying regulatory processes. These schemes can be further divided to deterministic and stochastic ones. While time-scales are fixed for each node in deterministic schemes, stochastic schemes randomly select a node to be updated or update all nodes in random order [16].

2.3. Bayesian Networks

Bayesian networks (BYNs) (also referred to as *probabilistic networks*) are a combination of probability calculus, i.e., they incorporate the stochastic nature of gene regulation, and graph theory [3]. BYNs describe the regulatory interaction in regulatory networks with *directed acyclic graph* presentation, where each vertex has a similar interpretation as in *directed graph* presentation (see section 2.1). Random variables that describe the expression levels correspond to each vertex. Conditional distribution is defined for each random variable, i.e., $p(X_i|parents(X_i))$, where i is the vertex, X_i random variable, that describes its expression level and $parents(X_i)$ random variables describing direct regulators of i. BYN description is therefore defined with the directed acyclic graph G and the conditional probability distributions Θ, which defines local conditional probability distributions for each vertex in graph [22]. An example of a BYN graph is presented in Fig. 3(a). Described approach is also known as *static Bayesian network* modelling. Its main disadvantages are in the fact that it is unable to capture feedback loops in gene regulation and can only be used to analyse dependencies between genes. It is therefore not suitable for simulation of system's dynamics.

Dynamic Bayesian networks (DBYNs) (also referred to as *dynamic probabilistic networks*) are a temporal extension of BYNs, that separate input nodes from output nodes. This modelling approach is therefore suitable for simulating the time-dependant dynamics of the system and is able to describe regulatory feedback loops with directed acyclic graphs. DBYN can be defined with a pair (B_0,B_1), where $B_0 = (G_0,\Theta_0)$ is initial BYN and $B_1 = (G_1,\Theta_1)$ a transition BYN which specifies transition probabilities, i.e., $P(\mathbf{X(t)}|\mathbf{X(t-1)})$ [26]. An example of a transition graph is presented in Fig. 3(b).

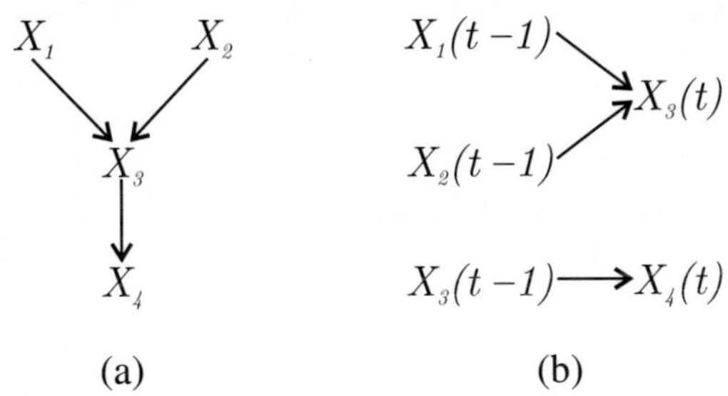

Figure 3. Directed acyclic graphs presenting an example of static Bayesian network (a) and a transition graph of the dynamic Bayesian network for the same regulatory network (b).

The main advantages of BYNs are in their capabilities to model large systems, but on the account of their coarse granularity. These networks are also easy to reconstruct, even if incomplete knowledge about the system is available or from the combination of different types of data. BYNs as such have many features which are suitable for the modelling of regulatory networks [3, 22]. While DBYN solve basic limitations of static BYNs, computational costs drastically increase when inferring these models from experimental data and are therefore not suitable for large regulatory networks.

2.4. Ordinary Differential Equations

Ordinary differential equation (ODE) models consider the concentrations of observed chemical species (such as mRNAs and proteins) as time-dependant variables [22]. Although ODE based models are mainly deterministic, their extensions with the introduction of Poisson random variables has also been reported [27]. These models mainly rely on mass action kinetics or on phenomenological representations of reaction mechanisms [17]. Let's presume, that state of the system in time t is defined as $\mathbf{x}(t) = [x_1(t), x_2(t),...,x_N(t)]^T$, where $x_i(t)$ is current abundance of chemical species i as non-negative real value. ODE model can therefore be defined with a set of ordinary differential equations of form:

$$\frac{dx_i}{dt} = f_i(\mathbf{x}(t),\mathbf{u}(t)); 1 \leq i \leq N, \quad (1)$$

where $\mathbf{u}(t)$ is a vector of concentrations of input components and $f_i(\cdot)$ in most cases a nonlinear function. Due to the nonlinearity of ODE models their solutions are usually obtained numerically (e.g., with Euler method). Qualitative properties of the system can also be derived from these models (see Section 4).

In the context of gene regulatory networks three main reaction types are described with the given set of differential equations, namely *transcription*, *translation* and *degradation*. While translation and degradation usually follow ordinary mass-action kinetics, transcription regulation, i.e., activation and repression, is in most cases modelled with the sigmoid shape functions, such as Hill functions [28]. Let's consider that an activator X increases the rate of transcription when it binds to its regulatory region. Promoter activity can be expressed as

$$f_A(X) = \frac{\beta \cdot X^n}{K_d^n + X^n}, \tag{2}$$

where β is maximal promoter activity, K_d dissociation constant and n nonlinearity (Hill) coefficient. Hill coefficient defines the steepness of the transitions and is often interpreted as cooperativity coefficient although it may also arise from other factors [29]. Let's presume that the expression of protein Y is activated by transcription factor X. Simplified model of its dynamics can be presented by

$$\frac{dY}{dt} = \frac{\beta \cdot X^n}{K_d^n + X^n} + \beta_0 - \delta Y, \tag{3}$$

where β_0 is basal transcription rate and δ protein degradation/dilution rate. Complementary regulation type of activation is repression of the transcription, where promoter activity can be expressed as

$$f_R(X) = \frac{\beta}{1 + \left(\frac{X}{K_d}\right)^n}. \tag{4}$$

Even though Hill equations are a relatively good approximation of transcriptional activity, they may reflect inappropriate results in some cases. They imply that the transcription factors are bound to their corresponding binding sites simultaneously when their cooperativity is larger than 1 [30]. If oligomerisation rates are on the same time-scales as other reaction rates, different approaches, such as classical mass-action kinetics, have to be used.

Several other approaches have been derived from ODE models [22]. *Piecewise linear differential equation models* are obtained if sigmoid curves are approximated with step functions and are used in order to simplify the qualitative analysis. ODE models can be extended with the transition to *delayed differential equations* (DDEs) [31] of the form

$$\frac{dx_i}{dt} = f_i(\mathbf{x}(t-\tau), \mathbf{u}(t)); 1 \le i \le N, \tag{5}$$

where τ is an N dimensional non-negative vector of discrete time delays. This approach allows us to model slow biochemical reactions, such as gene transcription and translation and protein diffusion, more precisely.

Models based on ODEs are relatively simple and therefore easy to construct when kinetic rates of observed reactions are available. On the other hand the quality and quantity of data needed to derive these rates makes them difficult to apply to poorly characterized or noisy systems. Even when precise data are available, number of parameters that need to be estimated may present major difficulties when inferring larger networks. If inference is successfully performed, high computational costs can present another problem when dealing with such networks. These models are therefore hard to scale up to more complex systems.

2.5. Thermodynamic Modelling

Thermodynamic (TD) or *fractional occupancy* modelling derives from the presumption, that gene expression is proportional to the level of bound activators and inversely proportional to the level of bound repressors [32]. It considers DNA binding and protein interactions in equilibrium condition [33]. The benefit of TD modelling approach is that we can predict occupancies of different binding site types very accurately, e.g., when transcription factors are competing for overlapping binding sites or cooperatively interacting at nearby binding sites. It can also account for very specific nonlinear regulatory responses, such as transcription synergy [34].

Given a set of binding sites, concentrations of transcription factors and their binding affinities, relative probabilities of each binding site configuration can be calculated in the first step. The probability of each configuration can be calculated according to its statistical weight which depends on the number and affinities of occupied binding sites in the configuration and interactions among bound transcription factors. Probabilities of binding sites that lead to transcription activation can be summed in *fractional occupancy*, which may also be expressed as a ratio of weights of binding site occupancies, that lead to transcription activation, to weights of all possible binding site occupancies.

In the second step gene expression level is calculated according to determined fractional occupancies, for which different techniques can be used. For example, calculation can be performed with the product among fractional occupancy of each promoter and its corresponding expression level.

TD models cannot describe dynamical nature of biological systems by themselves [33]. However, it is possible to combine them with the differential equation modelling in order to incorporate changes of gene expression over time. In each iteration fractional occupancies of binding sites and their corresponding expression levels are calculated in dependence of transcription factors concentrations. Concentration values of chemical species are on the other hand derived from the set of ordinary differential equations in each time step.

Let's again consider the same example as described in section 2.4, where protein X activates the transcription. Its fractional occupancy may be expressed with the following equation:

$$f_A = \frac{X^n}{{K_d}^n + X^n}, \tag{6}$$

where K_d describes dissociation constant and n nonlinearity coefficient. Gene expression can be therefore expressed as $f_A \cdot \beta$, where β corresponds to maximal promoter activity. If we combine this model with protein degradation and basal expression of promoter, equation 3 is obtained. If protein X would have a complementary, i.e., repressible role, fractional occupancy would be expressed as:

$$f_R = \frac{1}{1 + \left(\frac{X}{K_d}\right)^n + X^n}. \tag{7}$$

Even though fractional occupancy is able to precisely capture the dynamics when dealing with different types of binding site occupancies, models obtained in these two simple examples obtained with TD modelling are the same as ODE based models.

2.6. Single Molecule Level Models

Biological systems neither posses deterministic or continuous nature which is mostly presumed by ODE and thermodynamic models. *Single molecule level* models on the other hand describe their dynamics on the molecular level. Concentrations of observed species are thus presented with whole numbers. Moreover, reactions which affect their abundances are determined stochastically. When we are dealing with large systems or observing a system over longer periods of time the differences in response of different approaches are negligible [17]. But if the molecular population is small or if the system is sensitive to noise effects, continuous deterministic models may lead us to wrong conclusions [19].

Similar as with deterministic models state of the system can be defined with a vector $\mathbf{x}(t) = [x_1(t), x_2(t), ..., x_N(t)]^T$, where $x_i(t)$ is current abundance of chemical species S_i as non-negative whole number. Each of the chemical species from the set $\{S_1, S_2, ..., S_N\}$ interacts with others through so called reaction channels $\{R_1, R_2, ..., R_M\}$, which are defined by their *state change vectors* and their *propensities*. State change vector defines the species and their quantities that are produced and consumed by each reaction j in the form $\nu_j = (\nu_{1j}, \nu_{2j}, ..., \nu_{Nj})$. Propensities $(a_j(\mathbf{x}(t)))$ can be derived from reaction rates and system volume and are used to calculate the probability for each reaction to occur in a time step dt. Dynamics of the system can be analysed with the establishment of so called *Chemical Master Equation* [19]:

$$\dot{p}(\mathbf{x}(t)) = -p(\mathbf{x}(t)) \sum_{j=1}^{M} a_j(\mathbf{x}(t)) + \sum_{j=1}^{M} p(\mathbf{x}(t) - \nu_j; t) a_j(\mathbf{x}(t) - \nu_j). \tag{8}$$

This equation completely determines the probability of each state, but can be solved analytically only for very small systems. Various numerical approximations are therefore used instead, such as *Stochastic simulation algorithm* (SSA) [35], which can be described as:

1. Initialize the system.
2. Calculate current propensities $a_j(\mathbf{x}(t))$ and their sums $a_0(\mathbf{x}(t))$.
3. Generate random values r_1 and r_2.
4. Determine next time step: $\tau = \frac{1}{a_0(\mathbf{x}(t))} \ln\left(\frac{1}{r_1}\right)$.
5. Determine next reaction R_j to occur according to equation $\sum_{j'=1}^{j-1} a_{j'}(\mathbf{x}(t)) \leq r_2 a_0(\mathbf{x}(t)) < \sum_{j'=1}^{j} a_{j'}(\mathbf{x}(t))$.
6. Calculate state change: $\mathbf{x}(t+\tau) = \mathbf{x}(t) + \nu_j$.
7. Increase time: $t \leftarrow t + \tau$.
8. Return to step 2 unless conditions for stopping the simulation are fulfilled.

Computational time increases drastically with the number of observed chemical species. One of the solutions is to use another approximative approach, namely τ leaping, which does not calculate time steps in each iteration, but uses a fixed value through the whole simulation. In order to obtain valid results, time step must be chosen carefully (no propensity function should change its value significantly within defined time step).

Single molecule level models can be extended with the incorporation of time delays in a similar way as ODE models are extended to DDE models. Here, time delays can be assigned to each product of observed chemical reactions. Dynamics of such models can be analysed with *Delayed SSA*, which can be presented with the following steps [36]:

1. Initialize the system. Clear the queue L, which will contain the chemical species representing queued products and their designated times of appearance.

2. Determine next time step τ and next reaction R_j to occur in the same way as in basic SSA.

3. Calculate state change: Let t be current time and t_{min} be the lowest value in queue L. If $t + \tau < t_{\text{min}}$, calculate the state change in the same way as in SSA, with an exception of delayed products, which are inserted into L together with their time of appearance in the system. If $t + \tau \geq t_{\text{min}}$ release all the elements of the queue, with their designated times of appearance lower than $t+\tau$ and accordingly update the state vector.

4. Increase time:
$$t \leftarrow \begin{cases} t + \tau, & \text{if } t + \tau < t_{\text{min}} \\ t_{\text{min}}, & \text{otherwise} \end{cases}$$

5. Return to step 2 unless conditions for stopping the simulation are fulfilled.

The effectiveness of described approaches is drastically reduced, when the number of observed chemical reactions and chemical species increases. However, several improvements could be used to decrease the computational complexity of the algorithms based on CME. *Multi Time Scale* modelling approach [37] considers the fact that chemical reactions occur in different time scales. The rates of transcription processes for example are usually significantly higher than the rates of DNA and RNA binding reactions, such as scaffolding, dimerization, linking etc. The common kinetic-propensity approaches presented above may already take into consideration these differences. Nested simulations for each time scale reaction set may on the other hand improve the effectiveness of simulating the state changes [37]. A simple application of this concept can be easily implemented by using SSA specifically for the reactions that occur in time scales of minutes or hours and a nested SSA for the reactions that occur in time scales of seconds.

2.7. Petri Nets

Petri Nets (PNs) are in their most basic form used for modelling and analysis of various concurrent, asynchronous and distributed systems. The mathematical background of PNs enables us to analyse the system we are modelling, while a PN graph gives us its graphical representation. With recent development of different PN extensions they are becoming a powerful tool for describing biological systems. PNs were at first applied to metabolic networks [38], but they are, however, becoming increasingly popular for modelling gene regulatory networks.

PN is a directed-bipartite graph with two different types of vertices: *places* and *transitions*. When modelling biological systems, places correspond to chemical species and transitions to the events occurring in the system, namely chemical reactions that govern dynamics of the system. They are connected by *arcs* (directed edges) which represent how different chemical species interact in the system. At any time, places can hold zero or a positive number of *tokens*. Depending on what reaction we are modelling, these tokens can represent species concentration or simply presence or absence of a certain chemical compound. Distribution and allocation of tokens over places represents current state of the system which is called a *marking* of the PN. Marking of a PN changes when a transition *fires*. Transition can be fired only if all required conditions for that transition are met, e.g., all the chemical species needed for a reaction are present. Let's presume we have two chemical species x_1 and x_2 that can be combined with a chemical reaction t_1 which produces x_3. We can represent the model of this reaction and its different states with a PN as shown in Fig. 4.

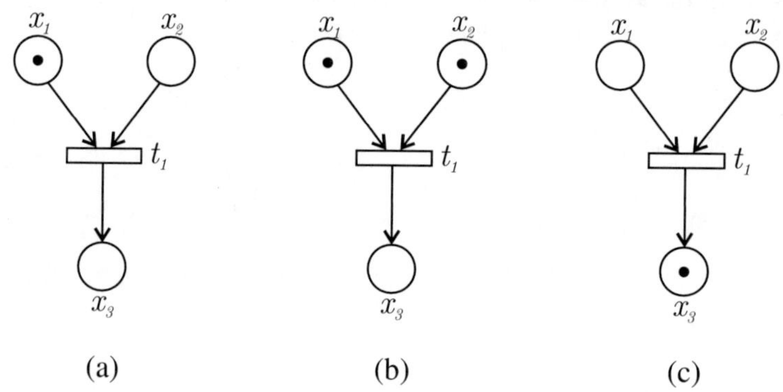

Figure 4. Example of a simple Petri Net of a chemical reaction $x_1 + x_2 \rightarrow x_3$. Fig. (a) presents a scenario where chemical reaction will not happen since one of the reactants is missing (x_2). Fig. (b) shows an enabled transition (chemical reaction can happen) and Fig. (c) the configuration after transition was fired.

We can construct a PN for any reaction or process which is a part of a larger biological system. By combining these basic parts, PNs presenting larger regulatory networks are constructed. Basic PNs support only strictly discrete modelling without the notion of time and are as such used as a framework for many different purely qualitative static modelling approaches [39]. However, with different extensions, PNs can be also used to construct dynamic continuous and discrete [40] models, while considering both deterministic [41] and stochastic [42] representations of the model. Additionally, possibilities to augment continuous deterministic PNs with fuzzy methods are currently being analysed [43]. The proposed solution aims to solve the problem of parameter sloppiness while maintaining a relatively good accuracy of established models at the same time.

Because PNs are a versatile tool for modelling biological systems, the size of the system we can efficiently analyse and model depends on the granularity and approach we aim for. While we can efficiently analyse large static qualitative models, different extensions increase the complexity of PN dynamics and can only be used to construct smaller models.

3. Parameter Estimation Techniques

A common problem of several computational modelling approaches is in their dependence on accurate kinetic data, such as kinetic rate constants or diffusion coefficients. It may be difficult, economically infeasible or even impossible to obtain these parameters experimentally in some cases. *Parameter estimation problem* often represents a serious obstacle when the modelling constraints require a high standard of reliability. Several computational techniques have been developed to overcome this problem. However, no standard has actually been defined, because of intrinsic uncertainty of the underlying biological systems. These methods have to face nonlinear constraints, which are implicit to such systems. The most relevant contributions in the field of parameter estimation have come from control theory of dynamical systems. Control theory is responsible for the development of several optimization and estimation methods used in automatic systems control. Many of these methods, such as *extended Kalman filter*, have already been successfully applied to computational biology [44, 45, 46].

Extended Kalman filtering approach presumes that an arbitrary nonlinear dynamic system may be approximated with a set of state change equations

$$\begin{aligned} \mathbf{x}_k &= f(\mathbf{x}_{k-1}, \mathbf{u}_{k-1}, \theta) + \mathbf{w}_k, \\ \mathbf{y}_k &= h(\mathbf{x}_k) + \mathbf{v}_k, \end{aligned} \tag{9}$$

where $\mathbf{x}$ is state vector of chemical species, i.e., it defines state of the system, $\mathbf{u}$ vector of system inputs, θ parameter vector which defines kinetic constants, $\mathbf{y}$ output vector of the system, $\mathbf{w}$ and $\mathbf{v}$ Gaussian noise vectors with zero mean and covariance matrices $\mathbf{R}$ and $\mathbf{Q}$ respectively; h is output and f transition function of the system, which completely defines its state change dynamics. Input vector $\mathbf{u}$ contains the parameters that define external influences on the system, such as temperature or pH variations. Output vector $\mathbf{y}$ usually contains experimentally obtained data. Function h describes these data, e.g., it can be interpreted as a response function that approximates the time course of a certain output protein concentrations.

Extended Kalman filter is able to estimate the state vector $\mathbf{x}$ on each discrete time step k, with its estimation vector $\hat{\mathbf{x}}_k$. In order to estimate unknown parameters at the same time state extension has to be performed:

$$\mathbf{x} = \begin{bmatrix} \mathbf{x} \\ \theta \end{bmatrix}. \tag{10}$$

The estimation is obtained in a two stage computation of the predictor-corrector form (see Fig. 5). In first step predicted state vector $\hat{\mathbf{x}}_{k|k-1}$ and covariance matrix $P_{k|k-1}$, which contains predicted variance changes of previously estimated state vector $\hat{\mathbf{x}}_{k-1}$, are evaluated. These predictions are used to construct gain K and to update the state and covariance matrix estimation, $\hat{\mathbf{x}}_k$ and P_k respectively. State estimation $\hat{\mathbf{x}}_k$ is evaluated by adjusting current (predicted) state estimation $\mathbf{x}_{k|k-1}$ with the difference between predicted and estimated output of the system, i.e., $\mathbf{y}_k$ and $h(\hat{\mathbf{x}}_k - 1)$ respectively, amplified or muted by the obtained gain K.

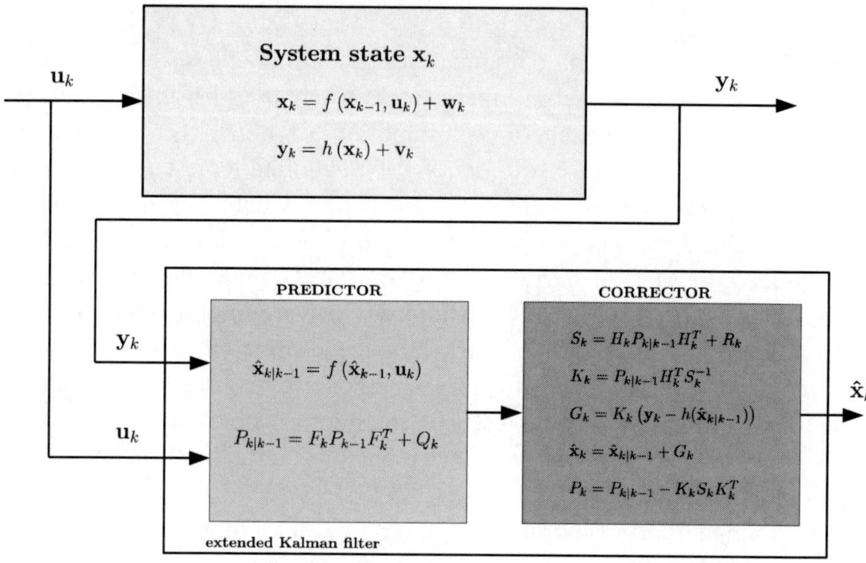

Figure 5. A schematic description of extended Kalman filtering, where F_k is Jacobian of function f, evaluated on the previous *a priori* state estimates $\hat{\mathbf{x}}_{k|k-1}$, formally denoted as $J_f^x(\hat{\mathbf{x}}_{k|k-1})$ and similarly, H_k is Jacobian of h, formally denoted as $J_h^x(\hat{\mathbf{x}}_{k|k-1})$. We refer to [47] for a complete derivation of extended Kalman filter equations.

Initial state estimation has to be performed before the filtering. Initial states are usually set to the mean values of all initial concentrations of chemical species $\bar{\mathbf{x}}$. Initial covariance matrix P_0 is on the other hand set to be a diagonal positive definite matrix containing initial mean variance of the state vector $\mathbf{x}_0$ [44]:

$$\hat{\mathbf{x}}_0 = \bar{\mathbf{x}}_0, \tag{11}$$

$$\hat{P}_0 = E\left\{(\mathbf{x} - \bar{\mathbf{x}}_0)(\mathbf{x} - \bar{\mathbf{x}}_0)^T\right\}. \tag{12}$$

The initial estimations are very important for the global convergence of the filter. Wrong estimation will cause the errors to be carried on during the entire filtering process, accumulating more and more noisy data. Once the initialization is performed, the evaluations of predictor and corrector equations can be performed in each time step (see Fig. 5). Computational complexity of extended Kalman filtering approach depends on the size of the state vector $\hat{\mathbf{x}}$ and on the filtering time, i.e., number of samples from which estimations are performed.

The main disadvantage of other state of the art parameter estimation methods is in their computational complexity, when applied to models with high numbers of unknown parameters. A model representation of GRN which implements a simple logic gate, such as AND or NOR, may hide several tens of unknown kinetic constants. Extended Kalman filtering approach seems to decrease computational complexity and increase the quality of estimations. Unfortunately, extended Kalman filter may sometimes diverge, which can result in several magnitudes of variations among numerical estimations. However, a validation of estimations based on statistical tests, such as χ^2, can be performed after the filtering [44] in order to confirm the reliability of estimates. Other approaches from control theory have also been successfully applied to parameter estimation problem, e.g., state estimation techniques by *state observers methods* [48, 49] and *particle filtering* [50]. We refer to [8, 44] for a comprehensive review of these approaches.

4. Advanced Model Analysis

Confidence in results obtained with established models can be increased with several validation techniques. Computational models are often based on hypothetical assumptions which are either well known or have to be confirmed. Computational approaches can be used for hypotheses confirmation, e.g., using statistical tests, such as χ^2 [44] or robustness criteria [51]. If confirmation is negative, the identification of erroneous model components and the review of the basic hypothesis of the model itself is vital. In worst case it is necessary to redesign the entire model including the principal hypothesis. If satisfactory accuracy of modelled dynamics is achieved further analytic approaches can be used in order to estimate the *performance*, *robustness* and *stability* of observed biological system.

4.1. Performance Evaluation of Biological Systems

Performance evaluation techniques are used to objectively evaluate the behaviour of biological systems by establishing various *objective functions*. In order to analyse specific performances, both modelling and experimental data can be included in the domain of these functions. A typical example of objective function, which is vastly used in reverse engineering, is *mean square error*:

$$E(Z,X) = \frac{1}{N}\sum_{i=1}^{N}(z_i - x_i)^2, \tag{13}$$

where $\{z_1,z_2,\ldots,z_N\} \in Z$ are samples reflecting the desired dynamics of the system and $\{x_1,x_2,\ldots,x_N\} \in X$ are samples obtained from modelling or experimental results. At glance function $E(Z,X)$ is a simple error measure. However, it can also be used as a naive approach to validate the model accuracy, e.g., by using the experimental results as Z and the modelling results as X in Eq. 13. We can thus quantitatively describe the similarities among the modelling and experimental results. Furthermore, mean square error function can be used to estimate unknown model parameters (see Section 3). The best model response may be obtained by the minimization of the function $E(Z,X)$ regarding the unknown parameters. Unfortunately, computational complexity of this approach makes it applicable only to problems with small numbers of unknown parameters. Improvements can be obtained with the use of special heuristics [8].

Different type of objective functions can be established with several metrics, such as *signal to noise ratio* or quantities which describe the results obtained with robustness, sensitivity or stability analysis (see Sections 4.2 and 4.3). In [13] various metrics were introduced and used to estimate the performance of modelled biological system from the information processing perspective. Characteristics that are similar to the ones used to describe the performance of digital electronic circuits, such as switching times and logical levels, were applied to biological systems. This work was extended in [52], where robustness and logical compatibility among different biological processing structures was also considered in order to automatize the construction of more complex information processing biological systems. While these metrics were only used on selected models of hypothetical biological processing structures, they could also be applied to data gathered from laboratory experiments.

4.2. Robustness and Sensitivity Analysis

Robustness is believed to be the key factor of adaptability in the evolutionary process of biological systems [53]. In cell biology, the robustness is the property of a biological system to remedy a substantial fluctuation in its homeostasis due to a sudden change in the conditions for its stability. Such changes can be provoked by external perturbations on the key parameters of the system. A robust system may respond with a counterbalance effects to these parameter changes, such as bacterial chemotaxis behaviour [28].

Although a general quantitative measure for robustness has still not been established, many efforts have come from various scientific disciplines, especially from the control theory. The use of bifurcation analysis, i.e., the Hopf bifurcation, was studied in [54] for evaluating the robustness of the Laub and Loomis model of cAMP oscillations in *Dictyostelium discoideum* cells. Similarly in [55] the same model was analysed, but with the prevalent use of μ-analysis. An interesting method for robustness analysis using *linear time logic* (LTL) was proposed in [56]. Despite its complexity, this methodology appears to offer a large spectrum of application, especially for synthetic gene networks.

Sensitivity analysis may also represent a metric to quantitatively evaluate the robustness of computational models [57]. This analysis aims to identify the parameters for which small input variations cause substantial variations in model response. Sensitivity analysis approaches can be divided in two categories, i.e., *local sensitivity analysis* and *global sensitivity analysis* approaches. Local sensitivity analysis refers to analytical methods that are capable to evaluate how much the variations in the model outputs can be apportioned to small variations in input parameter values [57]. On the other hand global sensitivity analysis aims to analyse large and even complex perturbations of parameter values with various numerical and statistical methods. State of the art sensitivity analysis approaches can be mainly applied to deterministic models only (for an application to stochastic models see [58]).

Local sensitivity can be estimated by evaluating first-order derivatives of the model output response relatively to input parameters. A quantitative measure can be mathematically represented by sensitivity coefficients of the form [57]:

$$S_i = \frac{\partial y_i}{\partial p} = \lim_{\Delta p \to 0} \frac{y_i(p + \Delta p) - y_i(p)}{\Delta p}. \tag{14}$$

Finite difference approximation, direct differential method and adjoint sensitivity analysis [57] evaluate sensitivity in terms of sensitivity coefficients as described in Eq. 14. In metabolic control analysis (MCA) [59] elasticity coefficients were developed for estimating the relationship between model output y_i and specific model parameter p.

Several global sensitivity analysis methods exist (see review in [57]). In order to analyse large parameter perturbations, effective sampling of all possible parameter values becomes crucial. Latin hypercube sampling is usually used in this context rather than Monte Carlo random sampling [57]. Global sensitivity analysis methods can be further divided in two subcategories, i.e., *variance based* and *variance non-based approaches*. Variance based methods, such as *Sobol sensitivity analysis* and *Fourier amplitude sensitivity test* (FAST), aim to estimate the global sensitivity as a relation between statistical variances of model outputs and chosen model parameters. These relations can assume a coefficient-like form such as in Sobol sensitivity analysis [57]:

$$S_{i_1 i_2 \ldots i_s} = \frac{D_{i_1 i_2 \ldots i_s}}{D}, \tag{15}$$

where $S_{i_1 i_2 \ldots i_s}$ are sensitivity coefficient, D is total variance of the system and $D_{i_1 i_2 \ldots i_s}$ are partial variances regarding the chosen parameter x_i. The main disadvantage of variance-based methods is their computational complexity.While the main advantage of non-variance based methods, such as *multi-parametric sensitivity analysis* (MPSA) and *Morris sensitivity analysis*, is their low computational complexity, they imply a certain grade of monotonicity in the model response. An additional disadvantage of the Morris method is in its unreliability when the model response exhibits negative values.

A closely related concepts to sensitivity and robustness are *reliability* and *scalability*. Robustness can be seen as a metric for evaluating reliability of a certain model. Reliable models might be used further for building scalable systems, for which the reliability of each component tends to be crucial for the entire structure. Hence a compatibility among basic modules may be required. Compatibility has already been analysed in [13] in the context of information processing biological structures.

4.3. Stability Analysis

The main goal of stability analysis is to obtain the insights into system's asymptotic behaviour, i.e., its behaviour after a long period of time ($t \rightarrow \infty$) without external perturbations, and dependencies of its asymptotic behaviour of given parameter set, also referred to as *bifurcation analysis*. Observed biological systems can in this context exhibit convergence towards *steady states*, which can be analysed with *steady state analysis*, or *stable oscillatory behaviour*, which can be analysed with *limit cycle analysis*. All these methods derive from the theory of *nonlinear dynamical systems* [60], which are usually described with a system of ordinary differential equations. Stability analyses of biological systems are therefore performed on their ODE models (see Section 2.4). We will presume that external inputs are fixed through the course of stability analysis and therefore omit the factor $\mathbf{u}(t)$ from the Equation 1.

The main goal of steady state analysis is to investigate the existence and types of steady states the observed biological system may reflect [61]. Existence of a steady state $\mathbf{x}^* = (x_1^*, x_2^*, \ldots, x_n^*)$ can be conditioned with the equation

$$\frac{dx_i}{dt} = f_i(\mathbf{x}^*) = 0; \forall i \in 1,...,N. \tag{16}$$

Steady states can be divided regarding their stability in two types, i.e., *stable* and *unstable*. The neighbourhood will always converge into a stable steady state, but diverge from an unstable steady state. Stability can be determined with the construction of Jacobian matrix of the form

$$J(\mathbf{x}(\mathbf{t})) = \begin{bmatrix} \frac{\partial f_1(\mathbf{x}(\mathbf{t}))}{\partial x_1} & \frac{\partial f_1(\mathbf{x}(\mathbf{t}))}{\partial x_2} & \cdots & \frac{\partial f_1(\mathbf{x}(\mathbf{t}))}{\partial x_n} \\ \frac{\partial f_2(\mathbf{x}(\mathbf{t}))}{\partial x_1} & \frac{\partial f_2(\mathbf{x}(\mathbf{t}))}{\partial x_2} & \cdots & \frac{\partial f_2(\mathbf{x}(\mathbf{t}))}{\partial x_n} \\ \vdots & \vdots & \ddots & \vdots \\ \frac{\partial f_n(\mathbf{x}(\mathbf{t}))}{\partial x_1} & \frac{\partial f_n(\mathbf{x}(\mathbf{t}))}{\partial x_2} & \cdots & \frac{\partial f_n(\mathbf{x}(\mathbf{t}))}{\partial x_n} \end{bmatrix}. \tag{17}$$

Steady state can be referred to as stable, when all real parts of eigenvalues of the Jacobian matrix are negative.

Existence of oscillatory behaviour on the other hand depends on the existence of a *stable limit cycle*. Limit cycle is an isolated simple oriented closed curve trajectory, which does not contain singular points (i.e., steady stable states) [61]. If the system converges to a limit cycle with time, i.e., $t \to \infty$, limit cycle is *stable*. The easiest way to analyse the existence of a limit cycle is with the state space investigation. When dealing with biological systems, this space is strictly limited by minimum and maximum concentrations of observed chemical species and its exhaust investigation does therefore not issue large computational complexities.

The type of behaviour system exhibits is strictly dependant on the parameter values used in its ODE description. Adjusting these values can therefore drastically change system's asymptotic behaviour, e.g., from a stable state convergence to self-sustained oscillations. Transitions among different types of behaviour are called *bifurcations* and the set of parameter values at which the transitions occur *bifurcation points* [62]. Different types of transitions exist, regarding the characteristics of behaviour that arises and characteristics of behaviour that ceases with the transition through the bifurcation point [60], e.g., *saddle-node* bifurcations, *Hopf* bifurcations and *pitchfork* bifurcations.

Stability analysis can not only be used when analysing the asymptotic behaviour of biological systems, but also when evaluating their robustness, e.g., if the distance of parameter values from the bifurcation point is large, the probability that the system will in reality reflect predicted behaviour is much higher and the system is therefore more robust.

5. Computational Design of Biological Systems

The most common approaches to *de novo* engineering of biological systems with desired behaviour are *directed evolution* [63] and *rational design* [64]. While directed evolution is an experimental method that performs artificial evolution on an initial biological system and therefore mimics natural evolution, but in a much shorter time scale, rational design uses engineering approaches to build novel biological systems and is as such a cornerstone of synthetic biology. These approaches combine modularization, rationalization and modelling [64]. Probably most famous results of rational design approach are toggle switch [65] and repressilator [66] circuits. Rational design combined with computer modelling can be also regarded as *computer aided design* of biological systems in which computational models of basic biological components are connected to each other rationally into more complex modules and systems. If analysed behaviour of these networks is appropriate, experimental realization can be conducted. The analyses of their behaviour can be performed with the methods described in Section 4.

The other computational approach in the design of biological systems is based on the unsupervised design with the investigation of search space of all possible solutions and optimization of objective functions which define the correlation among the desired and reflected behaviour of biological system [12]. Until recently this approach was only used for the design of protein and amino acids nucleotide sequences [67]. With the development of characterisation of basic genetic regulatory elements, mutation effects on their genetic functionalities and increased accuracy of their appropriate models, automatic design approaches can also be applied to the regulatory networks [68]. Given an input, which defines the specified behaviour, computational tools, such as AutoBioCAD [68], are able to find the nucleotide sequence and computational model of regulatory network with appropriate dynamics. Initial solution, on which the evolution is performed, is constructed randomly or with rational design from the basic parts characterized within appropriate libraries. In order to achieve the desired behaviour these solution is evolved with the employment of mutation operators, i.e., modifications of its topology (addition, deletion and replacement of basic parts) and kinetic rates (e.g., with the promoter mutations). Search space is therefore comprised of all possible combinations of elementary structures and their mutations that are provided by available libraries [69]. Each intermediate solution is evaluated with the calculation of objective (fitness) function and the best ones are selected for the next iteration of evolution. These mutation and selection operators are applied in accordance with various metaheuristics such as simulated annealing [68] and genetic algorithm [12]. Even though other automatic design approaches have also been reported, they will not be presented here on account of their several limitations in comparison with the described approach. These limitations include either genetic and functional diversity [70, 71] or requirements for predefinition of network topology [72].

Computational design approaches provide us with the results which rely on computational models of elementary structures and their mutual interactions. Experimental realization of these solutions can reflect the behaviour unpredicted within underlying models. On the other hand these solutions can be a basis for further optimization and fine-tuning with various experimental methods such as directed evolution.

6. Conclusion

While the complexity and the size of controllable biological systems rises, computational approaches gain more and more important role in their design and analysis. Here, we reviewed a collection of such approaches, which we find the most important, even though many others exist. Novel techniques are being developed on the account of many limitations of existent ones, such as incompatibility of accuracy of the modelling results with the size and complexity of modelled system. New methodologies are being introduced to the field also from other engineering disciplines, e.g., with the use of Kalman filtering or fuzzy logic methods. In the near future we expect the field of computational biology to evolve further on with the collaboration of new researchers from various scientific disciplines, until the final goal is achieved, i.e., to establish computational methods, that would allow high accuracy in the modelling, analysis and design of both, engineered and natural biological systems of arbitrary size and complexity.

References

[1] A. Kremling and J. Saez-Rodriguez, "Systems biology - an engineering perspective," *Journal of Biotechnology*, vol. 129, pp. 329–351, 2007.

[2] S. A. Benner and A. M. Sismour, "Synthetic biology," *Nature Reviews Genetics*, vol. 6, pp. 533–543, 2005.

[3] M. Hecker, S. Lambeck, S. Toepfer, E. van Someren, and R. Guthke, "Gene regulatory network inference: Data integration in dynamic models - a review," *BioSystems*, vol. 96, pp. 86–103, 2009.

[4] C. Sima, J. Hua, and S. Jung, "Inference of gene regulatory networks using time-series data: A survey," *Current Genomics*, vol. 10, pp. 416–429, 2009.

[5] P. B. Madhamshettiwar, S. R. Maetschke, M. J. Davis, A. Reverter, and M. A. Ragan, "Gene regulatory network inference: evaluation and application to ovarian cancer allows the prioritization of drug targets," *Genome Medicine*, vol. 4(41), pp. 1–15, 2012.

[6] C. A. Voigt, "Genetic parts to program bacteria," *Current Opinion in Biotechnology*, vol. 17(5), pp. 548–557, 2006.

[7] J. Cao and H. Zhao, "Estimating dynamic models for gene regulation networks," *Bioinformatics*, vol. 24(14), pp. 1619–1624, 2008.

[8] J. Sun, "Parameter estimation using metaheuristics in systems biology: A comprehensive review," *IEEE/ACM Transactions on Computational Biology and Bioinformatics*, vol. 9(1), pp. 185–202, 2012.

[9] R. Hamed, S. Ahson, and R. Parveen, "Designing genetic regulatory networks using fuzzy Petri nets approach," *International Journal of Automation and Computing*, vol. 7(3), pp. 403–412, 2010.

[10] J. Goutsias, "Quasiequilibrium approximation of fast reaction kinetics in stochastic biochemical systems," *The Journal of Chemical Physics*, vol. 122, pp. 1–15, 2005.

[11] G. Batt, B. Yordanov, R. Weiss, and C. Belta, "Robustness analysis and tuning of synthetic gene networks," *Bioinformatics*, vol. 23(18), pp. 2415–2422, 2012.

[12] G. Rodrigo, J. Carrera, T. E. Landrain, and A. Jaramillo, "Perspectives on the automatic design of regulatory systems for synthetic biology," *FEBS Letters*, vol. 586, pp. 2037–2042, 2012.

[13] M. Moškon and M. Mraz, "Modelling and analysing the information processing capabilities of simple biological systems," *Mathematical Modelling and Analysis*, vol. 17, pp. 467–484, 2012.

[14] V. Filkov, *Handbook of Computational Molecular Biology*. Chapman & Hall/CRC Press, 2005, ch. Identifying gene regulatory networks from gene expression data, pp. 27.1–27.29.

[15] G. Karlebach and R. Shamir, "Modelling and analysis of gene regulatory networks," *Nature*, vol. 9, pp. 770–780, 2008.

[16] R. S. Wang, A. Saadatpour, and R. Albert, "Boolean modeling in systems biology: an overview of methodology and applications," *Physical Biology*, vol. 9(5), pp. 1–14, 2012.

[17] M. Kaern, W. J. Blake, and J. Collins, "The engineering of gene regulatory networks," *Annual Review of Biomedical Engineering*, vol. 5, pp. 179–206, 2003.

[18] M. Kaern, T. C. Elston, W. J. Blake, and J. J. Collins, "Stochasticity in gene expression: From theories to phenotypes," *Nature Reviews Genetics*, vol. 6, pp. 451–464, 2005.

[19] H. El Samad, M. Khammash, L. Petzold, and D. Gillespie, "Stochastic modeling of gene regulatory networks," *International Journal of Robust and Nonlinear Control*, vol. 15, pp. 691–711, 2005.

[20] D. J. Wilkinson, "Stochastic modelling for quantitative description of heterogeneous biological systems," *Nature Reviews Genetics*, vol. 10(2), pp. 122–133, 2009.

[21] T. Schlitt and A. Brazma, "Current approaches to gene regulatory network modelling," *BMC Bioinformatics*, vol. 8(6), 2007.

[22] H. de Jong, "Modeling and simulation of genetic regulatory systems: A literature review," *Journal of Computational Biology*, vol. 9, pp. 67–103, 2002.

[23] G. Pavlopoulos, M. Secrier, C. Moschopoulos, T. Soldatos, S. Kossida, J. Aerts, R. Schneider, and P. Bagos, "Using graph theory to analyze biological networks," *BioData Mining*, vol. 4(1), 2011.

[24] I. A. Maraziotis, A. Dragomir, and D. Thanos, "Gene regulatory networks modelling using a dynamic evolutionary hybrid," *BMC Bioinformatics*, vol. 11(140), pp. 1–17, 2010.

[25] I. Shmulevich, E. R. Dougherty, S. Kim, and W. Zhang, "Probabilistic boolean networks: a rule-based uncertainty model for gene regulatory networks," *Bioinformatics*, vol. 18(2), pp. 261–274, 2002.

[26] H. Lähdesmäki, S. Hautaniemi, I. Shmulevich, and O. Yli-Harja, "Relationships between probabilistic boolean networks and dynamic bayesian networks as models of gene regulatory networks," *Signal Processing*, vol. 86(4), pp. 814–834, 2006.

[27] T. Tian and K. Burrage, "Stochastic models for regulatory networks of the genetic toggle switch," *Proceedings of the National Academy of Sciences*, vol. 103, pp. 8372–8377, 2006.

[28] U. Alon, *An Introduction to Systems Biology*. Chapman & Hall/CRC, 2007.

[29] M. Andrecut, J. D. Halley, D. A. Winkler, and S. H. S, "A general model for binary cell fate decision gene circuits with degeneracy: Indeterminacy and switch behavior in the absence of cooperativity," *PLoS ONE*, vol. 6(5), 2011.

[30] J. N. Weiss, "The hill equation revisited: uses and misuses," *FASEB Journal*, vol. 11, pp. 835–841, 1997.

[31] M. Barrio, K. Burrage, A. Leier, and T. Tian, "Oscillatory regulation of hes1: Discrete stochastic delay modelling and simulation," *PLoS Computational Biology*, vol. 2(9), 2006.

[32] H. M. Sauro, *Enzyme Kinetics for Systems Biology*. Future Skill Software (Ambrosius Publishing), 2012.

[33] A. Ay and D. N. Arnosti, "Mathematical modeling of gene expression: a guide for the perplexed biologist," *Critical Reviews in Biochemistry and Molecular Biology*, vol. 46(2), pp. 137–151, 2011.

[34] X. He, M. A. H. Samee, C. Blatti, and S. Sinha, "Thermodynamics-based models of transcriptional regulation by enhancers: The roles of synergistic activation, cooperative binding and short-range repression," *PLoS Computational Biology*, vol. 6(9), 2010.

[35] D. T. Gillespie, "A general method for numerically simulating the stochastic time evolution of coupled chemical reactions," *Journal of Computational Physics*, vol. 22, pp. 403–434, 1976.

[36] A. S. Ribeiro and J. Lloyd-Price, "SGN Sim, a stochastic genetic networks simulator," *Bioinformatics*, vol. 23, pp. 777–779, 2007.

[37] W. E, D. Liu, and E. Vanden-Eijnden, "Nested stochastic simulation algorithms for chemical kinetic systems with multiple time scales," *Journal of Computational Physics*, vol. 221(1), pp. 158–180, 2007.

[38] R. Hofestdt and S. Thelen, "Quantitative modeling of biochemical networks," *In Silico Biology*, vol. 1, pp. 39–53, 1998.

[39] C. Chaouiya, A. Naldi, E. Remy, and D. Thieffry, "Petri net representation of multi-valued logical regulatory graphs," *Natural Computing*, vol. 10(2), pp. 727–750, 2011.

[40] H. Matsuno, A. Doi, M. Nagasaki, and S. Miyano, "Hybrid Petri net representation of gene regulatory network." *Pacific Symposium On Biocomputing*, pp. 341–352, 2000.

[41] D. Gilbert and M. Heiner, "From Petri nets to differential equations - an integrative approach for biochemical network analysis," in *Petri Nets and Other Models of Concurrency - ICATPN 2006*, ser. Lecture Notes in Computer Science, S. Donatelli and P. Thiagarajan, Eds. Springer Berlin Heidelberg, 2006, vol. 4024, pp. 181–200.

[42] P. J. E. Goss and J. Peccoud, "Quantitative modeling of stochastic systems in molecular biology by using stochastic Petri nets," *Proceedings of the National Academy of Sciences*, vol. 95(12), pp. 6750–6755, 1998.

[43] J. Bordon, M. Moškon, and M. Mraz, "Modeling gene regulatory networks using Petri nets," in *Proceedings of 22nd International Electrotechnical and Computer Science Conference*, 2012.

[44] G. Lillacci and M. Khammash, "Parameter Estimation and Model Selection in Computational Biology," *PLoS Computational Biology*, vol. 6(3), 2010.

[45] X. Sun, L. Jin, and M. Xiong, "Extended Kalman Filter for Estimation of Parameters in Nonlinear State-Space Models of Biochemical Networks," *PLoS ONE*, vol. 3(11), 2008.

[46] X. Liu and M. Niranjan, "State and parameter estimation of the heat shock response system using Kalman and particle filters," *Bioinformatics*, vol. 28(11), pp. 1501–1507, 2012.

[47] D. Simon, *Optimal State Estimation: Kalman, H Infinity, and Nonlinear Approaches*. Wiley-Interscience, 2006.

[48] D. Dochain, "State and parameter estimation in chemical and biochemical processes : a tutorial," *Journal of Process Control*, vol. 13, pp. 801–818, 2003.

[49] M. Quach, N. Brunel, and F. Alche, "Estimating parameters and hidden variables in non-linear state-space models based on ODEs for biological networks inference," *Bioinformatics*, vol. 23(23), pp. 3209–3216, 2007.

[50] G. Lillacci and M. Khammash, "A distribution-matching method for parameter estimation and model selection in computational biology," *International Journal of Robust and Nonlinear Control*, 2012.

[51] D. Bates and C. Cosentino, "Validation and invalidation of systems biology models using robustness analysis," *IET Systems Biology Biology*, vol. 5, no. 4, pp. 229–244, 2011.

[52] M. Moškon, "Computer structures perspective on switching dynamics of simple biological systems," Ph.D. dissertation, University of Ljubljana, Faculty of Computer and Information Science, 2012.

[53] H. Kitano, "Biological robustness," *Nature Reviews Genetics*, vol. 5(11), pp. 826–837, 2004.

[54] R. Ghaemi and D. D. Vecchio, "Evaluating the robustness of a biochemical network model," in *Proceedings of the 46th IEEE Conference on Decision and Control*, 2007, pp. 615–620.

[55] J. Kim, D. G. Bates, I. Postlethwaite, L. Ma, and P. A. Iglesias, "Robustness analysis of biochemical network models," *IEE Proceedings Systems Biology*, vol. 153(3), pp. 96–104, 2006.

[56] A. Rizk, G. Batt, F. Fages, and S. Soliman, "A general computational method for robustness analysis with applications to synthetic gene networks," *Bioinformatics*, vol. 25, no. ISMB 2009, pp. i169–i178, 2009.

[57] Z. Zi, "Sensitivity analysis approaches applied to systems biology models," *IET Systems Biology*, vol. 5(6), pp. 336–346, 2011.

[58] A. Degasperi and S. Gilmore, "Sensitivity Analysis of Stochastic Models of Bistable Biochemical Reactions," *Lecture Notes in Computer Science*, no. 5016, pp. 1–20, 2008.

[59] C. Reder, "Metabolic control theory: A structural approach," *Journal of Theoretical Biology*, vol. 135, pp. 175–201, 1988.

[60] S. Strogatz, *Nonlinear Dynamics and Chaos: With Applications to Physics, Biology, Chemistry, and Engineering*, ser. Studies in nonlinearity. Perseus Books Group, 2008.

[61] L. Edelstein-Keshet, *Mathematical Models in Biology*. Siam, Philadelphia, 2005.

[62] P. Fall, E. S. Marland, J. M. Wagner, and J. J. Tyson, *Computational Cell Biology (Interdisciplinary Applied Mathematics)*. Springer, 2002.

[63] F. H. Arnold, "Design by directed evolution," *Accounts of Chemical Research*, vol. 31, pp. 125–131, 1998.

[64] A. S. Khalil and J. J. Collins, "Synthetic biology: Applications come of age," *Nature Reviews Genetics*, vol. 11(5), pp. 367–379, 2010.

[65] T. S. Gardner, C. R. Cantor, and J. J. Collins, "Construction of a genetic toggle switch in Escherichia coli," *Nature*, vol. 403, pp. 339–342, 2000.

[66] M. B. Elowitz and S. Leibler, "A synthetic oscillatory network of transcriptional regulators," *Nature*, vol. 403, pp. 335–338, 2000.

[67] G. Rodrigo, T. E. Landrain, and A. Jaramillo, "De novo automated deisgn of small RNA circuits for engineering synthetic riboregulation in living cells," *Proceedings of the National Academy of Sciences*, 2012.

[68] G. Rodrigo and A. Jaramillo, "AutoBioCAD: Full biodesign automation of genetic circuits," *ACS Synthetic Biology*, vol. 0, no. 0, p. null, 0.

[69] G. Rodrigo, J. Carrera, and A. Jaramillo, "Computational design of synthetic regulatory networks from a genetic library to characterize the designability of dynamical behaviors," *Nucleic Acids Research*, vol. 39, 2011.

[70] M. A. Marchisio and J. Stelling, "Automatic design of digital synthetic gene circuits," *PLoS Computational Biology*, vol. 7(2), 2011.

[71] J. Beal, R. Weiss, D. Densmore, A. Adler, E. Appleton, J. Babb, S. Bhatia, N. Davidsohn, T. Haddock, J. Loyall, R. Schantz, V. Vasilev, and F. Yaman, "An end-to-end workflow for engineering of biological networks from high-level specifications," *ACS Synthetic Biology*, vol. 1, no. 8, pp. 317–331, 2012.

[72] L. Huynh, J. Kececioglu, M. Koppe, and I. Tagkopoulos, "Automatic design of synthetic gene circuits through mixed integer non-linear programming," *PLoS ONE*, vol. 7(4), 2012.

In: Recent Advances in Systems Biology Research ISBN: 978-1-62948-736-6
Editors: A. X. C. N. Valente, A. Sarkar and Y. Gao

Chapter 9

Modelling Transcriptional Regulation with Gaussian Processes

Vinny Davies and Dirk Husmeier
School of Mathematics and Statistics, University of Glasgow, United Kingdom

Abstract

A challenging problem in systems biology is the quantitative modelling of transcriptional regulation. Transcription factors (TFs), which are the key proteins at the centre of the regulatory processes, may be subject to post-translational modification, rendering them unobservable at the mRNA level, or they may be controlled outside of the subsystem being modelled. In both cases, a mechanistic model description of the regulatory system needs to be able to deal with latent activity profiles of the key regulators. A promising approach to deal with these difficulties is based on using Gaussian processes to define a prior distribution over the latent TF activity profiles. Inference is based on the principles of non-parametric Bayesian statistics, consistently inferring the posterior distribution of the unknown TF activities from the observed expression levels of potential target genes. This chapter starts with a self-contained overview of the methodology. It then extends previous articles in four important aspects: it provides a complete set of derivations for all required integrations and inference equations; it discusses issues related to identifiability, a neglection of which can lead to unreliable inference; it evaluates the method on transcriptional profiles from a realistic simulation study, based on mass action and Michaelis-Menten kinetics; and it quantifies the effect of a violation of the assumption of additive Gaussian noise.

1. Introduction

Quantitative modelling of transcriptional regulation is a highly topical and challenging research problem in contemporary systems biology. Over the last few years, a whole plethora of methods have been developed that aim to infer transcriptional regulatory networks from gene expression and DNA sequence data; see work by Lin and Husmeier [9] for a concise review. A particular challenge is that transcription factors (TFs), the regulatory proteins at the heart of transcriptional regulation, are frequently subject to post-translational modification, which may affect their DNA binding capability. Consequently, gene expression levels of TFs contain only limited information about their actual activities. Various authors have applied latent variable and factor analysis models to infer genome-wide patterns of TF-gene interactions from high-throughput transcription profiles while allowing for the fact that the concentrations of active TFs may actually be unknown [9, 11, 13, 14].

An alternative and complementary research objective is to develop less abstract mechanistic models, based on differential equations, for smaller biochemical systems. Recent advances in computational statistics have demonstrated that kinetic parameters and interaction structures can be reliably inferred from completely observed concentration profiles of the molecular components involved [17]. However, the application of these methods to transcriptional regulation is impeded by the fact that concentrations of active TFs are often not directly observable, as discussed above. In addition, the dynamics of TF activities may be controlled outside of the subsystem being modelled, e.g. regulated by external protein signalling pathways. In both cases, the dynamics of TFs must be inferred indirectly, based on the response of the genes whose expression they control.

A promising approach to deal with these difficulties was proposed by Gao et al. [4], inspired by the work of Barenco et al. [1]. The authors advocate the use of Gaussian processes to define prior distributions over the latent TF activity profiles. Inference is soundly based on the principles of non-parametric Bayesian statistics, consistently inferring the posterior distribution of the unknown TF activities from the observed expression levels of potential target genes, and inferring regulatory network structures after marginalizing over the unknown TF activity profiles. The work has more recently been extended to incorporate a model of translation, which improves the quality of TF profile inference when TFs are primarily regulated at the transcriptional level [5], and to generalise the approach to multiple co-regulating transcription factors [16].

Problems with this method however come in the form of identifiability issues. Using a Gaussian process prior over the TF due to data being unobservable leads to a large level of flexibility within the estimates of the TF profile. Under these modelling assumptions some of the characteristic of the TF profile can be incorporated into the parameters of the model leading to a misestimation of the TF profile. We show more formally how this can occur and analyse how this effects the inference of the Gaussian process model.

The choice of a non-parametric prior distribution from the Gaussian process family is not a restrictive modelling assumption, as Gaussian processes have been shown to possess universal approximation capability [10]. Somewhat more restrictive is the assumption of additive Gaussian noise, which can be found in all previous applications [4, 5, 16]. A common approach is to assume that transcriptional data follows a log-normal distribution and to subject transcription profiles to a logarithmic transformation. However, Durbin et al. [3] and Huber et al. [6] showed that mRNA concentrations obtained from microarray experiments are only asymptotically log-normally distributed, in the limiting case of high transcription rates, and that the noise distribution for intermediate rates is of a more complex form. We look at how this can affect the inference of the model proposed by Gao et al. [4], which assumes additive Gaussian noise.

This chapter provides a self-contained review of the methodology of transcriptional modelling with Gaussian processes, which is based on Lawrence et al. and Gao et al. [7, 8, 4]. We then extend the previous articles in four important aspects. Firstly, we provide a complete set of derivations for all required integrations and inference equations, which were missing from the previous work. Secondly, we discuss issues related to identifiability, a neglection of which can lead to unreliable inference and spurious results. Thirdly, we evaluate the method on transcriptional profiles from a realistic simulation study, based on mass action and Michaelis-Menten kinetics. Fourthly, we quantify the effect of a violation of the assumption of additive Gaussian noise, on which all previous models are based, and which is known to be invalid for real gene expression data.

Table 1. Notation description

Symbols	Description
G	number of genes
T	number of time points
$x_i(t)$	true gene expression level of gene i
$y_i(t)$	noisy measurements of gene expression level of gene i
B_i	basal transcription rate of gene i
S_i	sensitivity to binding of TF of gene i
D_i	decay rate of gene i
$\boldsymbol{\theta}_c$	hyperparameters of the Gaussian process prior
$\boldsymbol{\theta}'$	parameters in covariance function $\mathbf{K}_{\mathbf{f},\mathbf{f}}$
$\boldsymbol{\theta}$	parameters in eq. (5)
$\mathbf{f}$	TF activity

2. Methodology

An overview of the notation used in the following exposition can be found in Table 1. A linear model of gene expression was proposed by Barenco et al. [1]

$$\frac{dx_i(t)}{dt} = B_i + S_i f(t) - D_i x_i(t) \tag{1}$$

where $i \in \{1, \ldots, G\}$ is a set of genes regulated by the same transcription factor TF, $x_i(t)$ are the (unknown) true gene expression levels at time point t, $f(t)$ is the (unknown) TF activity, B_i is the basal transcription rate of gene i, S_i is the sensitivity to binding of TF, and D_i is a decay rate. We assume that (noisy) measurements of $x_i(t)$ can be obtained, and $f(t)$ is unobservable. Eq. (1) has the analytic solution

$$x_i(t) = \frac{B_i}{D_i} + S_i \int_0^t \exp(-D_i(t-u)) f(u) du \tag{2}$$

where transient terms have been ignored.

2.1. GP Inference

Lawrence et al. [8] and Gao et al. [4] proposed a non-parametric Bayesian approach to inference in this model by placing a Gaussian process prior on the unknown TF activities $\mathbf{f} = (f(t_1), \ldots, f(t_T))$ at time points $\mathbf{t} = (t_1, \ldots, t_T)$

$$p(\mathbf{f}) = \mathcal{N}(\mathbf{f}|\mathbf{0}, \mathbf{K}_{\mathbf{f},\mathbf{f}}) \tag{3}$$

that is, the prior probability of the TF activities, $p(\mathbf{f})$, is a zero-mean multivariate Gaussian distribution with covariance matrix $\mathbf{K_{f,f}}$. The linear form of eq. (2) implies that the joint prior distribution of the expression profiles of all regulated genes

$$\mathbf{x}_i = [x_i(t_1), \ldots, x_i(t_T)]; i = 1, \ldots, G \tag{4}$$

is described by a Gaussian process prior with a covariance matrix, $\mathbf{K}$, that depends on the hyperparameters $\boldsymbol{\theta}_c$ and the parameters that characterise the transcriptional regulation processes via eq. (1):

$$\begin{aligned} p(\mathbf{x}_i|\boldsymbol{\theta}') &= \mathcal{N}(\mathbf{B}./\mathbf{D}, \mathbf{K})\,;\ \mathbf{K} = \mathbf{K}(\boldsymbol{\theta}') \\ \boldsymbol{\theta}' &= (\boldsymbol{\theta}_c, B_1, \ldots, B_G, S_1, \ldots, S_G, D_1, \ldots, D_G) \end{aligned} \tag{5}$$

where $\mathbf{B} = \{B_1, \ldots, B_1, \ldots, B_G, \ldots, B_G\}$ for example and $\mathbf{B}./\mathbf{D}$ is a point wise vector division.

In order to derive an expression for the covariance matrix $\mathbf{K}$, we firstly define the linear operator relating to the latent function f and the gene expression level $x_j(t)$ as shown by Lawrence et al. [8]

$$L_j[f](t) = S_j \exp(-D_j t) \int_0^t f(u) \exp(D_j u) du. \tag{6}$$

This contains all the time varying elements of eq. (2) and implies a covariance of

$$\text{cov}(L_j[f](t), L_k[f](t')) = L_j \otimes L_k[k_{f,f}](t,t'). \tag{7}$$

which represents the tensor product of two vectors. This results in the elements of the covariance matrix being expressed as

$$\begin{aligned} k_{x_j,x_k}(t,t') = S_j S_k \exp(-D_j t - D_k t') \int_0^t \exp(D_j u) \\ \times \int_0^{t'} \exp(D_k u') k_{f,f}(u,u') du' du \end{aligned} \tag{8}$$

where $k_{f,f}(t,t')$ is the covariance matrix associated with $f(t)$. If we then let this be the Radial Basis Function (RBF) kernel;

$$k_{f,f}(t,t') = a^2 \exp\left(-\frac{(t-t')^2}{l^2}\right) \tag{9}$$

where l is the length scale and a^2 is the amplitude parameter, then the integral becomes tractable and can be solved using Laplace transformations (see Appendix A).

From Appendix A, the resulting covariance elements are given by

$$k_{x_j,x_k}(t,t') = a^2 S_j S_k \frac{\sqrt{\pi}l}{2}[h_{kj}(t',t) + h_{jk}(t,t')] \tag{10}$$

where

$$h_{kj}(t',t) = \frac{\exp(\gamma_k^2)}{D_j + D_k}\Bigg\{\exp[-D_k(t'-t)]\left[\mathrm{erf}\left(\frac{t'-t}{l} - \gamma_k\right) + \mathrm{erf}\left(\frac{t}{l} + \gamma_k\right)\right]$$
$$- \exp[-(D_j t + D_k t')]\left[\mathrm{erf}\left(\frac{t'}{l} - \gamma_k\right) + \mathrm{erf}(\gamma_k)\right]\Bigg\}.$$

Here we define $\gamma_k = \frac{D_k l}{2}$ and $\mathrm{erf}(x) = \frac{2}{\sqrt{\pi}}\int_0^x \exp(-y^2)dy$. This then allows us to compute the likelihood dependent on parameters $\boldsymbol{\theta}'$.

In order to infer the unknown TF activity, a covariance matrix must be found relating $x_j(t)$ and $f(t)$. This is similarly given as

$$k_{x_j,f}(t,t') = S_j \exp(-D_j t)\int_0^t \exp(D_j u) k_{f,f}(u,t')du. \tag{11}$$

Again when the RBF kernel is used the integral is solvable with a Laplace transformation (see Appendix B), with the covariance matrix given as

$$k_{x_j,f}(t,t') = a^2 S_j \frac{\sqrt{\pi}l}{2}\exp(\gamma_j^2)\exp[-D_j(t-t')]$$
$$\times \left[\mathrm{erf}\left(\frac{t-t'}{l} - \gamma_j\right) + \mathrm{erf}\left(\frac{t'}{l} + \gamma_j\right)\right].$$

Standard regression techniques, such as shown by Rasmussen and Williams [12], then lead to f being modelled such that

$$f|\mathbf{x} \sim \mathcal{N}(K_{f,\mathbf{x}}K_{\mathbf{x},\mathbf{x}}^{-1}(\mathbf{x} - \mathbf{B}./\mathbf{D}), K_{f,f} - K_{f,\mathbf{x}}K_{\mathbf{x},\mathbf{x}}^{-1}K_{\mathbf{x},f}) \tag{12}$$

where $\mathbf{x}$ denotes the vector of true gene expression levels for all genes, $\mathbf{K}$ is the matrix which arises from evaluating the covariance functions between all time points and $\mathbf{B}./\mathbf{D}$ is the point wise division of vectors $\mathbf{B}$ and $\mathbf{D}$.

To relate the unknown true gene expression profiles $\mathbf{x}_i$ to noisy measurements $\mathbf{y} = (\mathbf{y}_1, \ldots, \mathbf{y}_G)^\top$, $\mathbf{y}_i = [y_i(t_1), \ldots, y_i(t_T)]; i = 1, \ldots, G$, the standard approach (e.g. [2], Section 6.4.2) assumes additive Gaussian noise σ^2:

$$p(\mathbf{y}|\mathbf{x}, \sigma^2) \;=\; \mathcal{N}(\mathbf{y}|\mathbf{x}, \sigma^2\mathbf{I}) \tag{13}$$

where $\mathbf{I}$ is the identity matrix. The marginalisation over $\mathbf{y}$ is analytically tractable and gives, with the definition $\boldsymbol{\theta} = (\boldsymbol{\theta}', \sigma^2)$:

$$p(\mathbf{y}|\boldsymbol{\theta}) \;=\; \int p(\mathbf{y}|\mathbf{x}, \sigma^2)p(\mathbf{x}|\boldsymbol{\theta}')d\mathbf{x} \;=\; \mathcal{N}(\mathbf{y}|\mathbf{0}, \mathbf{C}(\boldsymbol{\theta}))$$
$$\mathbf{C}(\boldsymbol{\theta}) = \mathbf{K}(\boldsymbol{\theta}') + \sigma^2\mathbf{I} \tag{14}$$

giving a log likelihood of

$$\ln(\mathbf{y}|\boldsymbol{\theta}) = -\tfrac{T}{2}\ln(2\pi) - \tfrac{1}{2}\ln|\mathbf{C}(\boldsymbol{\theta})| - (\mathbf{y} - \mathbf{B}./\mathbf{D})^\top\mathbf{C}(\boldsymbol{\theta})^{-1}(\mathbf{y} - \mathbf{B}./\mathbf{D}). \tag{15}$$

Inference of the parameters $\boldsymbol{\theta}$ can then be achieved in a maximum likelihood or Bayesian framework, as described in standard textbooks on Gaussian process modelling [2, 12]. For non-linear optimisation methods, the gradient of the log likelihood with respect to the parameters $\boldsymbol{\theta}$ helps to improve the optimisation and these are derived in Appendix C.

Standard Gaussian process regression, as described in [2, 12], yields the mean and covariance function of the posterior process on f as

$$\langle f \rangle_{post} = \mathbf{C}_{f\mathbf{y}} \mathbf{C}_{\mathbf{yy}}^{-1} \mathbf{y} \tag{16}$$

$$\sigma_{post}^2 = C_{ff} - \mathbf{C}_{f\mathbf{y}} \mathbf{C}_{\mathbf{yy}}^{-1} \mathbf{C}_{\mathbf{y}f}. \tag{17}$$

Implementing these methods is often harder than suggested in the original papers from which they were proposed. For instance when inferring parameters within either a maximum likelihood or Bayesian framework, problems can occur in the form of ill-conditioned matrices. This may lead to spikes in the likelihood in the tails of the distribution, which may require the exploration of various numerical stabilisation techniques.

2.2. Identifiability Issues

The identifiability issues within the model come from the original make up of eq. (1) proposed by Barenco et al. [1] and the fact that data from the TF is unobservable. With only data from the downstream genes and no data from the TF activity, the Gaussian process effectively learns the shape of the TF, but can fail to accurately estimate the baseline activity and exact concentration levels of the profile. This can lead to problems as the method will still continue to give accurate estimates for the gene expression profiles, so it may be erroneously expected that the TF profile is similarly accurate. Following Barenco et al. [1], let $\tilde{f}(t)$ denote the unscaled profile of the TF, and consider the linear transformation

$$f(t) = \alpha + \beta \tilde{f}(t). \tag{18}$$

where additive and multiplicative terms, α and β respectively, have been introduced. We can show how this can be manipulated into the form of eq.(1), where $\tilde{f}(t)$ can be seen as the TF and incorrect parameters can be learnt for the ODEs:

$$\begin{aligned} \frac{dx_i(t)}{dt} &= B_i + S_i f(t) - D_i x_i(t) \\ &= B_i + S_i\{\alpha + \beta \tilde{f}(t)\} - D_i x_i(t) \\ &= \{B_i + S_i \alpha\} + \{S_i \beta\} \tilde{f}(t) - D_i x_i(t) \\ &= \tilde{B}_i + \tilde{S}_i \tilde{f}(t) - D_i x_i(t). \end{aligned} \tag{19}$$

As can be seen, the derivatives of the gene expression values are invariant with respect to an affine transformation of the chemical kinetic parameters: $\tilde{B}_i = B_i + S_i \alpha$, $\tilde{S}_i = S_i \beta$, and two constraints are required to fix their values. In practice, this requires measurements of two independent quantities directly related to the chemical kinetic parameters, e.g. measuring one of the base transcription rates, say B_1, and an affinity or interaction strength parameter, say S_1 (this will be assumed in our simulation studies).

2.3. Noise Assumptions

Previous publications, such as [4, 5, 16], have assumed white additive Gaussian noise of the form of eq. (13). This assumption leads to a closed-form solution of the convolution integral given in eq. (14). The result is a simple and straightforward modification of the noise-free scenario, in which the noise variance is added to the diagonal elements of the covariance matrix.

Unfortunately it has been shown by both Durbin et al. [3] and Huber et al. [6] that transcriptional data are subject to noise of a more complex form;

$$\begin{aligned} y_i(t) &= c + x_i(t)\exp(\mu_t) + \varepsilon_t \\ \mu_t &\sim \mathcal{N}(0, \sigma_\mu^2);\ \varepsilon_t \sim \mathcal{N}(0, \sigma_\varepsilon^2). \end{aligned} \tag{20}$$

Maximising the log likelihood under the assumption of white additive Gaussian noise is equivalent to minimising the residual sum of squares difference between the observations and the interpolant, which in expectation is equivalent to learning the expectation value $\mathbb{E}(y_i(t))$. However,

$$\begin{aligned} y_i(t) &= x_i(t)\exp(\mu_t) + \varepsilon_t \\ \mathbb{E}[y_i(t)] &= x_i(t)\mathbb{E}[\exp(\mu_t)] + \mathbb{E}[\varepsilon_t] \\ &= x_i(t)\mathbb{E}[\exp(\mu_t)] \\ &\geq x_i(t)\exp(\mathbb{E}[\mu_t]) = x_i(t)\exp(0) = x_i(t) \end{aligned} \tag{21}$$

by Jensen's inequality. Replacing the true signal $x_i(t)$ in eq. (21) by the smoothed signal $\tilde{x}_i(t) = \mathbb{E}[y_i(t)]$ from the Gaussian interpolation will lead to a systematic distortion as a consequence of the overestimation $\tilde{x}_i(t) \geq x_i(t)$. This implies that

$$\frac{d\tilde{x}_i(t)}{dt} = B_i + S_i f(t) - D_i\tilde{x}_i(t) \leq B_i + S_i f(t) - D_i x_i(t) = \frac{dx_i(t)}{dt} \tag{22}$$

will be systematically underestimated. We look at how in practise this incorrect assumption of additive Gaussian noise affects the predictions of the model parameters and the profiles of both the downstream genes and also the TF.

3. Evaluation on Simulated Data

To assess the model performance, we carried out evaluation procedures on two known forms of TF. The first TF profile is in the form of a Gaussian mixture model while the second is based on a more biologically realistic TF based on mass action and Michaelis-Menten kinetics. Here, we exploit the fact that the ground truth is known and that the performance of the model can therefore be objectively assessed.

3.1. Data Generation

We tested the performance of the proposed scheme on data simulated from the regulatory networks shown in Figure 1. A single TF regulates three downstream genes linearly according to eq.(2). We generated data with both additive Gaussian noise, such as in Gao et

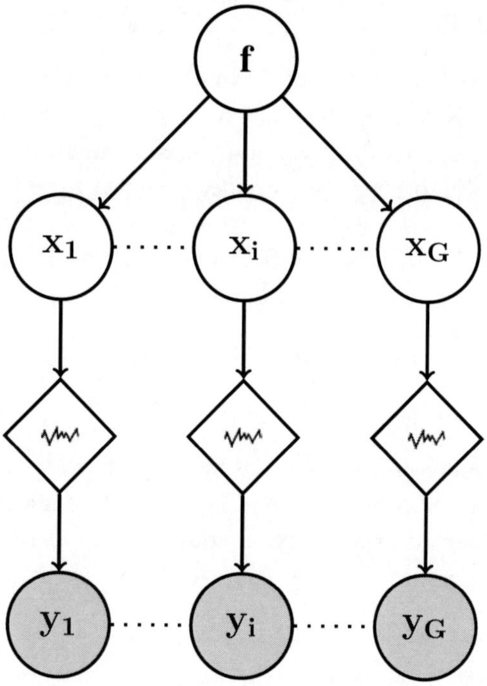

Figure 1. **TF simulated from a synthetic system.** The active form of the TF is given by f and regulates G genes with expression values $x_1,\dots,x_i,\dots,x_G$, whose noisy measurements are $y_1,\dots,y_i,\dots,y_G$.

al. [4], and also with more realistic non-Gaussian noise from eq. (20), i.e. using the model proposed in Durbin et al. [3]. The simulations require the specification of the latent TF profile.

We approached this in two different ways. In the first approach, we followed Lawrence et al. [7] and chose the TF profile to have the following fixed functional form:

$$f(t) = \sum_{j=1}^{4} a_j \exp\left(-\frac{(t-\mu_j)^2}{\sigma^2}\right). \tag{23}$$

For the parameters, we chose the same values as Lawrence et al. [7], namely $\sigma = 1.5, a_1 = a_2 = 1.5, a_3 = a_4 = 0.5, \mu_1 = 4, \mu_2 = 6, \mu_3 = 8.5$ and $\mu_4 = 10.5$. For our empirical evaluation studies, we generated the TF signal between time units 0 and 100, out of which we sampled 18 values regularly spaced in $[0, 100]$.

In the second approach, we chose a more realistic simulation, where we explicitly modelled the post-translational modification of the TF. To this end, we combined the regulatory network of the previous figure, Figure 1, with a simple protein signalling pathway taken from work by Vyshemirsky and Girolami [17]; see Figure 2. The active form of the TF is Rpp, which is derived from the native form of the protein, R, via phosphorylation. The phosphorylation is catalysed by the enzyme S via formation of the dimer RS, where the enzyme S can irreversibly decay into its degraded form dS. The kinetics of these processes are described by the following system of mass action and Michaelis-Menten type differential equations:

$$\begin{aligned}
\frac{d[S]}{dt} &= -k_1 \cdot [S] - k_2 \cdot [S] \cdot [R] + k_3 \cdot [RS] \\
\frac{d[dS]}{dt} &= k_1 \cdot [S] \\
\frac{d[R]}{dt} &= -k_2 \cdot [S] \cdot [R] + k_3 \cdot [RS] + \frac{V \cdot [Rpp]}{k_m + [Rpp]} \\
\frac{d[RS]}{dt} &= k_2 \cdot [S] \cdot [R] - k_3 \cdot [RS] - k_4 \cdot [RS] \\
\frac{d[Rpp]}{dt} &= k_4 \cdot [RS] - \frac{V \cdot [Rpp]}{k_m + [Rpp]}
\end{aligned} \tag{24}$$

where the square brackets, [.], denote concentrations, and the kinetic parameters were set in the same way as Vyshemirsky and Girolami [17]: $k_1 = 0.07, k_2 = 0.6, k_3 = 0.05, k_4 = 0.3, V = 0.017, k_m = 0.3$.

In this way we have emulated the scenario described in the Introduction, where the TF activity profiles are unobservable due to post-translational modification, and the processes leading to the formation of active TFs are controlled outside of the subsystem being modelled (which only consists of the TF and the three downstream genes). Again, we generated the TF signal between time units 0 and 100, and sampled 18 values regularly spaced in $[0, 100]$.

From the TF profiles, the three downstream genes were generated by solving eq. (2) with the kinetic parameter values given in Table 2. All values were the same for both TFs with the exception of D_1 which was changed in order to give more reasonable profiles in the case of the second TF. In the case of the Gaussian mixture model, eq. (2) was solved explicitly using similar techniques to those used for finding the covariance matrix. For the TF based on mass action and Michaelis-Menten kinetics a numerical approach was used, in this case Simpson's method, where only grid values of the TF were required to find the true gene expression profiles. Data points were then simulated for the downstream genes with either additive Gaussian or non-Gaussian noise of different sizes.

Table 2. Values for the kinetic parameters used to generate the downstream genes. All parameter values are used for both TFs, with the exception of D_1 where a different value, $D_1 = 0.2$, was used in the more biologically realistic TF based on mass action and Michaelis-Menten kinetics.

	B_i	S_i	D_i
$i = 1$	0.0	1.0	$1.0\ (0.2)$
$i = 2$	7.5×10^{-2}	0.4	5×10^{-2}
$i = 3$	2.5×10^{-3}	0.4	1×10^{-3}

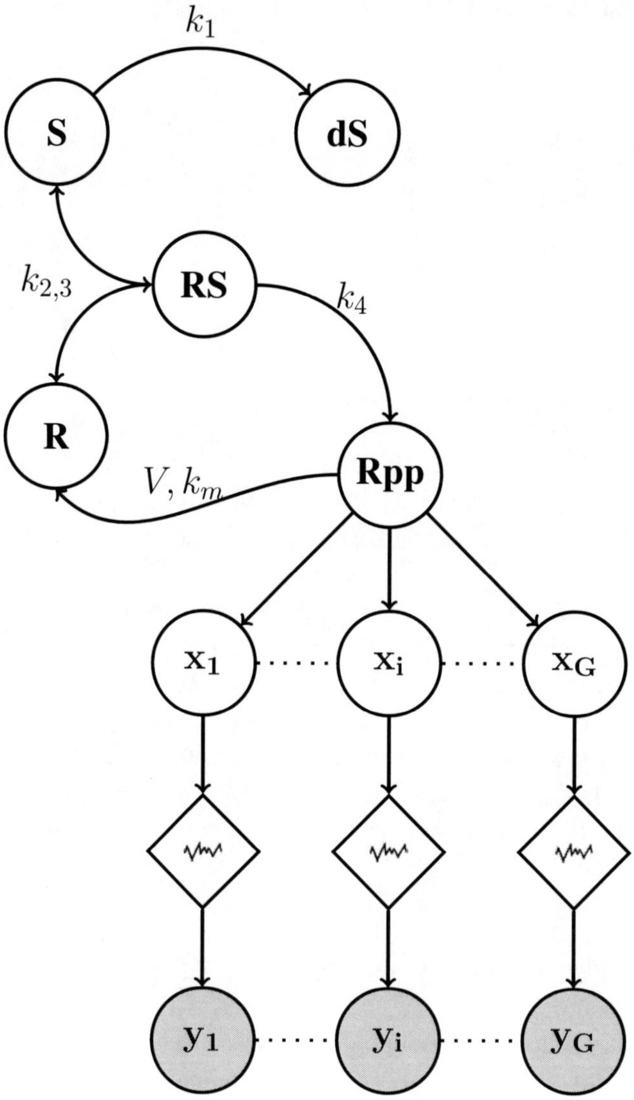

Figure 2. **Transcriptional regulation by a TF that is subject to post-translation modification.** The active form of the TF is Rpp, which is derived from the native form of the protein, R, via phosphorylation. The phosphorylation is catalysed by the enzyme S via formation of the dimer RS, where the enzyme S can irreversibly decay into its degraded form dS. The TF regulates G genes with expression values $x_1,\ldots,x_i,\ldots,x_G$, whose noisy measurements are $y_1,\ldots,y_i,\ldots,y_G$.

3.2. Evaluation Procedure

We emulated a real experiment, in which only the expression time series of the downstream genes are observed, and the objective is to infer both the unknown TF activities as well as the kinetic parameters, i.e. the set of $\{B_i, S_i, D_i\}_{i=1,2,\ldots,G}$ with $G = 3$ in eq. (1). For the TF activity levels represented in Figure 2 and defined by eq. (24) we simulated data sets based on a variety standard deviations. 10 independent data sets were generated in each case, where the standard deviations were chosen on a roughly log scale such that $\sigma_\mu, \sigma_\varepsilon = \{0.01, 0.03, 0.1, 0.3\}$. In each case we chose values for the equivalent standard deviation of additive Gaussian noise, σ, so to give a fair comparison between methods. We have represented the results in the form of box plots of the errors in the parameter and profile estimates.

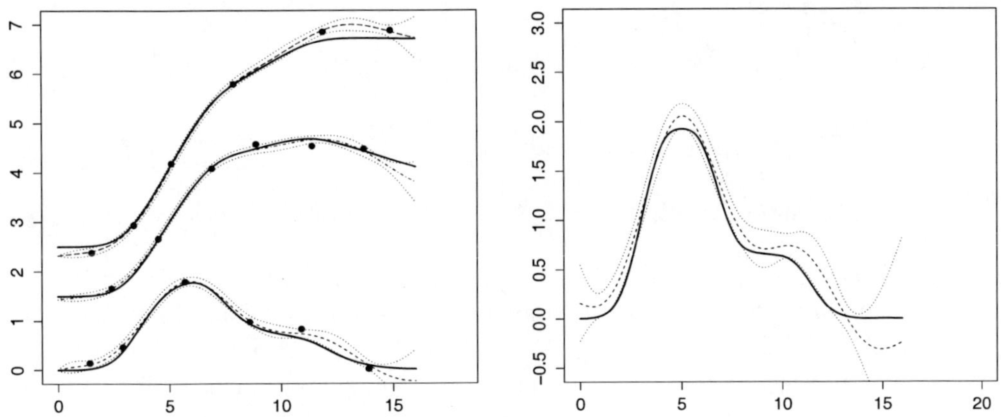

(a) Downstream Genes (b) Transcription Factor

Figure 3. True and estimated concentration profiles for (a) downstream genes and (b) transcription factor, based on the Gaussian mixture model TF profile of eq. (23). Solid lines indicate the true values of the profiles, the dashed lines give the predicted posterior mean, from eq. (16), and the dotted lines indicate the region within two standard deviations, using eq. (17). Solid points indicate the observations from the gene expression profiles modelled with additive Gaussian noise.

4. Simulation Results

We firstly explored the model for data with additive Gaussian noise similarly to Lawrence et al. [8] and Gao et al. [4], comparing this for both of the TF profiles and looking at how the predictions were affected by the issues with identifiability. We then used data simulated with the more biologically realistic noise from eq. (20) to look at how this affected our ability to make predictions on the more realistic TF activity represented in Figure 2 and defined in eq. (24). We analysed data sets generated from a variety of different values for σ_μ and σ_ε with the aim of quantifying the effect of the size of these standard deviations.

4.1. Additive Gaussian Noise

Figure 3a shows an example of the available observations from the gene expression profiles (in the form of black dots), with no data being directly observed from the TF profile in Figure 3b. Using the additive Gaussian noise model as in Lawrence et al. [8] leads to an

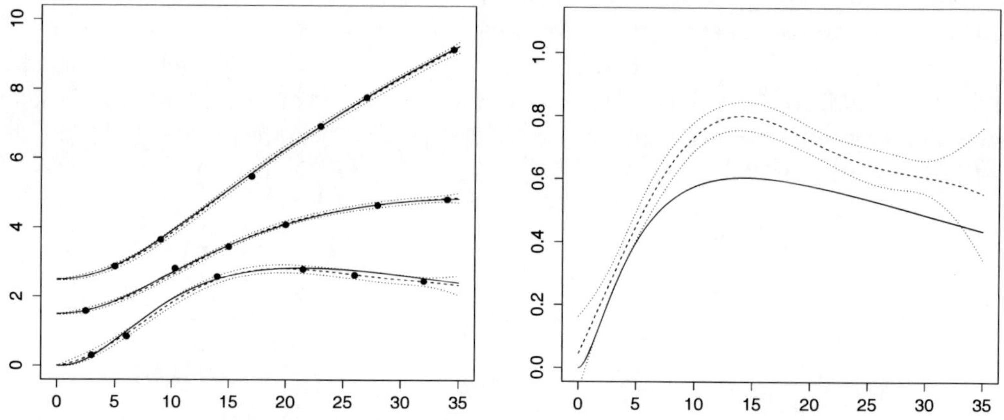

(a) Downstream Genes (b) Transcription Factor

Figure 4. True and estimated concentration profiles for (a) downstream genes and (b) transcription factor, based on the TF profile from the mass action and Michaelis-Menten kinetics of eq. (24). Solid lines indicate the true values of the profiles, the dashed lines give the predicted posterior mean, from eq. (16), and the dotted lines indicate the region within two standard deviations, using eq. (17). Solid points indicate the observations from the gene expression profiles modelled with additive Gaussian noise.

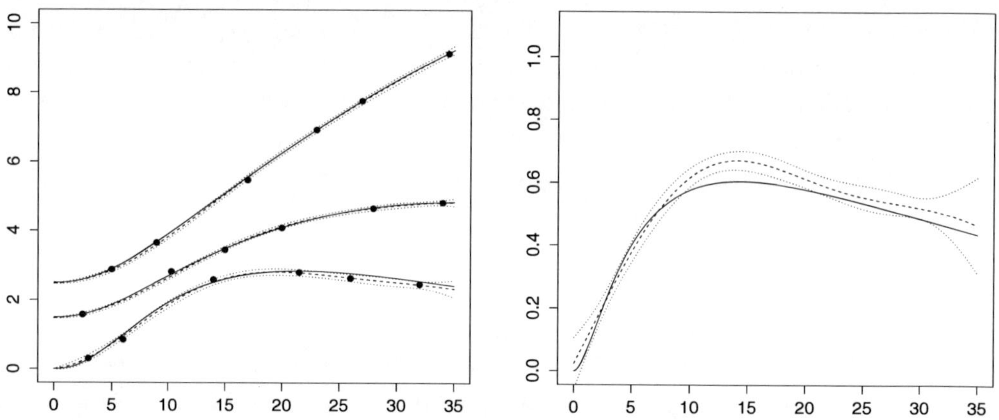

(a) Downstream Genes (b) Transcription Factor

Figure 5. True and estimated concentration profiles for (a) downstream genes and (b) transcription factor, based on the TF profile from the mass action and Michaelis-Menten kinetics of eq. (24). Solid lines indicate the true values of the profiles, the dashed lines give the predicted posterior mean, from eq. (16), and the dotted lines indicate the region within two standard deviations, using eq. (17). Solid points indicate the observations from the gene expression profiles modelled with additive Gaussian noise and in this case the parameters $S_1 = 1$ and $B_1 = 0$ are fixed.

accurate reconstruction of the unobserved TF activity profile from only a limited number of observations.

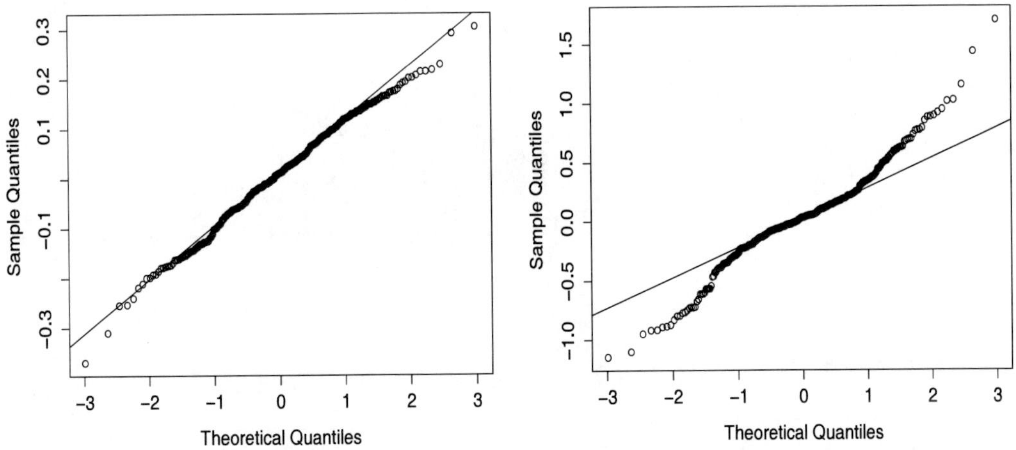

(a) Standard noise

(b) Realistic noise of the form (20)

Figure 6. QQ-plots showing the deviations from a normal distribution for data generated with the (a) standard noise assumption and (b) more realistic noise of the form of eq.(20).

For a more realistic emulation of true transcriptional regulatory processes, we simulated TF activity profiles from the model shown in Figure 2, and mathematically described in eq. (24). The results are shown in Figures 4 and 5. Figure 4 was obtained from a simulation in which all chemical kinetic parameters B_i, D_i, and S_i had to be inferred from the data. Although the profiles of the downstream gene expression profiles were estimated accurately, the TF activity profiles differ noticeably from the true profiles. The reason for this discrepancy is the identifiability problem discussed in Section 2.2: starting from the original form of the ODEs given in eq. (1), part of the size and shape of $f(t)$ can be easily absorbed into the parameters, as seen in eq. (19). To address this issue, we chose to keep two arbitrarily selected chemical kinetic parameters, B_1 and S_1, fixed at their true values, corresponding to accurate measurements. As can be seen from Figure 5, this leads to a considerable improvement in the TF activity profile reconstruction.

4.2. Non-Gaussian Noise

Durbin et al. [3] and Huber et al. [6] showed that transcriptional data is subject to noise of the form given in eq. (20). However previous work, [4, 5, 16], has been completed under the assumption of additive Gaussian noise to make the method tractable. Intuitively there is a significant difference between these types of noise and this can be seen in the form of QQ-plots given in Figure 6. In Figure 6b the non-Gaussian noise from eq. (20) shows considerable deviations from additive Gaussian noise, violating the modelling assumptions.

To investigate the influence of the noise, we simulated noise-free TF profiles from eq. (24), and then generated noisy data by either adding iid Gaussian noise, or subjecting the noise-free data to the more realistic noise process of eq. (20). We repeated this process for 10 independent noise instantiations, giving 10 noisy data sets. For each of these data

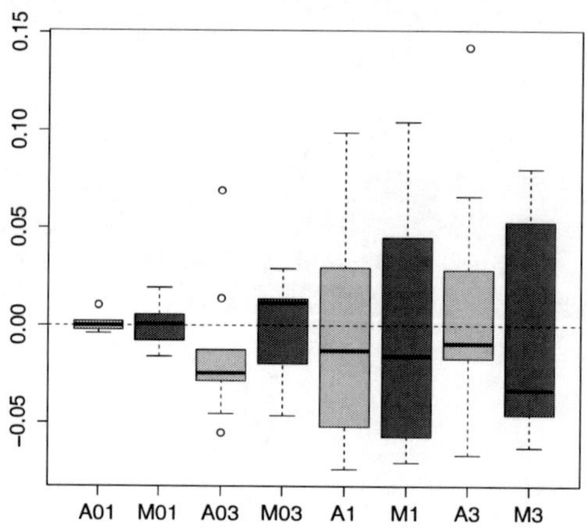

(a) Basel Transcription Rate, **B**

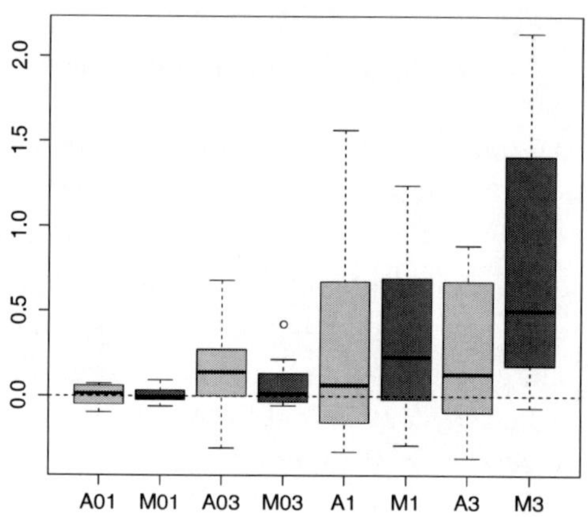

(b) Sensitivity, **S**

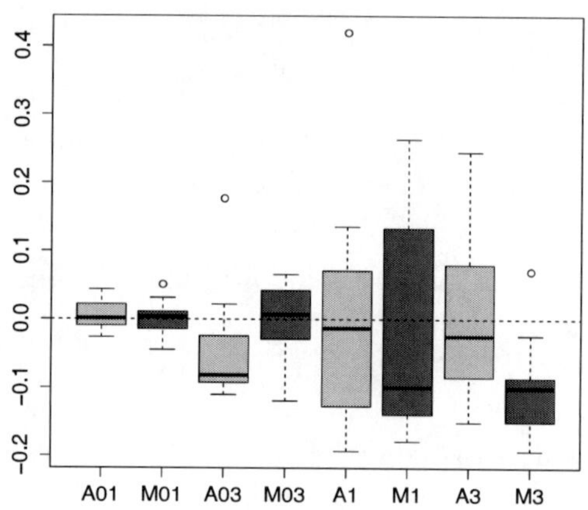

(c) Decay Rate, **D**

Figure 7. Box plots of the error of ODEs parameter predictions, indicating a distribution obtained from 10 independent data instantiations. Box plots for the additive Gaussian and non-Gaussian noise are given in light and dark grey respectively, also indicated by ‘A’ and ‘M’ relating to where the noise is additive or multiplicative. The standard deviations used for σ_μ and σ_ε are given under each box plot and represent the values $\{0.01, 0.03, 0.1, 0.3\}$.

sets, we inferred the chemical kinetic parameters and reconstructed the TF activity profiles. The results are shown in Figures 7 and 8.

Figure 7 gives box plots of the errors in the estimation of the parameters from eq. (1). It shows that there was little difference between the estimates based on data with additive Gaussian and non-Gaussian noise when the standard deviation is low. However when the standard deviation increases, the non-Gaussian noise begins to have a stronger deteriorating influence on the inference. Often the estimates have a bias away from the true parameter values and there also appears to be an increase in the variance of the estimates over when there is additive Gaussian noise.

Figure 8 shows box plots illustrating the accuracy of estimates for the gene and TF pro-

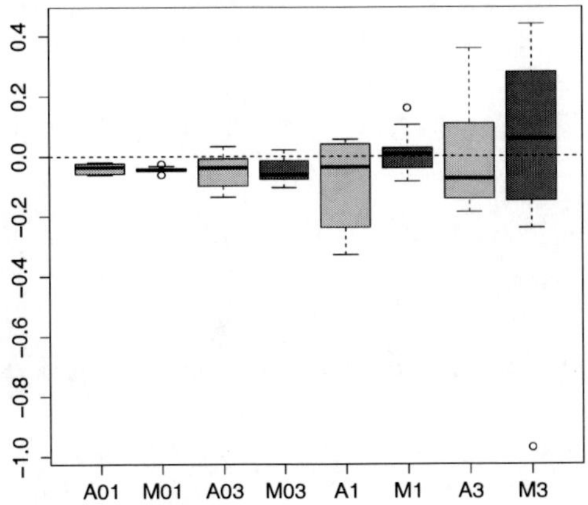

(a) Gene expression profile, $\mathbf{x}_1$

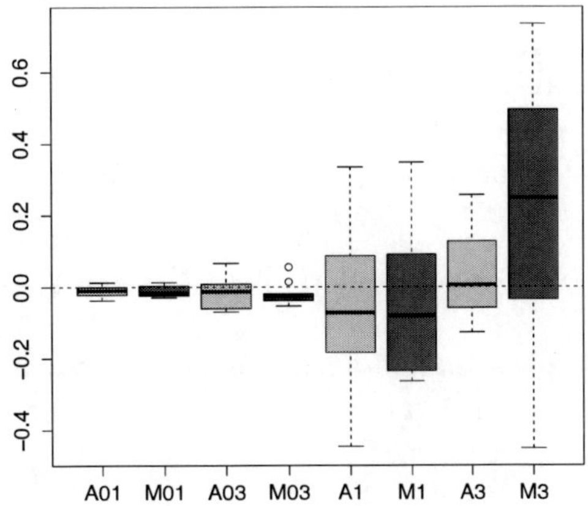

(b) Gene expression profile, $\mathbf{x}_2$

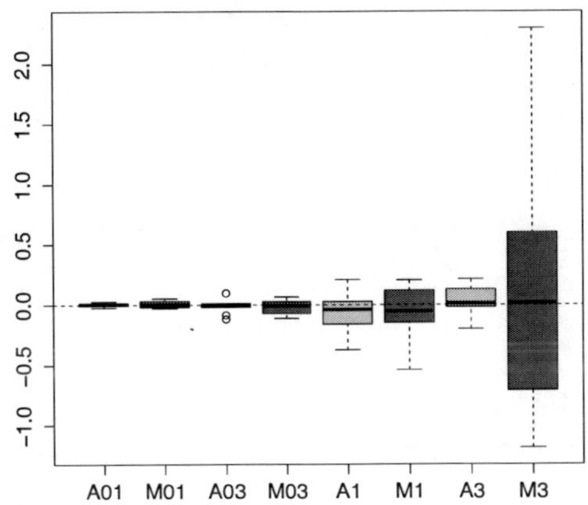

(c) Gene expression profile, $\mathbf{x}_3$

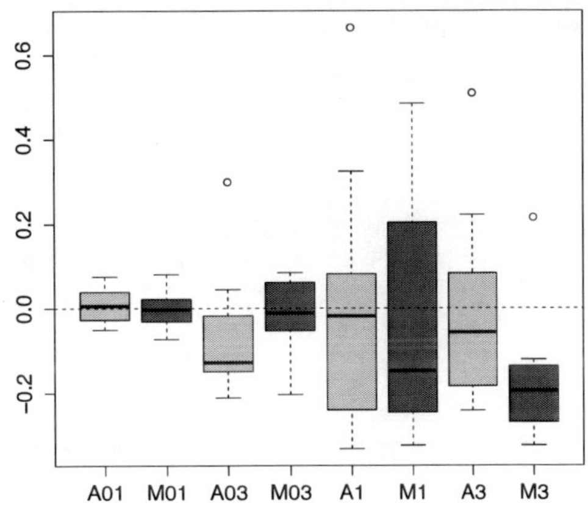

(d) TF Profile, $\mathbf{f}$

Figure 8. Box plots of the error of gene and TF profile predictions, indicating a distribution obtained from 10 independent data instantiations. Box plots for the additive Gaussian and non-Gaussian noise are given in light and dark grey respectively, also indicated by 'A' and 'M' relating to where the noise is additive or multiplicative. The standard deviations used for σ_μ and σ_ε are given under each box plot and represent the values $\{0.01,0.03,0.1,0.3\}$

files generated from the model based on mass action and Michaelis-Menten kinetics. The results show that for the gene expression profile, $\mathbf{x}_1$ in Figure 8a, where the expression levels are low, estimates are very similar and the differences occur relatively randomly. For the genes with higher expression levels however, the differences become more pronounced and our results show that the non-Gaussian noise can cause a strong bias within the estimates, where this effect is exaggerated when the standard deviation gets larger.

Figure 7 and 8 give a visual interpretation of how realistic non-Gaussian noise affected the inference of the model compared to when only additive Gaussian noise was added. To quantify this, we have compared the absolute values of the medians given in the box plots. For a relatively small number of data sets, 10, we have found that results were inconclusive for $\sigma_\mu, \sigma_\varepsilon = \{0.01, 0.03\}$ showing little negative effect in the inference. For larger standard deviations the results show a consistent deterioration when non-Gaussian noise was added, but this cannot be easily quantified until $\sigma_\mu, \sigma_\varepsilon = \{0.3\}$. For this final level of variance the median error in the parameter estimates shows a four fold increase, and similar increases are found in the TF activity profile estimates.

5. Extension to Nonlinear Regulation

To allow for non-linear regulation and saturation, the transcriptional regulation model can be extended to

$$\frac{dx_i(t)}{dt} = B_i + S_i g_i(f(t)) - D_i x_i(t) \tag{25}$$

where $g(.)$ is a non-linear function of the concentration of the TF activity. A common choice for transcriptional activation is

$$g_i(f(t)) = \frac{\exp(f(t))}{\exp(f(t)) + \gamma_i} \tag{26}$$

where the additional parameter γ_i is the so-called Michaelis-Menten constant. For a transcriptional repressor, the non-linear function has the form

$$g_i(f(t)) = \frac{1}{\exp(f(t)) + \gamma_i} \tag{27}$$

where again γ_i is the Michaelis-Menten constant. The equations have the formal solution

$$x_i(t) = \alpha_i \exp(-D_i t) + \frac{B_i}{D_i} + S_i \int_0^t \exp(-D_i(t-u)) g_i(f(u)) du. \tag{28}$$

However, as opposed to the linear case, the induced distribution of $x_i(t)$ is no longer a Gaussian process. In order to maintain analytical tractability, one can pursue a MAP-Laplace approximation; for details, see Lawrence et al. [7, 8].

6. Conclusion

This chapter has reviewed Gaussian processes for modelling transcriptional regulation, based on the work of Lawrence et al. [7, 8] and Gao et al. [4]. As opposed to previous articles, we have provided a complete set of the closed-form solutions to the integrations on which the method is based, and the reader has access to all the update equations that are required for a practical implementation of the framework. We have further evaluated the method on data from a realistic simulation of transcriptional regulation and post-translational modification, and we have discussed identifiability problems inherent in the inference scheme. A central part of the chapter has been a closer inspection of the influence of the observational noise. The previous articles have assumed the noise to be additive Gaussian, so as to maintain analytical tractability, despite the fact that its functional form is more complex. We have assessed the effect of ignoring the deviation from additive Gaussianity, and quantified the amount of deterioration of the inference results in dependence on the noise level. Our future work will look into ways to integrate more complex forms of noise into the model, based on the idea of warped Gaussian processes, as proposed by Snelsen et al. [15].

Appendices

A. Cross-Covariance between Target Genes

The expression levels of the target genes over time, $\{x_i(t)\}_{i=1}^G$, are given in terms of the transcription factor in eq. (2). This gives the formula for the cross-covariance shown in eq. (8) and repeated below:

$$k_{x_j,x_k}(t,t') = S_j S_k \exp(-D_j t - D_k t') \int_0^t \exp(D_j u) \times \int_0^{t'} \exp(D_k u') k_{f,f}(u,u') du' du$$

where the transient terms have been ignored. As stated, when using an RBF kernel this integral becomes tractable and can be solved using a Laplace transformation. The solution to the integral can then be derived in the following manner:

$$\begin{aligned} k_{x_j,x_k}(t,t') &= S_j S_k \exp(-D_j t - D_k t') \int_0^t \exp(D_j u) \\ &\times \int_0^{t'} \exp(D_k u') a^2 \exp\left(-\tfrac{1}{l^2}[u-u']^2\right) du' du \\ &= a^2 S_j S_k \exp(-D_j t - D_k t') \int_0^t \exp(D_j u) \\ &\times \int_0^{t'} \exp\left(-\frac{1}{l^2}[u^2 - 2uu' + u'^2 - D_k l^2 u']\right) du' du. \end{aligned}$$

We can then complete the square and move any factors that are constant out of the integral to give:

$$\begin{aligned} k_{x_j,x_k}(t,t') &= a^2 S_j S_k \exp(-D_j t - D_k t') \int_0^t \exp\left(\frac{1}{l^2}[u + \tfrac{1}{2}l^2 D_k]^2\right) \\ &\times \exp\left(D_j u - \frac{u^2}{l^2}\right) \int_0^{t'} \exp\left\{-\frac{1}{l^2}[u' - (u + \tfrac{1}{2}l^2 D_k)]^2\right\} du' du \\ &= a^2 S_j S_k \exp(-D_j t - D_k t') \exp(\gamma_k^2) \int_0^t \exp((D_j + D_k)u) \\ &\times \int_0^{t'} \exp\left\{-\frac{1}{l^2}[u' - (u + \tfrac{1}{2}l^2 D_k)]^2\right\} du' du. \end{aligned}$$

where $\gamma_k = \frac{D_k l}{2}$. At this point the manipulation becomes clearer if a substitution is used, in this case let $s = \frac{1}{l}[u' - (u + \frac{1}{2}l^2 D_k)]$ such that

$$k_{x_j,x_k}(t,t') = a^2 S_j S_k l \exp(-D_j t - D_k t')\exp(\gamma_k^2) \\ \times \int_0^t \exp((D_j + D_k)u) \int_{s'}^{s''} \exp(-s^2) ds du$$

where the l comes from the change of variable and the upper and lower bounds of the second integral are given as $s'' = \frac{1}{l}[t' - (u + \frac{1}{2}l^2 D_k)]$ and $s' = \frac{1}{l}[-(u + \frac{1}{2}l^2 D_k)]$ respectively. Defining the error function to be $\text{erf}(x) = \frac{2}{\sqrt{\pi}}\int_0^x \exp(-y^2)dy$ and splitting the integral gives:

$$\begin{aligned} k_{x_j,x_k}(t,t') &= \frac{\sqrt{\pi}}{2} a^2 S_j S_k l \exp(-D_j t - D_k t')\exp(\gamma_k^2) \\ &\times \int_0^t \exp((D_j + D_k)u)[\text{erf}(s'') - \text{erf}(s')]du \\ &= \frac{\sqrt{\pi}}{2} a^2 S_j S_k l \exp(-D_j t - D_k t')\exp(\gamma_k^2) \\ &\times \int_0^t \exp((D_j + D_k)u)\left[\text{erf}\left(\frac{t'-u}{l} - \gamma_k\right) + \text{erf}\left(\frac{u}{l} + \gamma_k\right)\right] du. \end{aligned}$$

Solving the first integral still leads to a further integral needing to be solved. In order to make this calculation visibly easier to follow, the integral is derived separately from the main formula in two separate parts (see Appendix A.1). Once this has been solved, the solutions can be put back into the formula to give the final solution:

$$\begin{aligned} k_{x_j,x_k}(t,t') = \frac{\sqrt{\pi} a^2 S_j S_k l}{2(D_j + D_k)} \exp(-D_j t - D_k t')\exp(\gamma_k^2)\Bigg[\\ \exp((D_j + D_k)t)\,\text{erf}\left(\frac{t'-t}{l} - \gamma_k\right) - \text{erf}\left(\frac{t'}{l} - \gamma_k\right) - \exp(-\gamma_k^2) \\ \times \exp(\gamma_j^2)\exp((D_j + D_k)t')\left\{\text{erf}\left(\frac{t-t'}{l} - \gamma_j\right) + \text{erf}\left(\frac{t'}{l} + \gamma_j\right)\right\} \\ + \exp((D_j + D_k)t)\,\text{erf}\left(\frac{t}{l} + \gamma_k\right) - \text{erf}(\gamma_k) \\ - \exp(-\gamma_k^2)\exp(\gamma_j^2)\left\{\text{erf}\left(\frac{t}{l} - \gamma_j\right) + \text{erf}(\gamma_j)\right\}\Bigg] \end{aligned}$$

where $\gamma_j = \frac{D_j l}{2}$ and $\gamma_k = \frac{D_k l}{2}$. This can then be rearrange to give the solution stated in eq. (10).

A.1. Integral Solutions

In order to calculate the cross-covariance between two target genes, the following integral must be solved:

$$\int_0^t \exp((D_j + D_k)u)\left[\text{erf}\left(\frac{t'-u}{l} - \gamma_k\right) + \text{erf}\left(\frac{u}{l} + \gamma_k\right)\right] du.$$

This can derived most simply by firstly splitting it into two and then integrating by parts. In this way the first part of the integral is given by;

$$I_1 = \int_0^t \exp((D_j + D_k)u)\,\mathrm{erf}\left(\frac{t' - u}{l} - \gamma_k\right) du$$

which can be solved in the following manner:

$$\begin{aligned}
I_1 &= \frac{1}{D_j + D_k}\left[\exp((D_j + D_k)u)\,\mathrm{erf}\left(\frac{t' - u}{l} - \gamma_k\right)\right.\\
&\left.+ \frac{2}{l\sqrt{\pi}}\int \exp((D_j + D_k)u)\exp\left(-\left(\frac{t' - u}{l} - \gamma_k\right)^2\right) du\right]_0^t\\
&= \frac{1}{D_j + D_k}\left[\exp((D_j + D_k)t)\,\mathrm{erf}\left(\frac{t' - t}{l} - \gamma_k\right) - \mathrm{erf}\left(\frac{t'}{l} - \gamma_k\right)\right.\\
&\left.+ \frac{2}{l\sqrt{\pi}}\int_0^t \exp((D_j + D_k)u)\exp\left(-\frac{1}{l^2}[t' - u - \tfrac{1}{2}l^2 D_k]^2\right) du\right].
\end{aligned}$$

Again to make the formula more manageable we make a substitution, so let $P = \exp((D_j + D_k)t)\,\mathrm{erf}(\frac{t'-t}{l} - \gamma_k) - \mathrm{erf}(\frac{t'}{l} - \gamma_k)$. Then as before we work towards a situation where we can complete the square:

$$\begin{aligned}
I_1 &= \frac{1}{D_j + D_k}\left[P + \frac{2}{l\sqrt{\pi}}\int_0^t \exp((D_j + D_k)u)\right.\\
&\left.\times \exp\left(-\frac{1}{l^2}[u^2 - 2u(t' - \tfrac{1}{2}l^2 D_k) + (t' - \tfrac{1}{2}l^2 D_k)^2]\right) du\right]\\
&= \frac{1}{D_j + D_k}\left[P + \frac{2}{l\sqrt{\pi}}\exp\left(-\frac{1}{l^2}[t' - \tfrac{1}{2}l^2 D_k]^2\right)\right.\\
&\left.\times \int_0^t \exp\left(-\frac{1}{l^2}[u^2 - 2u(t' + \tfrac{1}{2}l^2 D_j - \tfrac{1}{2}l^2 D_k + \tfrac{1}{2}l^2 D_k)]\right) du\right]\\
&= \frac{1}{D_j + D_k}\left[P + \frac{2}{l\sqrt{\pi}}\exp(\gamma_j^2 - \gamma_k^2)\exp(t'(D_j + D_k))\right.\\
&\left.\times \int_0^t \exp\left(-\frac{1}{l^2}[u - (t' + \tfrac{1}{2}l^2 D_j)]^2\right) du\right].
\end{aligned}$$

We then once again make a substitution to simplify, $z = \frac{1}{l}[u - (t' + \frac{1}{2}l^2 D_j)]$. This then gives the upper and lower bounds as $z'' = \frac{1}{l}[t - (t' + \frac{1}{2}l^2 D_j)]$ and $z' = \frac{1}{l}[-(t' + \frac{1}{2}l^2 D_j)]$ respectively. Then splitting these as before gives the solution

$$\begin{aligned}
I_1 &= \frac{1}{D_j + D_k}\left[P + \exp(\gamma_j^2 - \gamma_k^2)\exp(t'(D_j + D_k))\{\mathrm{erf}(z'') - \mathrm{erf}(z')\}\right]\\
&= \frac{1}{D_j + D_k}\left[\exp((D_j + D_k)t)\,\mathrm{erf}\left(\frac{t' - t}{l} - \gamma_k\right) - \mathrm{erf}\left(\frac{t'}{l} - \gamma_k\right)\right.\\
&\left.+ \exp(\gamma_j^2 - \gamma_k^2)\exp(t'(D_j + D_k))\left\{\mathrm{erf}\left(\frac{t - t'}{l} - \gamma_j\right) + \mathrm{erf}\left(\frac{t'}{l} + \gamma_j\right)\right\}\right].
\end{aligned}$$

The second part of the integral is given as

$$I_2 = \int_0^t \exp((D_j + D_k)u) \operatorname{erf}\left(\frac{u}{l} + \gamma_k\right) du$$

which can again be solved through integration by parts:

$$\begin{aligned} I_2 &= \frac{1}{D_j + D_k} \left[\exp((D_j + D_k)u) \operatorname{erf}\left(\frac{u}{l} + \gamma_k\right) \right. \\ &\quad \left. - \frac{2}{l\sqrt{\pi}} \int \exp((D_j + D_k)u) \exp\left(-\left(\frac{u}{l} + \gamma_k\right)^2\right) du \right]_0^t \\ &= \frac{1}{D_j + D_k} \left[\exp((D_j + D_k)t) \operatorname{erf}\left(\frac{t}{l} + \gamma_k\right) - \operatorname{erf}(\gamma_k) \right. \\ &\quad \left. - \frac{2}{l\sqrt{\pi}} \int_0^t \exp((D_j + D_k)u) \exp\left(-\frac{1}{l^2}[u + \tfrac{1}{2}l^2 D_k]^2\right) du \right]. \end{aligned}$$

Again for simplicity, we denote $Q = \exp((D_j + D_k)t) \operatorname{erf}\left(\frac{t}{l} + \gamma_k\right) - \operatorname{erf}(\gamma_k)$ leading to

$$\begin{aligned} I_2 &= \frac{1}{D_j + D_k} \left[Q - \frac{2}{l\sqrt{\pi}} \int_0^t \exp((D_j + D_k)u) \right. \\ &\quad \left. \times \exp\left(-\frac{1}{l^2}[u^2 + l^2 D_k u + \frac{l^4 D_k^2}{4}]\right) du \right] \\ &= \frac{1}{D_j + D_k} \left[Q - \frac{2}{l\sqrt{\pi}} \exp(-\gamma_k^2) \right. \\ &\quad \left. \times \int_0^t \exp\left(-\frac{1}{l^2}[u^2 - 2u(\tfrac{1}{2}D_j l^2 - \tfrac{1}{2}D_k l^2 + \tfrac{1}{2}D_k l^2)]\right) du \right] \\ &= \frac{1}{D_j + D_k} \left[Q - \frac{2}{l\sqrt{\pi}} \exp(\gamma_j^2 - \gamma_k^2) \int_0^t \exp\left(-\tfrac{1}{l^2}\left[u - \tfrac{1}{2}D_j l^2\right]^2\right) du \right]. \end{aligned}$$

Again we use a similar substitution to before, letting $v = \frac{1}{l}[u - \frac{1}{2}D_j l^2]$. This gives the upper and lower bounds respectively as $v'' = \frac{1}{l}[t - \frac{1}{2}D_j l^2]$ and $v' = \frac{1}{l}[-\frac{1}{2}D_j l^2]$. Splitting the integral then leads to the following solution to I_2:

$$\begin{aligned} I_2 &= \frac{1}{D_j + D_k} \left[Q - \exp(\gamma_j^2 - \gamma_k^2)\{\operatorname{erf}(v'') - \operatorname{erf}(v')\} \right] \\ &= \frac{1}{D_j + D_k} \left[\exp((D_j + D_k)t) \operatorname{erf}\left(\frac{t}{l} + \gamma_k\right) - \operatorname{erf}(\gamma_k) \right. \\ &\quad \left. - \exp(\gamma_j^2 - \gamma_k^2) \left\{ \operatorname{erf}\left(\frac{t}{l} - \gamma_j\right) + \operatorname{erf}(\gamma_j) \right\} \right]. \end{aligned}$$

B. Cross-Covariance between Target Genes and the Transcription Factor

In a similar manner to finding $k_{x_j,x_k}(t,t')$ we must also find $k_{x_j,f}(t,t')$. In order to do this we must solve the integral in eq. (11), given below:

$$k_{x_j,f}(t,t') = S_j \exp(-D_j t) \int_0^t \exp(D_j u) k_{f,f}(u,t') du$$

where the transient terms have again been ignored. Again when an RBF kernel is chosen we can use Laplace transformation to solve the integral as follows:

$$\begin{aligned} k_{x_j,f}(t,t') &= S_j \exp(-D_j t) \int_0^t \exp(D_j u) a^2 \exp\left(-\frac{1}{l^2}[u-t']^2\right) du \\ &= a^2 S_j \exp(-D_j t) \int_0^t \exp\left(-\frac{1}{l^2}[u^2 - 2ut' + t'^2 - D_j l^2 u]\right) du. \end{aligned}$$

We can then complete the square and move all of the factors that are constant outside of the integral

$$\begin{aligned} k_{x_j,f}(t,t') &= a^2 S_j \exp(-D_j t) \exp\left(\frac{[t' + \frac{1}{2} D_j l^2]^2}{l^2}\right) \exp\left(-\frac{t'^2}{l^2}\right) \\ &\times \int_0^t \exp\left(-\frac{1}{l^2}[u - (t' + \tfrac{1}{2} D_j l^2)]^2\right) du \\ &= a^2 S_j \exp(-D_j(t-t')) \exp(\gamma_j^2) \\ &\times \int_0^t \exp\left(-\frac{1}{l^2}[u - (t' + \tfrac{1}{2} D_j l^2)]^2\right) du \end{aligned}$$

where $\gamma_j = \frac{1}{2} D_j l$. A substitution of $s = \frac{1}{l}[u - (t' + \frac{1}{2} D_j l^2)]$ then makes the manipulation easier to manage

$$k_{x_j,f}(t,t') = a^2 S_j l \exp(-D_j(t-t')) \exp(\gamma_j^2) \int_{s'}^{s''} \exp(-s^2) ds$$

where l comes from the change of variable and the upper and lower bounds are given by $s'' = \frac{1}{l}[t - t' - \frac{1}{2} D_j l^2]$ and $s' = \frac{1}{l}[-t' - \frac{1}{2} D_j l^2]$ respectively. Defining the error function $\mathrm{erf}(x)$ as before and splitting the integral allows the derivation of the final solution

$$\begin{aligned} k_{x_j,f}(t,t') =& a^2 S_j l \exp(-D_j(t-t')) \exp(\gamma_j^2) \\ &\times \left[\int_0^{s''} \exp(-s^2) ds - \int_0^{s'} \exp(-s^2) ds\right] \\ =& \frac{\sqrt{\pi}}{2} a^2 S_j l \exp(-D_j(t-t')) \exp(\gamma_j^2) [\mathrm{erf}(s'') - \mathrm{erf}(s')] \\ =& \frac{\sqrt{\pi}}{2} a^2 S_j l \exp(-D_j(t-t')) \exp(\gamma_j^2) \\ &\times \left[\mathrm{erf}\left(\frac{t-t'}{l} - \gamma_j\right) + \mathrm{erf}\left(\frac{t'}{l} + \gamma_j\right)\right]. \end{aligned}$$

C. Gradients of the Log Likelihood for GP

In order to learn the hyperparameters, $\boldsymbol{\theta}$, of the model it is necessary to maximise the likelihood function $p(\mathbf{y}|\boldsymbol{\theta})$ obtained from eq. (14). This is given in the form of a multivariate Gaussian distribution, given as

$$\ln p(\mathbf{y}|\boldsymbol{\theta}) = -\frac{1}{2}\ln|\mathbf{C}(\boldsymbol{\theta})| - \frac{1}{2}(\mathbf{y} - \mathbf{B}./\mathbf{D})^\top \mathbf{C}(\boldsymbol{\theta})^{-1}(\mathbf{y} - \mathbf{B}./\mathbf{D}) - \tfrac{N}{2}\ln(2\pi).$$

where $\mathbf{B}./\mathbf{D}$ is a point wise vector division. For non-linear optimisation techniques, such as the conjugate gradients, the gradient of the log likelihood is helpful for giving accurate and consistent estimates. Using some basic results for the derivation of matrices, see [2], the gradient is given as follows for the parameters $a^2, l, S_1, S \ldots, S_G$ and σ^2

$$\begin{aligned}\frac{\partial}{\partial\theta_i}\ln p(\mathbf{y}|\boldsymbol{\theta}) &= -\frac{1}{2}\operatorname{Tr}\left(\mathbf{C}(\boldsymbol{\theta})^{-1}\frac{\partial\mathbf{C}(\boldsymbol{\theta})}{\partial\theta_i}\right) \\ &+ \frac{1}{2}(\mathbf{y} - \mathbf{B}./\mathbf{D})^\top \mathbf{C}(\boldsymbol{\theta})^{-1}\frac{\partial\mathbf{C}(\boldsymbol{\theta})}{\partial\theta_i}\mathbf{C}(\boldsymbol{\theta})^{-1}(\mathbf{y} - \mathbf{B}./\mathbf{D}) \end{aligned} \tag{29}$$

where the trace function is defined as to be the sum of the elements of the main diagonal of the matrix. The result is also useful for $B_1, \ldots, B_G$ and $D_1, \ldots, D_G$, although not directly applicable due to the parameter values also occurring outside of $\mathbf{C}(\boldsymbol{\theta})$.

For the linear model $\boldsymbol{\theta} = (a^2, l, B_1, \ldots, B_G, S_1, \ldots, S_G, D_1, \ldots, D_G, \sigma^2)$ where $G =$ 3 in the example and a gradient function must be found for each of these hyper parameters. All of these gradient functions can be found through differentiation by parts, although the expression often becomes very large. The expression for the derivative of l is given by

$$\frac{\partial}{\partial l}k_{x_j,x_k}(t,t') = \frac{1}{l}k_{x_j,x_k}(t,t') + l_{dif1} + l_{dif2}$$

where $k_{x_j,x_k}(t,t')$ are elements of $\mathbf{K}(\boldsymbol{\theta})$, so have no noise, and where

$$\begin{aligned} l_{dif1} = \frac{\sqrt{\pi}la^2S_jS_k}{2(D_j+D_k)}&\left[\frac{1}{2}D_k^2 l\exp(\gamma_k^2)\right. \\ &\times\left\{\exp(D_k(t-t'))\left[\operatorname{erf}\left(\frac{t'-t}{l}-\gamma_k\right)+\operatorname{erf}\left(\frac{t}{l}+\gamma_k\right)\right]\right. \\ &\left.-\exp(-(D_jt+D_kt'))\left[\operatorname{erf}\left(\frac{t'}{l}-\gamma_k\right)+\exp(\gamma_k)\right]\right\} \\ &+\frac{1}{2}D_j^2 l\exp(\gamma_j^2) \\ &\times\left\{\exp(D_j(t'-t))\left[\operatorname{erf}\left(\frac{t-t'}{l}-\gamma_j\right)+\operatorname{erf}\left(\frac{t'}{l}+\gamma_j\right)\right]\right. \\ &\left.\left.-\exp(-(D_jt+D_kt'))\left[\operatorname{erf}\left(\frac{t}{l}-\gamma_j\right)+\exp(\gamma_j)\right]\right\}\right] \end{aligned}$$

and

$$
\begin{aligned}
l_{dif2} = \frac{la^2 S_j S_k}{D_j + D_k} \Bigg[& \exp(D_k(t-t')) \exp(\gamma_k^2) \\
& \times \left\{ \exp\left(-\left(\frac{t'-t}{l} - \gamma_k \right)^2 \right) \left(-\frac{D_k}{2} - \frac{t'-t}{l^2} \right) \right. \\
& \left. + \exp\left(-\left(\frac{t}{l} + \gamma_k \right)^2 \right) \left(\frac{D_k}{2} - \frac{t}{l^2} \right) \right\} \\
& - \exp(-(D_j t + D_k t')) \exp(\gamma_k^2) \\
& \times \left\{ \exp\left(-\left(\frac{t'}{l} - \gamma_k^2 \right)^2 \right) \left(-\frac{D_k}{2} - \frac{t'}{l^2} \right) + \exp(-\gamma_k^2) \frac{D_k}{2} \right\} \\
& + \exp(D_j(t'-t)) \exp(\gamma_j^2) \\
& \times \left\{ \exp\left(-\left(\frac{t-t'}{l} - \gamma_j \right)^2 \right) \left(-\frac{D_j}{2} - \frac{t-t'}{l^2} \right) \right. \\
& \left. + \exp\left(-\left(\frac{t'}{l} + \gamma_j \right)^2 \right) \left(\frac{D_j}{2} - \frac{t'}{l^2} \right) \right\} \\
& - \exp(-(D_j t + D_k t')) \exp(\gamma_j^2) \\
& \times \left\{ \exp\left(-\left(\frac{t}{l} - \gamma_j^2 \right)^2 \right) \left(-\frac{D_j}{2} - \frac{t}{l^2} \right) + \exp(-\gamma_j^2) \frac{D_j}{2} \right\} \Bigg].
\end{aligned}
$$

This can be put into eq. (29) to give the function for the gradient for l.

The gradient functions for B_1 is given by

$$
\begin{aligned}
\frac{\partial}{\partial B_1} \ln p(\mathbf{y}|\boldsymbol{\theta}) = -\frac{1}{2} \big[& (\mathbf{y} - \mathbf{B}./\mathbf{D})^\top \mathbf{C}(\boldsymbol{\theta})^{-1} (-\mathbf{1}./\mathbf{D_1}, \mathbf{0}, \mathbf{0}) \\
& + (-\mathbf{1}./\mathbf{D_1}, \mathbf{0}, \mathbf{0})^\top \mathbf{C}(\boldsymbol{\theta})^{-1} (\mathbf{y} - \mathbf{B}./\mathbf{D}) \big]
\end{aligned}
$$

where the vector $(-\mathbf{1}./\mathbf{D_1}, \mathbf{0}, \mathbf{0})$ has the same length as $\mathbf{y}$ with each part having a length corresponding to the number of observations from the related gene. Similar gradient functions are available for B_2, B_3, etc., where the above vector is replaced by $(\mathbf{0}, -\mathbf{1}./\mathbf{D_2}, \mathbf{0})$ for B_2 for example when $G = 3$.

The gradient function for $S_1, \ldots, S_G$ and $D_1, \ldots, D_G$ is more complicated due to the varying occurrence of the $\mathbf{S_i}$ and $\mathbf{D_i}$ within the covariance matrix. This can be illustrated if we think of the covariance matrix as being made up of G^2 equal sized smaller matrices. Thinking in this manner means that for the parameter S_1 for example, the small matrix at the top left of the larger matrix has $S_j = S_k = S_1$. Similarly the small matrices in the same row or column as this have either $S_j = S_1$ and $S_k \neq S_1$ or $S_j \neq S_1$ and $S_k = S_1$. For all of the other matrices $S_j \neq S_1$ and $S_k \neq S_1$, which ultimately means that different elements of the matrix have different equations for calculating the gradient.

Looking specifically at the parameters, $S_1, \ldots, S_G$, where $S_j = S_i$ and $S_k = S_i$ the differential is given by

$$\frac{\partial}{\partial S_i} k_{x_i,x_i}(t,t') = \frac{2}{S_i} k_{x_i,x_i}(t,t').$$

Similarly for $S_j = S_i$ and $S_k \neq S_i$ or $S_j \neq S_i$ and $S_k = S_i$ the differential is given by

$$\frac{\partial}{\partial S_i} k_{x_j,x_k}(t,t') = \frac{1}{S_i} k_{x_j,x_k}(t,t').$$

Finally where $S_j \neq S_i$ and $S_k \neq S_i$, the differential is equal to zero. Combining these various elements gives the matrix for the differential and this can be put into eq. (29) to give the function for the gradient for S_i.

The gradient function for D_1 is given by

$$\begin{aligned}
\frac{\partial}{\partial D_1} \ln p(\mathbf{y}|\boldsymbol{\theta}) = &-\frac{1}{2} \operatorname{Tr}\left(\mathbf{C}(\boldsymbol{\theta})^{-1} \frac{\partial \mathbf{C}(\boldsymbol{\theta})}{\partial D_1}\right) \\
&- \frac{1}{2}(\mathbf{B_1}./\mathbf{D_1^2}, \mathbf{0}, \mathbf{0})^\top \mathbf{C}(\boldsymbol{\theta})^{-1}(\mathbf{y} - \mathbf{B}./\mathbf{D}) \\
&- \frac{1}{2}(\mathbf{y} - \mathbf{B}./\mathbf{D})^\top \mathbf{C}(\boldsymbol{\theta})^{-1}(\mathbf{B_1}./\mathbf{D_1^2}, \mathbf{0}, \mathbf{0}) \\
&+ \frac{1}{2}(\mathbf{y} - \mathbf{B}./\mathbf{D})^\top \mathbf{C}(\boldsymbol{\theta})^{-1} \frac{\partial \mathbf{C}(\boldsymbol{\theta})}{\partial D_1} \mathbf{C}(\boldsymbol{\theta})^{-1}(\mathbf{y} - \mathbf{B}./\mathbf{D}). \qquad (30)
\end{aligned}$$

Again similarly to $B_1, \ldots, B_G$, this can be adapted for D_2, D_3, etc., where the vector $(\mathbf{B_1}./\mathbf{D_1^2}, \mathbf{0}, \mathbf{0})$ is replaced $(\mathbf{0}, \mathbf{B_2}./\mathbf{D_2^2}, \mathbf{0})$ for example for D_2 when $G = 3$.

We are now required to work out the differential of D_i to put into eq. (30). Again different elements of the matrix have varying differentials, so firstly where $D_j = D_i$ and $D_k = D_i$ we have

$$\frac{\partial k_{x_i,x_i}(t,t')}{\partial D_i} = (\tfrac{1}{2} D_i l^2 - \tfrac{1}{D_i}) k_{x_i,x_i}(t,t') + D_{i,i,dif1} + D_{i,i,dif2}$$

where

$$\begin{aligned}
D_{i,i,dif1} = &\frac{\sqrt{\pi} l a^2 S_i^2}{4 D_i} \exp(\gamma_i^2) \\
&\times \Bigg[(t - t') \exp(-D_i(t' - t)) \left\{\operatorname{erf}\left(\frac{t' - t}{l} - \gamma_i\right) + \operatorname{erf}\left(\frac{t}{l} + \gamma_i\right)\right\} \\
&+ (t + t') \exp(-D_j(t + t')) \left\{\operatorname{erf}\left(\frac{t'}{l} - \gamma_i\right) + \operatorname{erf}(\gamma_i)\right\} \\
&+ (t' - t) \exp(-D_i(t - t')) \left\{\operatorname{erf}\left(\frac{t - t'}{l} - \gamma_i\right) + \operatorname{erf}\left(\frac{t'}{l} + \gamma_i\right)\right\} \\
&+ (t + t') \exp(-D_j(t + t')) \left\{\operatorname{erf}\left(\frac{t}{l} - \gamma_i\right) + \operatorname{erf}(\gamma_i)\right\}\Bigg]
\end{aligned}$$

and

$$D_{i,i,dif2} = \frac{l^2 a^2 S_i^2}{4D_i} \exp(\gamma_i^2)$$
$$\times \Bigg[\exp(-D_i(t'-t)) \Bigg\{ -\exp\left(-\left(\frac{t'-t}{l} - \gamma_i\right)^2\right) + \exp\left(-\left(\frac{t}{l} + \gamma_i\right)^2\right) \Bigg\}$$
$$- \exp(-D_i(t+t')) \Bigg\{ -\exp\left(-\left(\frac{t'}{l} - \gamma_i\right)^2\right) - \exp(-\gamma_i^2) \Bigg\}$$
$$+ \exp(-D_i(t-t')) \Bigg\{ -\exp\left(-\left(\frac{t-t'}{l} - \gamma_i\right)^2\right) + \exp\left(-\left(\frac{t}{l} + \gamma_i\right)^2\right) \Bigg\}$$
$$- \exp(-D_i(t+t')) \Bigg\{ -\exp\left(-\left(\frac{t}{l} - \gamma_i\right)^2\right) + \exp(-\gamma_i^2) \Bigg\} \Bigg]$$

Similarly when $D_j = D_i$ and $D_k \neq D_i$, then the differential is given by

$$\frac{\partial k_{x_i,x_k}(t,t')}{\partial D_i} = -\frac{1}{D_i + D_k} k_{x_i,x_k}(t,t') + D_{i,k,dif1} + D_{i,k,dif2} + D_{i,k,dif3}$$

where

$$D_{i,k,dif1} = \frac{\sqrt{\pi} l a^2 S_i S_k}{2(D_i + D_k)} \Bigg\{ t \exp(\gamma_k^2) \exp(-(D_i t + D_k t'))$$
$$\times \left[\operatorname{erf}\left(\frac{t'}{l} - \gamma_k\right) + \operatorname{erf}(\gamma_k) \right] + \tfrac{1}{2} D_i l^2 \exp(\gamma_i^2)$$
$$\Bigg\{ \exp(-D_i(t-t')) \left[\operatorname{erf}\left(\frac{t-t'}{l} - \gamma_i\right) + \operatorname{erf}\left(\frac{t'}{l} + \gamma_i\right) \right]$$
$$- \exp(-(D_i t + D_k t')) \left[\operatorname{erf}\left(\frac{t}{l} - \gamma_i\right) + \operatorname{erf}(\gamma_i) \right] \Bigg\} ,$$

$$D_{i,k,dif2} = \frac{\sqrt{\pi} l a^2 S_i S_k}{2(D_i + D_k)} \exp(\gamma_i^2) \Bigg\{ (t'-t) \exp(-D_i(t-t'))$$
$$\times \left[\operatorname{erf}\left(\frac{t-t'}{l} - \gamma_i\right) + \operatorname{erf}\left(\frac{t'}{l} + \gamma_i\right) \right] + t \exp(-(D_i t + D_k t'))$$
$$\times \left[\operatorname{erf}\left(\frac{t}{l} - \gamma_i\right) + \operatorname{erf}(\gamma_i) \right] \Bigg\}$$

and

$$D_{i,k,dif3} = \frac{l^2 a^2 S_i S_k}{2(D_i + D_k)} \exp(\gamma_i^2) \Bigg\{ \exp(-D_i(t-t'))$$
$$\times \left[-\exp\left(-\left(\frac{t-t'}{l} - \gamma_i\right)^2\right) + \exp\left(-\left(\frac{t'}{l} + \gamma_i\right)^2\right) \right]$$
$$+ \exp(-(D_i t + D_k t')) \left[\exp\left(-\left(\frac{t}{l} - \gamma_i\right)^2\right) - \exp(-\gamma_i^2) \right] \Bigg\}$$

There is also a similar expression also available for $D_j \neq D_i$ and $D_k = D_i$, where t and t' are swapped and D_j replaces D_k. Then similarly to **S**, when $D_j \neq D_i$ and $D_k \neq D_i$, then this part of the differential is equal to zero. Once calculated, all parts of the differential can be combined and used within eq. (30).

Finally expressions must be found for the gradient of σ^2 and a^2. By eq. (14) it immediately become obvious that the derivative for σ^2 is simply the identity matrix of the appropriate length. This can then be easily be used in eq. (29) to give the gradient function as required. The derivative for a^2 can also be simply used eq. (29) and this is given by

$$\frac{\partial}{\partial a^2} k_{x_j,x_k}(t,t') = \frac{1}{a^2} k_{x_j,x_k}(t,t').$$

References

[1] M. Barenco, D. Tomescu, D. Brewer, R. Callard, J. Stark, and M. Hubank. Ranked prediction of p53 targets using hidden variable dynamic modeling. *Genome Biology*, 7(3):R25, 2006.

[2] C.M. Bishop. *Pattern Recognition and Machine Learning*. Springer, Singapore, 2006.

[3] B.P Durbin, J.S. Hardin, D.M. Hawkins, and D.M. Rocke. A variance-stabilizing transformation for gene-expression microarray data. *Bioinformatics*, 18(suppl1):S105–S110, 2002.

[4] P. Gao, A. Honkela, M. Rattray, and N.D. Lawrence. Gaussian process modelling of latent chemical species: applications to inferring transcription factor activities. *Bioinformatics*, 24:i70–i75, 2008.

[5] A. Honkela, C. Girardot, E.H. Gustafson, Y.-H. Liu, E.E.M. Furlong, N.D. Lawrence, and M. Rattray. Model-based method for transcription factor target identification with limited data. *PNAS*, 107(17):7793–7798, 2010.

[6] W. Huber, A. von Heydebreck, H. Sultmann, A. Poustka, and M. Vingron. Variance stabilization applied to microarray data calibration and to the quantification of differential expression. *Bioinformatics*, 18(suppl1):S96–104, 2002.

[7] N.D. Lawrence, M. Rattray, P. Gao, and M. Titsias. Gaussian processes for missing species in biochemical systems. In N.D. Lawrence, M.A. Girolami, M. Rattray, and G. Sanguinetti, editors, *Learning and Inference in Computational Systems Biology*. MIT Press, 2010.

[8] N.D. Lawrence, G. Sanguinetti, and M. Rattray. Modelling transcriptional regulation using Gaussian processes. *Advances in Neural Information Processing Systems (NIPS)*, 21:785–792, 2008.

[9] K. Lin and D. Husmeier. Modelling transcriptional regulation with a mixture of factor analyzers and variational bayesian expectation maximization. *EURASIP Journal on Bioinformatics and Systems Biology*, 2009. Article ID 601068.

[10] R.M. Neal. *Bayesian Learning for Neural Networks*, volume 118 of *Lecture Notes in Statistics*. Springer, New York, 1996. ISBN 0-387-94724-8.

[11] I. Pournara and L. Wernisch. Factor analysis for gene regulatory networks and transcription factor activity profiles. *BMC Bioinformatics*, 8, 2007.

[12] C.E. Rasmussen and C.K.I. Williams. *Gaussian Processes for Machine Learning*. MIT Press, 2006.

[13] C. Sabatti and G.M. James. Bayesian sparse hidden components analysis for transcription regulation networks. *Bioinformatics*, 22(6):739–46, 2006.

[14] G. Sanguinetti, M. Rattray, and N.D. Lawrence. A probabilistic dynamical model for quantitative inference of the regulatory mechanism of transcription. *Bioinformatics*, 22(14):1753–9, 2006.

[15] E. Snelsen, C.E. Rasmussen, and Z. Ghahramani. Warped Gaussian processes. *Neural Information Processing Systems (NIPS)*, 16, 2004.

[16] M.K. Titsias, A. Honkela, N.D. Lawrence, and M. Rattray. Identifying targets of multiple co-regulating transcription factors from expression time-series by Bayesian model comparison. *BMC Systems Biology*, 6(53), 2012.

[17] V. Vyshemirsky and M.A. Girolami. Bayesian ranking of biochemical system models. *Bioinformatics*, 24(6):S833–839, 2008.

In: Recent Advances in Systems Biology Research
Editors: A. X. C. N. Valente, A. Sarkar and Y. Gao
ISBN: 978-1-62948-736-6

Chapter 10

Cancer Systems Biology

***Rachael Hageman Blair*[1,*], *David L. Tritchler*[1,2] *and Daniel P. Gaile*[1]**
[1]Department of Biostatistics,
State University of New York at Buffalo, Buffalo, NY, US
[2]Department of Biostatistics,
University of Toronto, Toronto, ON, Canada

Keywords: Cancer, omics, metabolism, ODEs, steady-state, dynamic, graphical models, prediction, databases, sensitivity

Abstract

Cancer is a complex and deadly disease. Staggering mortality and recurrence rates reflect the robust nature of the cancer cell for survival. These survival mechanisms manifest in the form of drug resistance, which has proven to be a major obstacle for emerging targeted therapies. Massive experimental efforts have been fueled by the advances and availability of high-throughput technologies. However, gained knowledge is not translating into treatment. Consequently, experts and funding organizations are calling for a paradigm shift in the field of Cancer Biology, which moves the focus from single molecular targets to pathways and systems. Approaches in Systems Biology will be critical for this endeavor. Mathematical and Statistical models are central to Systems Biology approaches, and hold tremendous promise for advancing our understanding about the disease. In this Chapter, we describe several modeling paradigms and their applications to Cancer Biology. The methods described fall into two complementary categories: deterministic models and statistical graphical models. Strengths of these approaches include: flexibility of the modeling paradigms to generalize to different data and applications, and the ability of the model to make predictions about quantities, which may be difficult or impossible to measure *in vivo*. Current challenges regarding inference and computation are detailed. A rich spectrum of applications are

*E-mail address: hageman@buffalo.edu

provided, including, predicting system-wide effects of drug treatment, disease prognosis, tumor classification, forecasting treatment outcomes, and predicting survival for individual patients.

1. Introduction

Declining costs, coupled with advances in the design and application, of high-throughput biotechnologies, have fueled the growth of open source *omics* databases. The *omics* designation refers to the fact that the collected data correspond to assays designed to interrogate, under observed or experimentally designed conditions, sample genomes, proteomes, metabolomes, etc.. Unfortunately, our ability to interpret "Big Data" of the *omics* variety for knowledge and discovery has not kept pace with data collection and archiving efforts. Systems Biology approaches will play a crucial role in addressing the challenge of large-scale data interpretation and integration. Central to Systems Biology paradigms are Mathematical and Statistical models, which can serve as a tool to drive experimental advances in terms of predication, classification, and hypothesis generation. These approaches hold great promise for systems-based interpretations of omics data, which may provide novel insights into complex disease pathologies.

Omics technologies have guided detailed investigations and characterizations for a number of different cancers [18]. Despite tremendous progress in understanding the molecular mechanisms underlying the disease, gained knowledge is not translating into effective cancer treatments [71]. This message was echoed in NCI's recent funding opportunity addressing *provocative questions*, which pose game-changing scientific questions to drive progress against cancer (RFA-CA-11-01) [111]. The questions were intended to inspire researchers to "...step back from the momentum of these discoveries and make sure we have left no stone unturned and no important but perhaps not obvious question left unexplored". Cancer system biology is especially relevent to provocative question 17, which asks: "Since current methods to assess potential cancer treatments are cumbersome, expensive and often inaccurate, can we develop other methods to rapidly test interventions for cancer treatment or prevention?".

The slow translation of experimental advances into clinical practice is evidenced by the staggering mortality rates and lack of new and effective cancer treatments. The most common forms of clinical treatment remain surprisingly primitive in light of the knowledge that has been obtained through massive experimental efforts. For example, chemotherapy was discovered in chemical warfare during World War I; it was first used to treat cancer in the 1940s when little was understood about the disease [45], and remains the most common form of treatment for most types of cancers. Chemotherapy drugs target rapidly-dividing cells; as a result, normal tissues with high growth rates suffer and patients often experience adverse and sometimes deadly side effects. Over the past 15 years, drugs have emerged that target cancer metabolism, either directly through enzymes that facilitate metabolic reactions or indirectly through signaling pathways [50, 120]. Targeted therapy is typically less damaging to normal cells than chemotherapy. However, cancer cells are extremely robust for survival and often completely insensitive to perturbations or develop resistance over time. Drug resistance occurs when non-targeted genes or proteins kick in to *rescue* the cancer cell by rerouting growth requirements through alternative mechanisms and pathways. Drug resistance is a major limitation to targeted therapies. For this reason, targeted therapies are most effective when used in combination with chemotherapy treatments.

It is becoming apparent that in order to develop effective targeted therapies that combat resistance, the drug development paradigm must shift from single molecular targets to pathways [4, 110]. Systems biology approaches, which utilize computational models of cancer networks will play a pivotal role in this endeavor. Computational models can be used as an exploratory tool to complement and guide experimental work. Simulations performed with these models are known as *in silico* experiments, and can be used as a cost-effective avenue for hypothesis generation and to make predictions about quantities that are difficult or currently impossible to measure *in vivo*. Predictions can provide much-needed insight into the pathways driving cancer progression, and the robust compensatory mechanisms that protect cancer cells from drug intervention. Model simulations can also be used to predict the system-wide effects of molecular targets, e.g., side-effects in specific genetic populations. In addition, they can be used as an important clinical tool for the classification of benign and malignant tumors, prediction of disease prognosis for individual patients, and prediction treatment outcomes.

In this chapter, we describe complementary and widely-used systems-level modeling paradigms: deterministic models of cellular systems and graphical modeling (Figure 1). These modeling approaches require the solution to unique sets of inverse problems, which are necessary for the characterization of the computational model. Deterministic models of cellular metabolism are constructed in a *bottom-up* approach from known stoichiometry, principles of mass balance, and physiological constraints. These models require inference of statistical parameters or molecular constants in order to characterize the computational model, and predictions can be made in terms of molecular profiles, e.g., fluxes and time course data. In contrast, graphical models are developed from data in a *top-down* fashion. These models can require up two levels of inference for both the model parameters and for the network structure itself. Novel associations and causal pathways can be revealed through the inferred network structure.

This chapter is organized as follows. We begin by describing deterministic bottom-up approaches that are rooted in ordinary differential equations. These mechanistic models are used to simulate steady-state and dynamic cellular mechanisms. We follow this discussion with graphical models that are statistical in nature, and rooted in the concept of modularity. A spectrum of models ranging from clustering, to directed and undirected graphical models, and hybrid models are discussed. Cancer applications using deterministic and probabilistic model systems are highlighted throughout, and genome-scale modeling advances are also described. Finally, we highlight data resources which can be leveraged for model construction and integration, and discuss some of the current challenges and opportunities in the field.

1.1. Deterministic Models of Cellular Metabolism and Cell Signaling

Cancer cells undergo profound alterations to their metabolic and signaling pathways, which distinguish them from normal cells. Many drugs that are either available or in the development phase target proteins or enzymes in these pathways in order to slow or halt cancer growth [6]. Proliferation, motility, and survival are tightly controlled by normal cells,

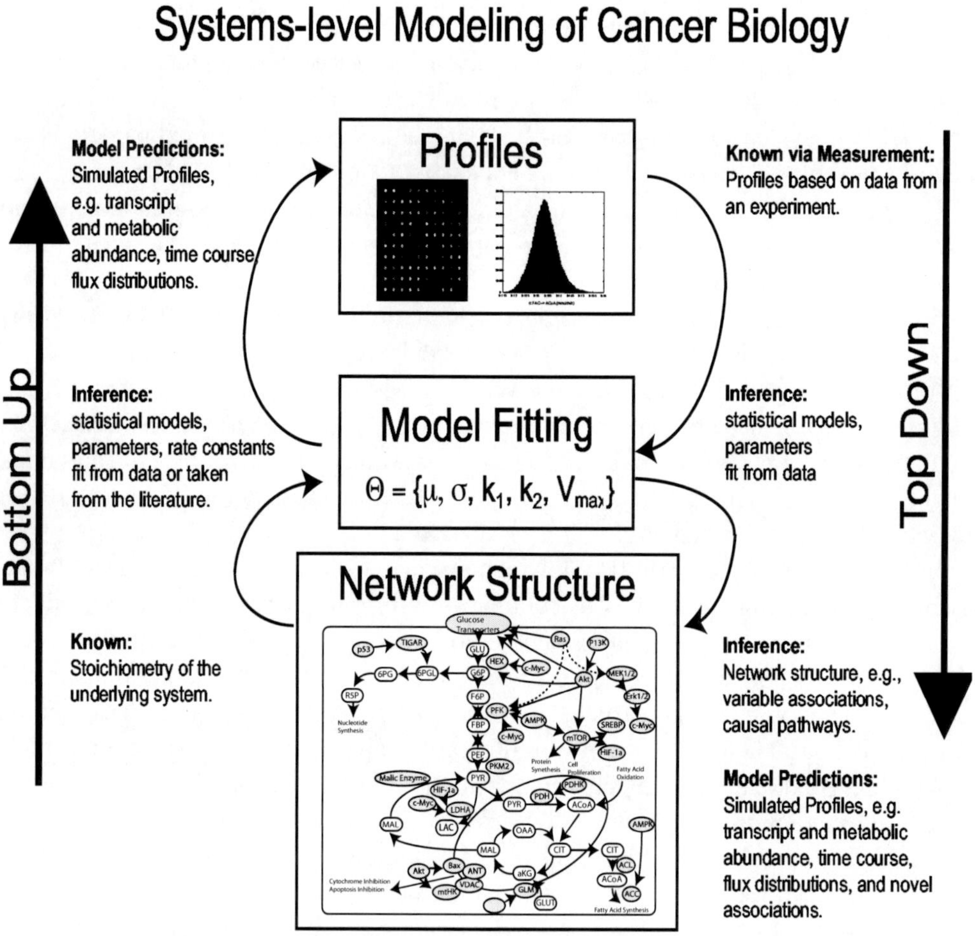

Figure 1. Outline of complimentary approaches to systems-level modeling of Cancer Biology. The *bottom-up* approach utilizes known stoichiometry of pathways and uses inference to fit the computational model. The *top-down* approach aims to infer both network parameters and structure directly from the data. These approaches characterize different aspects of the biological system and can be used to make predictions about system-wide behavior under different model perturbations.

whereas cancer cells leverage coordinated adjustments to enable proliferation independent of exogenous signals, disrupt apoptosis, and elicit tumor angiogenesis and metastasis to surrounding tissues and vessels [58, 74]. In contrast to their normal counterparts, cancer cells use aerobic glycolysis instead of oxidative phosphorylation for energy production [113]. Glutamine is central to cancer cell protein and nucleotide biosynthesis, and is critical to the TCA cycle for anabolic processes [72]. Fatty acid biosynthesis occurs at high rates and the majority of fatty acids are produced *de novo* regardless of nutrition [77, 88]. These metabolic and signaling signatures are common to most forms of cancer.

Ordinary differential equations (ODEs) represent the most widely-used approach for modeling the cellular dynamics of signaling and metabolic pathways. Reactions are typically assumed to occur under well-mixed conditions and to involve reactants that are present in at least minimal abundance. The differential equations are derived from the laws of mass balance and describe the rate of change of a species (dC/dt) in terms of species production and utilization, i.e., $dC/dt = \text{production} - \text{utilization}$ (Figure 2A). In many cases, the stoichiometry of pathways are well understood from the basic principals of biochemistry, and the topology of the system can be modeled easily with ODEs. However, the underlying processes, e.g., reaction fluxes and transport rates, rely on parameters that are often unknown and may be computationally difficult to estimate. For example, estimation might be complicated by an under-determined inverse problem inherent to the time course data (Figure 2B) [30] under consideration. Additionally, systems can exhibit sharp transients on different time scales (stiffness), which requires computationally intensive numerical integration [73, 105]. The scale of dynamic models which can be investigated is limited by practical concerns such as computational feasibility and model parameter identifiability. Consequently, they are most often used in focussed metabolic simulations from small subsets of reactions and pathways.

ODE models have been used extensively to simulate the dynamic properties of cancer signaling pathways. For example, mathematical models of the MAPK (mitogen-activated protein kinase) pathway have shown great promise. As of 2005, 35 models of the MAPK pathway had been developed in different levels of detail and granularity [89]. Despite differences between models, computational simulations have been able to explain experimental data reasonably well and have provided insightful system-wide predictions about the pathway dynamics. The ODE modeling process and the underlying biological system are, in combination, fairly robust [89] as the differences between model outputs can mostly be attributed to model boundaries and simplifications.

Other notable examples of ODE modeling of cancer signaling pathways include p53 and NF-κB. A model of tumor suppressor p53 and oncogene Mdm2 revealed high variability in the oscillatory behaviors of cells after DNA damage [43]. NF-κB signaling plays a central role in intracellular signaling, apoptosis, and resistance to chemotherapy. A computational model was developed and simulated to distinguish the roles of NF-κB kinase isoforms, which regulate NF-κB through coordinated system dynamics [51]. Extensions of this model have been used to characterize feedback loops in the system and identify the activation of downstream pathways [16, 20, 116].

Advances in high-throughput technologies have spurred the development of comprehensive genome-scale metabolic models [85]. These models have developed from extensive curation of the data and literature. The metabolic system is described by hundreds of metabolic reactions, multiple compartments, and highly-interconnected pathways. Constraint Based Analysis (CBA) has been used to investigate the *steady-state* behavior of these systems under a variety of conditions. In the steady-state, metabolites are assumed to be stable and not exhibit change in concentration levels. Adopting such an assumption reduces a complex dynamical system of ODEs to a linear system, eliminating the need for large-scale parameter estimation (Figure 2C). The resulting linear system is described by reaction stoichiometry and the sets of genes and proteins that facilitate the reactions in the model.

Gene-Protein-Reaction (GPR) mapping has enabled the association of metabolic reactions with the genes that encode their catalyzing enzymes within the model [55]. Several types of associations exist, which accommodate different scenarios of regulation. For example, a reaction catalyzed by a protein complex depends in turn on all of the genes which

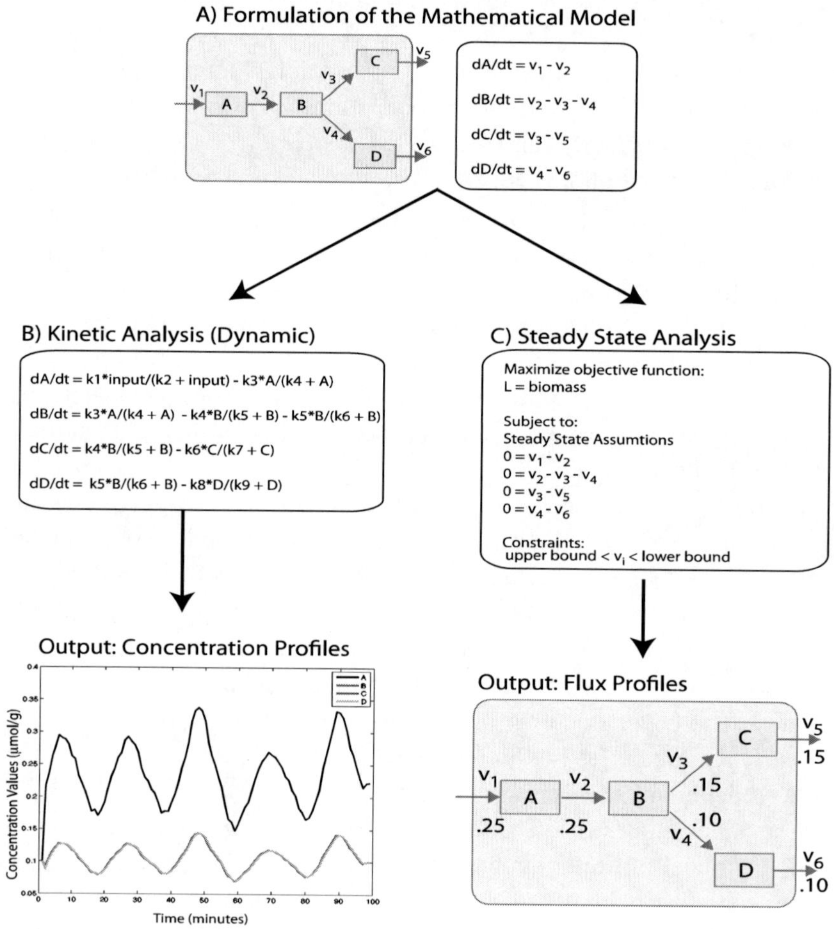

Figure 2. Simplified schematic depicting two different deterministic approaches to modeling cellular systems with ODEs. (**A**) The cellular system is translated into a mathematical model. The system of ODEs is governed by the mass balance of the cell. (**B**) Analyzing the dynamics of the cell requires specification of fluxes as non-linear function, which depend on a number of unknown parameters. Solving the ODE system results time course profiles of concentrations. (**C**) Steady-state analysis of the cellular system requires an objective function to be specified along with constraints. These simplifications reduce the dynamical system to a linear system of equations. The output of the analysis are optimal flux values which maximize the objective function.

encode the complex. Several isozymes can catalyze the same reaction, and may also catalyze other reactions within the system. These scenarios, and others, rely on careful curation of the biochemistry and literature to accurately depict the hierarchy of regulation that underlies the mass balance. Despite the detailed complexity and tedious mapping involved, taking this information into account through the GPR mappings has been shown to offer a considerable advantages in terms of accuracy and flexibility of the modeling process. In terms of model scope, the GPR mappings allow for the integration of tissue or disease specific data to influence the steady-state cellular dynamics. GPR mapping improves the CBA approach by making it more applicable to tissue and disease specific applications.

Steady-state analyses typically infer flux values that satisfy the steady-state assumptions and maximize an objective function describing the physiological objectives of the cell [68]. The solution space is bounded with system constraints, e.g., stoichiometric, thermodynamic, and enzyme capacity. In single cell organisms, such as *Escherichia coli* and *Saccharomyces cerevisiae*, the cellular objective is to proliferate. Critical reactions and pathways are included in a biomass function that is maximized to predict optimal steady-state flux values [27, 33]. In such cases, optimizing cellular growth is analogous to maximizing the likelihood of survival.

Defining cellular objectives is less straightforward in mammalian and human systems, which consist of interacting tissues and cells [25, 70, 104]. However, given their propensity for proliferation, cancer cells exhibit biomass requirements which can be leveraged in CBA modeling approaches. The Warburg effect and other common features exhibited across cancerous cells has been characterized by a generic genome-scale model which was validated against the full panel of NCI-60 cancer cell line mutation data [107]. The model demonstrated that the Warburg effect is a consequence of necessary metabolic adaptation that cancer cells undergo in order to increase biomass production. Model based predictions of flux profiles provided novel insights into phases of altered metabolism through cancer progression. The model also verified the utility of incorporating a biomass function which conveys the proliferation requirements of the cancer cell. A smaller and more simplified model was developed around a core set of known critical enzymes and coding genes for the purpose of predicting drug targets. The simplified model was also validated using the full panel of NCI-cell lines and known cancer growth supporting genes, and used to predict novel therapeutic targets [31]. Drug targets can be predicted by simply inhibiting the fluxes facilitated by the target of interest and then re-estimating a new set of optimal fluxes according to the new constrained model. The resulting perturbed model can be compared to the original to determine the overall system-wide effects on the metabolic system. A model perturbation can be performed to any parameter or combinations of parameters in the model. Potential candidate targets are those that repress cancer growth and do not exhibit undesirable metabolic consequences in the form of side-effects.

Promising steps have been taken toward integrating cancer specific data into the generic genome-scale model. For example, the model was used in connection with kidney cells from a mouse model of fumarate hydratase (FH) deficient cells [34]. The model predicted a novel pathway involved in the production and degradation of haem, which shifts the metabolism of the cell to enable TCA intermediates to accumulate and contribute to mitochondrial NADPH production. Further characterization of the pathway identified novel targets within the pathway as potential theraputic intervention points that would minimize undesirable impacts upon normal cells. In another application, novel methodology known as Metabolic Phenotypic Analysis (MPA) was developed to infer potential cancer-specific biomarkers based on high-throughput data from clinical data [56]. Briefly, the method estimates the potential of cells to produce essential metabolites in a specific context, e.g. cancer or non-cancerous cells, from high-throughput data. The inference relies on the estimation of exchange fluxes in the model, which are responsible for shuttling metabolites in and out of the cell as well as between cellular compartments. Two-group comparisons are made to identify significant metabolites, which are then flagged as potential biomarkers. MPA was used to predict known biomarkers for breast cancer, which were validated with amino acid profiles from breast cancer patients. Once validated, the model was used to predict novel biomarkers, including a number of choline containing metabolites which were previously identified as candidate positron emission tomography (PET) markers for breast cancer screening [87].

Due to their relative simplicity, ODEs are the most widely implemented "bottom up" modeling technique. Several other modeling paradigms that vary in complexity have been applied to study cancer cellular metabolism, signaling, and response to treatment. Boolean models have been used to simplify the representation of reactions as logical gates with two states: on and off [65, 79]. Partial differential equations (PDEs) are substantially more complex than ODEs and require detailed information about spatial dynamics, boundary conditions, and interactions between components [35, 99, 108]. Perturbation-response modeling approaches are based on fundamental linear response rules, which leverage flux conservation for parameter estimation. This approach has been used to characterize toll like receptor (TLR) signaling and tumor necrosis factor related apoptosis inducing ligand (TRAIL) resistance [95, 103]. Pharmacokinetic modeling has also been used to describe the time-dependent distribution of drugs in the system, a critical step toward drug development [41, 97, 101].

1.2. Graphical Models

Experimental data can be used to suggest possible network structures and associations when there is an insufficient level of understanding and inability to exploit the biological pathway in the context of deterministic systems. Descriptions of omics data can be given by graphical or network models which represent and visualize relationships among the variables. The intuitive notion that interactions among biological elements can depicted by connections between them motivates the study of graphical models for biological systems. Graphs consist of nodes and links. Nodes can represent quantities such as gene expression, genotype or protein abundance and links in the graph indicate the binary presence or absence of an interaction between a pair of nodes, or express a relevant measure of pairwise node association such as correlation (for which the association is symmetric), or a regression parameter (where causes and effects are distinguished by asymmetric connections).

Graphs constructed from prevailing knowledge contain graphical properties which can be viewed as descriptive of the underlying biological system. For example, the average degree of node connectivity conveys variable interdependence, central nodes may be interpreted as driving the function of the system, or a connected component of the graph may represent a distinct biological process. A compendium of useful graphical descriptors is given in [52]. *Small worlds* models also provide global descriptive measures of network connectivity which can give insight into the emergence and evolution of the network [114]. A popular organizing concept in systems biology is *modularity*, where a module is a part of an organism that is integrated with respect to a certain kind of process and relatively autonomous with respect to other parts of the organism. The modularity concept has gained popularity more-or-less simultaneously in molecular biology, systems biology, developmental biology and evolutionary biology [112]. Modular system structure corresponds to a partitioning of the associated graph.

Visualization is important for the analysis and interpretation of graphical models. Once a graph is constructed, it is desirable to have a clear and uncluttered display of the graph. In small-scale problems, the visualization task is manageable, interactions can be depicted and interpreted with relative ease. For graphs with many nodes the *ball of yarn* phenomena may occur and it may be necessary to display neighborhoods of nodes separately. In *omics*-scale data, visualizing the data and associations in a clear and concise graphical display is a major challenge. There has been considerable research effort in this direction, which has spurred the development of a number of visualization software tools, see reviews, [39, 92]. GRAPHVIZ is one of the more popular platforms for obtaining viewable static and dynamic graphs [28, 29]. The integration of ontology tools and R software [96] compatibility, makes GRAPHVIZ a powerful and attractive choice for visualization [13, 14, 119].

Cluster Analysis

Cluster analysis is widely used to describe large-scale biological systems and consists of a multitude of methodologies [23, 46, 57]. The method can be viewed as module identification since variables in the same cluster are considered to be *connected* according to some measure of statistical association, and different clusters are viewed as relatively *disconnected* from each other to varying degrees. This can be regarded as a partitioning of the graph into connected components. Relatedness of cluster members is typically given by a nonegative measure called a *similarity*, or by its complement, *distance*. Only cluster membership of the variables is inferred from the data. Therefore, a cluster-based model is relatively simple in that it does not deduce the detailed relationships or mechanisms among cluster members. To give the flavor of cluster analysis we give a general description of the widely used hierarchical clustering method. Hierarchical clustering proceeds in bottom-up fashion by initially considering each mode to be a cluster and then repeatedly consolidating similar clusters until some stopping rule is satisfied. Variations are based on an assortment of definitions of cluster similarity and stopping criteria. For example, the similarity of two clusters might be the minimum correlation of pairs of elements (one from each cluster), and the procedure might stop when a maximum cluster size is exceeded.

Statistical graphical models

Building a graph that accurately represents a biological system is achieved through the use of probability models, which guide in both the process of construcing the graph and clarify in its interpretation. Such a graph is referred to as a probabilistic graphical model (PGM). These models clearly define the absence of a link connecting two variables in terms of a conditional independence statement. In basic terms, this means that links in the graph imply a direct association between the variables which cannot be attributed to the influence of other variables in the system. Our discussion of PGMs will focus on general, statistically correct methods which can be applied to describe almost any biological system. There are several important problem-specific considerations for choosing and applying modeling methodologies and software. Specifically, the variables in the model (node type: continuous, discrete, or mixed), the associations between them (link type: undirected, directed, or both), the overall size of the problem (number of nodes), and other assumptions (e.g., causality), all influence the choice of graphical modeling technique. There is an assortment of general purpose graphical software associated with the popular statistical system R (http://cran.r-project.org/web/views/gR.html), which can apply to many problems [100]. We will also mention some application-based software which has been used for a specific problem.

Undirected graphs [66, 117] portray symmetric relationships (Figure 3A). A link in this model is present if the linked nodes are associated after controlling for the influence of all other nodes in the graph (conditional association). There is no notion of the directionality in this model; the links in undirected graphs express direct association between variable pairs, but no causal relationship is implied. Learning undirected graphical structures tends to be less demanding computationally, and recent methods enable inference of the links for very large scale systems [66, 117]. For graphs whose nodes are all continuous or all discrete, parameter estimation and interpretation of the graph is straightforward. The basic framework has been extended to mix continuous and discrete variables, which gives rise to subtleties in specifying permissible models [67]. The challenge is model selection, the task of determining which links belong in the graph. For 30 nodes the number of possible graphs is the same magnitude as the estimated number of atoms in the observable universe. Clearly, efficient computing algorithms are required.

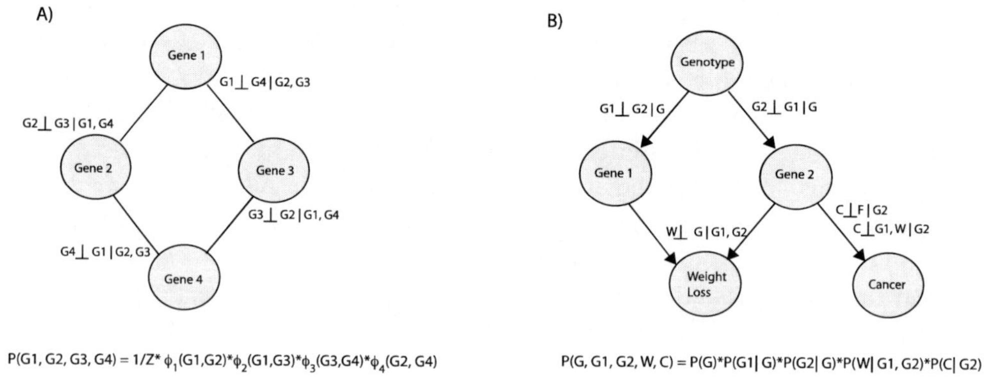

Figure 3. Probabilistic graphical models can have links that are **(A)** unidrected or **(B)** directed. Relationships between the variables in the model can be expressed using conditional independencies, allowing for compact representations of the joint distribution.

Causal direction is represented in Bayesian Networks (BNs) or directed acyclic graph models (DAGs), which have directed edges between nodes with no feedback permitted. In a directed graph, an edge $A \longrightarrow B$ implies that independent variable A (parent node) is upstream of the dependent variable B (child node) in the underlying causal process. Further, the directed edge implies a causal effect of A after the influence of the remaining nodes upstream of B (ancestors of B) have been controlled for or removed (conditional independence) (Figure 3B). The precise interpretation of these graphs in terms of conditional independence statements is given by the rules of d-separation [93]. The field of Bayesian networks originated when Pearl [94] showed how a consistent probability model could be assigned to a DAG using local distributions, and further gave algorithms for accumulating information or changing states of the system while maintaining consistency of the probabilities. The model required prior specification of the nodes, links, and directions of the links. The required local distributions can be estimated from data. This enabled valid decision support applications, and made so-called expert systems statistically valid. General purpose software to implement these systems is available in R and is also available in HUGIN software. HUGIN is commercial software but has a free full functionality version that can handle problems of limited size.

The structure of BNs can be inferred from data, and their are several methods for structural inference. For example, methods such as PC-algorithm [91] and hill-climbing greedy search algorithms [53], are used to infer the most likely structure. Usually multiple DAGs will give exactly the same fit to the observed data so it is important to compare equivalent models for plausibility. When the direction of some links is known beforehand that information should be input to guide the model search. Structural learning of BNs is an *NP*-hard problem [17]. Consequently, computation and scalability are major issues that limit BN utility and performance for large-scale data. Several algorithms use directed graphs to estimate the network of causal relationships among phenotypes and genotypes in segregating populations [47, 83]. They rely on computationally intensive Markov Chain Monte Carlo (MCMC) methods to sample networks from a posterior distribution. As a result, applications to high-throughput data requires variable selection to restrict model size (typically less than 30 variables), and there is risk of selection bias. A more cautious interpretation is required for methods that use graphical model as a base from which heuristic methods are applied to modify the network structure, e.g. direct or trim [19, 82].

BNs were used to integrate clinical and microarray data for the classification of breast cancer patients into good and poor prognosis groups [44]. Kahn et al. developed a BN called MammoNet for radiological decision support in distinguishing malignant and benign mammary tumors. The highly accurate classifier (88% correct diagnosis in test cases) was constructed from observational data, patient history, and expert advice from experienced radiologists [60]. This group later developed a similar BN classifier called OncOs to differentiate among bone lesions on the appendicular skeleton [59]. A practical value of these models is to provide a probabilistic guide for a clinician to diagnose and treat different cancers. Another use of PGMs is to sort out the underlying mutations which put individuals at high-risk. Conjunctive Bayesian Networks (CBNs) [7], which describe an accumulation of events, have been used to model the accumulation of mutations using CGH mutation data from the Progentix database [42]. The inference problem is to identify CGNs, which represent the dependencies among accumulating mutations in renal cell carcinoma, breast, and colorectal cancers. The models identified multiple independent mutations, which triggered downstream complex pathways.

A strength of PGM frameworks is the flexibility to integrate across diverse data types. Recently, a PGM methodology based on factor graphs known as PARADIGM (Pathway Recognition Algorithm using Data Integration on Genomic Models) was proposed, which integrates multiple high-throughput data sets together to identify perturbed molecular pathways [84]. This method was applied to breast cancer using gene expression data, and glioblastoma using gene expression and copy-number data, to identify pathways and disease subclasses which correlate with survival. PARADIGM was recently applied to the same task using a more comprehensive set of breast cancer data in the CGA, including, mRNA, copy number alterations, micro RNAs and methylation data. The method revealed disease subclasses and specific class signatures, which would not have been identified without leveraging the different data sources. Specific perturbations in immune response and interleukin signaling (IL-4, IL-6, IL-12, and IL-23) were also shown to be drivers of the classification and to have promising prognostic value. For example, patients with gene signature that favors high-T helper 1 cytotoxic T-lymphocyte response and represses Th2 driven humoral immunity, are more likely to have a better survival outcome.

Chain graphs [81] and block recursive models [21] are a directed/undirected hybrid. In these models variables are classified into causally ordered blocks such that variables in a block are potential causes of variables in later blocks but are potential effects of the variables in earlier blocks, and the links between blocks are directed to reflect this order. Within blocks the variables are unordered and links are undirected. Such an organization is quite common. For example, genotypes would precede gene expression, which would also preceed patient outcome. These models are easily interpreted and understood, but in general the statistical interpretation of graphs with both directed and undirected links is more complex [37].

An interesting application is the perturbation experiment. In these experiments, a single variable such as the expression of a specified gene is perturbed and the system is measured to determine the implied network connections. It can be shown that the methods that have been introduced to analyze these experiments are mathematically equivalent to a simple simultaneous model analysis used to study economic systems. Simultaneous models have been studied and refined by econometricians for decades and this field is potentially useful for biological applications. These models are qualitatively different from PGMs in that they are not required to be acyclic graphs. The construction of probability models for systems including cycles, where a node can both influence and be influenced by another node is an unresolved problem.

Genome-scale graphical models

Hypothesis generation from omics data is difficult due to the large-scale nature of the data, coupled with the vastness of the genome and the lack of preliminary information. One approach is to reduce the variable set to contain only those variables shown to be relevant. When the objective is cluster analysis or network estimation [22] and [2] propose methods for identifying the subset of variables that potentially participate in a pathway or network. When the objective is to describe the system which generates an outcome of interest such as survival, a useful first step is to isolate those variables demonstrating an effect on the outcome before elucidating the system properties of those variables. However, variable selection using classical regression analysis is not applicable to genome-scale variable selection since parameter estimates cannot be computed when the number of variables exceeds the number of experimental units studied. Traditional methods for dimension reduction such as principal component analysis and partial least squares are not appropriate. These methods reduce dimension by selecting linear combinations of the full set of predictors, thereby prohibiting explicit identification of relevant predictors.

Over the last 15 years, great strides have been made in large scale model selection by methods such as the lasso or elastic net, which are summarized by [90]. These methods can be used to address quite specific structural aspects of the genome such as identifying protein or transcription factor binding sites or DNA splice sites. Such information can be of great use in identifying and understanding biological systems. These methods exist both for outcomes on the form of continuous measurements or for categorical outcomes such as disease class or response/nonresponse. In contrast to lasso regression methods, the elastic net has the advantage of including correlated variables in the selected set, whereas lasso selects only one out of a correlated set. Since pathways consist of correlated variables the elastic net is strongly preferred as a preliminary step in network analysis.

There has been recent progress in sparse genome-scale models for undirected graphs, with applications that include protein signaling, breast cancer gene expression and the genetics of gene expression [15,26,36]. Sparse models can be estimated when the number of variables greatly exceeds the sample size. Approaches using techniques related to lasso regression can handle thousands of variables [36]. Some examples of modelling high-dimensional undirected graphs using sparse lasso-related techniques are the modeling of 324 gene expression nodes for 294 subjects [49] and 1001 gene expression nodes for 250 subjects [78]. Using a different approach [26] give an efficient method and demonstrate it on a network of 50,000 gene expression variables and SNPs. This is an active area of statistical research and is promising for inferring genomic-scale system structure.

1.3. Data Resources

There has never been a better time to study cancer in a systems biology framework. The last decade has seen a rapid increase in the number and scope of publicly accessible genomic databases, the National Center for Biotechnology Information (NCBI) Gene Expression Omnibus (GEO) [5], is a prime example. While the databases differ with respect to content, they share fundamental goals with respect to maintaining public utility. For example, most databases mirror the stated goals of GEO [1] which are to (1) "provide a robust, versatile database in which to efficiently store high-throughput functional genomic data", (2) "offer simple submission procedures and formats that support complete and well-annotated data deposits from the research community", and (3) "provide user-friendly mechanisms that allow users to query, locate, review and download studies and gene expression profiles of interest." To illustrate the size of such databases, as of January, 2013, GEO contained data from $10,937$ platforms, $861,862$ samples, $35,270$ series and $2,720$ manually curated data sets. The integration and unification of the languages of the ever increasing number of such database resources poses a major challenge moving forward [24].

Reactome and Kyoto Encyclopedia of Genes and Genomes (KEGG) [75,86] are currently the most widely used resources for the aggregation and dissemination of species specific topologies associated with signaling and metabolic pathways. Reactome, a manually curated resource, contains curated pathways for 28 different species, ranging from single cell organisms to humans, and is cross-referenced to bioinformatics databases such as Ensembl and NCBI. The KEGG resource integrates data from 15 primary databases [62], each with specialized content such as pathway maps, functional hierarchies, biochemical reactions, and reactant pairs. The aggregated data includes, for example, extensive pathway information for several thousand species, currently 176 eukaryotes and 2168 prokaryotes. The KEGG DISEASE and KEGG DRUG databases contain drug compounds and their known targets and interactions for a number of monogenic and multifactorial diseases [61]. KEGG DRUG catalogs the chemical structures and/or chemical components of a large collection of drugs which have been approved for use in the USA, Europe and Japan. KEGG DISEASE represents a notable effort to develop a resource that will allow genetic and infectious diseases to be modeled and analyzed in the context of perturbed states of molecular networks. At present, 17 different types of cancer (e.g., melanoma and prostate cancer) are represented and annotated with pathways of known disease genes and environmental triggers. KEGG provides a more general map of cancer cell metabolism, combining features across cancer specific disease pathways and differing organism types. Since its inception, the primary focus of the KEGG resource has been the molecular network. Moving forward, the architects of the KEGG resource have recently highlighted the importance of including "the association of network states and phenotypic features, such as network perturbations associated with diseases, drug effects, and other health conditions." [62].

The Cancer Genome Atlas contains data relating to a variety of cancer subtypes, which was generated through the interrogation of tissue samples using multiple high throughput biotechnologies, including, array comparative genomic hybridization (aCGH), expression array, methylation profiling, SNP genotyping, and microRNA profiling platforms. Notable publications have emerged from these efforts for glioblastomas [76], ovarian cancer [8], lung cancer [48], and breast cancer [64]. Additionally, several subsequent publications have used the data from TCGA data portal for secondary data analysis for methodology development and comparative studies. Another important resource is The International Cancer Genome Consortium, which was established "to generate comprehensive catalogues of genomic abnormalities (somatic mutations, abnormal expression of genes, epigenetic modifications) in tumors from 50 different cancer types and/or subtypes which are of clinical and societal importance across." The Sanger Cancer Genome Project has generated both data and bioinformatic resources, including a a catalog of somatic mutations in cancer [32] and a cancer gene census [38]. Array Express is a database of functional genomics experiments containing, as of January, 2013, data from 34, 149 experiments, 8, 488 of which are related to cancer.

The challenges associated with querying and locating data of potential value have spawned numerous data portals and search engines. A useful comparison of functionality across genomic search engines can be found in [118]. Federated databases such as Biomart [106] are available which are "data agnostic" and "platform independent", and are designed to provide biologists convenient access to an ever increasing quantity of "disparate, geographically distributed data sources". In fact, the number of *omics*-related databases has increased to the point that multiple online "databases of databases" have emerged in recent years. The growth of the number of database related publications and some of the major efforts to catalog them are conveniently summarized in [9].

The growth of publicly available datasets has created a demand for data management and analytical resources. That demand has been partially met by open source efforts such as Bioconductor [40], which is an extensive collection of 610 open source software libraries (as of January, 2013) for the R programming language. Bioconductor supports all major microarray, sequence and general high throughput platforms, and enables visualization, annotation (e.g., KEGG, Biomart, etc.). Packages such as GEOquery [102], connect to publicly available databases and allow for data interrogation and retrieval conveniently from the R workspace.

2. Conclusion

The ability to take full advantage of *omics* data for knowledge and discovery remains limited by challenges involving data integration [63, 109]. Comparisons and integration *within* omics data types are complicated by several factors, including differences in coverage and in the underlying biotechnologies which have been employed. Differences may also exist with respect to sample quality, array processing, the organism under investigation, tissue type, and observational or experimental conditions (e.g., diet). The integration *between* data types presents an even larger challenge (Figure 4). It is important to understand how different biological domains connect and give rise to a phenotype such as disease. Methods that integrate between and across different types of *omics* data have started to emerge [84], and are enabled by the flexibility of mathematical modeling. In Cancer Biology, the importance of data integration cannot be overemphasized because it presents the means to formally address the complex interplay between genetics, cell signaling, and metabolic pathways.

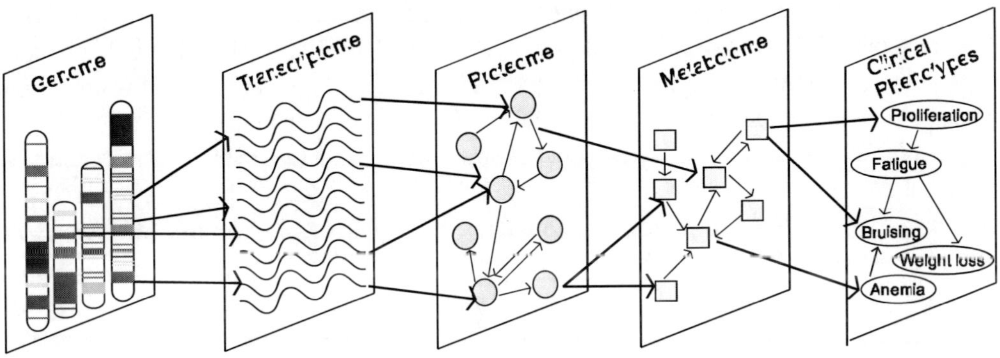

Figure 4. Understanding how biological domains form networks and interact to give rise to complex phenotypes and disease is a central goal of systems biology.

Often model performance can be improved when biological knowledge that is outside of the observational data can somehow be appropriately integrated into the analysis. To that end, the use of Bayesian methods that integrate *a priori* knowledge into the model have been shown to significantly improve model behaviors and predictive output. For example, we have described applications of BNs which incorporate expert advice from radiologists, which can be viewed as a prior for the predictive model [60]. In metabolic modeling, flexible Bayesian priors have been used to guide the parameter inference. In that context, priors favor parameter estimates which give rise to models that support the known physiology of the system, e.g., steady-state, dynamic trends, feasible bounds on concentration levels, and fluxes [11, 12]. In graphical models, priors based on entropy principals and energy functions have been used to guide network inference [54]. Notably, priors have also been used to encode known relational information from databases such as KEGG into the network inference process [80, 115], enforce sparsity in the network structure, and prevent over-fitting [47].

Improving and assessing the utility of mathematical models in the context of systems biology will continue to be an active area of research. Experimental design and perturbations have been shown to have major influence on parameter estimation, and subsequently the output and accuracy of the computational model [3]. Graphical model network inference can also be subject to a large proportion of false positive edges [69]. Environmental and experimental design factors that are not accounted for in the model can further misguide models [98]. Taken together, these modeling challenges can be problematic in achieving robust and physiologically relevant and accurate models.

George Box famously stated, “all models are wrong, but some are useful” [10]. Placing too much value on simulation and computational modeling can be detrimental. Rather, a continuous cycle between computational modeling and the wet-bench is critical to move systems biology forward. This cycle can be viewed as iterative refinement of both the computational model, and the scope and design of experiments. Sensitivity analysis should routinely be performed, as part of the iterative refinement process, in order to assess how sensitive the model output (predictions) are to small changes in the model parameters and input (data). Sensitivity analysis can also guide model reductions and expansions, e.g., marginalizing over parameters and variables which have no impact on the output. Over specifying models in terms of parameters is commonplace, but often these parameters are not physical, i.e. impossible to measure *in vivo*. In these cases, elimination in terms of model reductions to form a parsimoneous model would be desirable. Unfortunately, this type of analysis is far from routine. Mathematical models can provide, through model driven predictions and hypotheses generation, a cheap and fast catalyst for experimental advances in cancer systems biology. On the other hand, models which are more “wrong” than “useful” can slow progress by supporting the design and execution of experiments, which are likely to fail. Contrary to *in silico* studies, this can be a major waste of both time and money, and promote skepticism in such approaches.

In summary, computational network modeling techniques can be used to describe a wide range biological processes associated with cancer and other complex diseases. Developing models which mimic the true underlying biological mechanisms and physiology is a central goal of systems biology. We have described two complementary modeling approaches which proceed in a *bottom-up* and *top-down* fashion. Deterministic modeling of cellular metabolism and graphical modeling offer vastly different insights into biological systems. These models can be simulated to make a wide-range of predictions about pathway behavior, drug targets, side-effects, survival, and the clinical classification of cancer subtypes, to name only a few. The modeling paradigms we have described have been instrumental in progressing the field of Cancer Systems Biology. However, they remain far from mainstream. At present, there is an overwhelming demand to view cancer as a complex network in order to combat drug resistance, and develop viable treatment options. In the coming years, we believe the massive experimental effort will continue, hopefully in connection, and not in parallel to, the field of systems biology and modeling. We also predict a renewed interest in flexible computational modeling paradigms, such as the ones described here, for the interpretation and integration of *omics* data for knowledge and discovery.

References

[1] Geo overview: *http://www.ncbi.nlm.nih.gov/geo/info/overview.html*, March 2012.

[2] A. Hero, B. R. Large-scale correlation screening. *Journal of the American Statistical Association 106* (2011), 1540–52.

[3] Apgar, J. F., Witmer, D. K., White, F. M., and Tidor, B. Sloppy models, parameter uncertainty, and the role of experimental design. *Mol Biosyst 6* (2010), 1890–1900.

[4] Astsaturov, I., Ratushny, V., Sukhanova, A., Einarson, M., Bagnyukova, T., Zhou, Y., Devarajan, K., Silverman, J., Tikhmyanova, N., Skobeleva, N., Pecherskaya, A., Nasto, R., Sharma, C., Jablonski, S., Serebriiskii, I., Weiner, L., and Golemis, E. Synthetic lethal screen of an egfr-centered network to improve targeted therapies. *Science Signaling*, 140 (2010), ra67.

[5] Barrett, T., Troup, D. B., Wilhite, S. E., Ledoux, P., Rudnev, D., Evangelista, C., Kim, I. F., Soboleva, A., Tomashevsky, M., and Edgar, R. NCBI GEO: mining tens of millions of expression profiles–database and tools update. *Nucleic Acids Res. 35* (Jan 2007), D760–765. [PubMed Central:PMC1669752] [DOI:10.1093/nar/gkl887] [PubMed:17099226].

[6] Bates, S., Amiri-Kordestani, L., and Giaccone, G. Drug development: portals of discovery. *Clinical Cancer Research 18*, 1 (2012), 23–32.

[7] Beerenwinkel, N., Eriksson, N., and Sturmfels, B. Conjunctive bayesian networks. *Bernoulli 13*, 4 (2007), 893–909.

[8] Bell, D., Berchuck, A., Birrer, M. ., Ozenberger, B., Peterson, J., and Thomson, E. Integrated genomic analyses of ovarian carcinoma. *Nature 474* (Jun 2011), 609–615. [PubMed Central:PMC3163504] [DOI:10.1038/nature10166] [PubMed:21720365].

[9] Bolser, D. M., Chibon, P. Y., Palopoli, N., Gong, S., Jacob, D., Del Angel, V. D., Swan, D., Bassi, S., Gonzalez, V., Suravajhala, P., Hwang, S., Romano, P., Edwards, R., Bishop, B., Eargle, J., Shtatland, T., Provart, N. J., Clements, D., Renfro, D. P., Bhak, D., and Bhak, J. MetaBase–the wiki-database of biological databases. *Nucleic Acids Res. 40*, Database issue (Jan 2012), D1250–1254. [PubMed Central:PMC3245051] [DOI:10.1093/nar/gkr1099] [PubMed:22139927].

[10] Box, G. E. P., and Draper, N. R. *Empirical Model-building and Response Surfaces*. Wiley, 1987.

[11] Calvetti, D., Hageman, R., and Somersalo, E. Large-scale bayesian parameter estimation for a three-compartment cardiac metabolism model during ischemia. *Inverse Problems* (2006), 1797–1816.

[12] Calvetti, D., and Somersalo, E. Large-scale statistical parameter estimation in complex systems with an application to metabolic models. *Multiscale Modeling & Simulation 5* (2006), 1333.

[13] Carbon, S., Ireland, A., Mungall, C., Shu, S., Marshall, B., Lewis, S., et al. Amigo: online access to ontology and annotation data. *Bioinformatics 25*, 2 (2009), 288–289.

[14] Carey, V., Gentleman, R., Huber, W., and Gentry, J. Bioconductor software for graphs. *Bioinformatics and Computational Biology Solutions Using R and Bioconductor* (2005), 347–368.

[15] Carvalho, C., Chang, J., Lucas, J., Nevins, J., Wang, Q., and West, M. High-dimensional sparse factor modeling: applications in gene expression genomics. *Journal of the American Statistical Association 103*, 484 (2008), 1438–1456.

[16] Cheong, R., Hoffmann, A., and Levchenko, A. Understanding nf-κb signaling via mathematical modeling. *Molecular systems biology 4*, 1 (2008).

[17] Chickering, D., Geiger, D., and Heckerman, D. Learning bayesian networks is np-hard.

[18] Cho, W. Omics approaches in cancer research. *An Omics Perspective on Cancer Research* (2010), 1–9.

[19] Chu, J., Weiss, S., Carey, V., and Raby, B. A graphical model approach for inferring large-scale networks integrating gene expression and genetic polymorphism. *BMC systems biology 3*, 1 (2009), 55.

[20] Covert, M., Leung, T., Gaston, J., and Baltimore, D. Achieving stability of lipopolysaccharide-induced nf-κb activation. *Science 309*, 5742 (2005), 1854.

[21] D. R. Cox, N. W. *Multivariate Dependencies*. Chapman and Hall, 1996.

[22] D. Tritchler, E. Parkhomenko, J. B. Filtering genes for cluster and network analysis. *BMC Bioinformatics 10* (2009), 193.

[23] De Souto, M., Costa, I., De Araujo, D., Ludermir, T., and Schliep, A. Clustering cancer gene expression data: a comparative study. *BMC bioinformatics 9*, 1 (2008), 497.

[24] Demir, E., Cary, M., Paley, S., Fukuda, K., Lemer, C., Vastrik, I., Wu, G., D'Eustachio, P., Schaefer, C., Luciano, J., et al. The biopax community standard for pathway data sharing. *Nature biotechnology 28*, 9 (2010), 935–942.

[25] Duarte, N., Becker, S., Jamshidi, N., Thiele, I., Mo, M., Vo, T., Srivas, R., and Palsson, B. Global reconstruction of the human metabolic network based on genomic and bibliomic data. *Proceedings of the National Academy of Sciences 104*, 6 (2007), 1777.

[26] Edwards, D., De Abreu, G., and Labouriau, R. Selecting high-dimensional mixed graphical models using minimal aic or bic forests. *BMC bioinformatics 11*, 1 (2010), 18.

[27] Edwards, J., and Palsson, B. Metabolic flux balance analysis and the in silico analysis of escherichia coli k-12 gene deletions. *BMC bioinformatics 1*, 1 (2000), 1–1.

[28] Ellson, J., Gansner, E., Koutsofios, E., North, S., and Woodhull, G. Graphviz and dynagraphstatic and dynamic graph drawing tools. *Graph Drawing Software* (2004), 127–148.

[29] Ellson, J., Gansner, E., Koutsofios, L., North, S., and Woodhull, G. Graphviz: open source graph drawing tools. In *Graph Drawing* (2002), Springer, pp. 594–597.

[30] Erguler, K., and Stumpf, M. Practical limits for reverse engineering of dynamical systems: a statistical analysis of sensitivity and parameter inferability in systems biology models. *Mol. BioSyst.* (2011).

[31] Folger, O., Jerby, L., Frezza, C., Gottlieb, E., Ruppin, E., and Shlomi, T. Predicting selective drug targets in cancer through metabolic networks. *Molecular Systems Biology 7* (2011), 501.

[32] Forbes, S. A., Tang, G., Bindal, N., Bamford, S., Dawson, E., Cole, C., Kok, C. Y., Jia, M., Ewing, R., Menzies, A., Teague, J. W., Stratton, M. R., and Futreal, P. A. COSMIC (the Catalogue of Somatic Mutations in Cancer): a resource to investigate acquired mutations in human cancer. *Nucleic Acids Res. 38* (Jan 2010), D652–657. [PubMed Central:PMC2808858] [DOI:10.1093/nar/gkp995] [PubMed:19906727].

[33] Förster, J., Famili, I., Fu, P., Palsson, B., and Nielsen, J. Genome-scale reconstruction of the saccharomyces cerevisiae metabolic network. *Genome research 13*, 2 (2003), 244–253.

[34] Frezza, C., Zheng, L., Folger, O., Rajagopalan, K., MacKenzie, E., Jerby, L., Micaroni, M., Chaneton, B., Adam, J., Hedley, A., et al. Haem oxygenase is synthetically lethal with the tumour suppressor fumarate hydratase. *Nature 477*, 7363 (2011), 225–228.

[35] Friedman, A., Bellomo, N., and Maini, P. Mathematical analysis and challenges arising from models of tumor growth. *Mathematical Models and Methods in Applied Sciences 17* (2007), 1751–1772.

[36] Friedman, J., Hastie, T., and Tibshirani, R. Sparse inverse covariance estimation with the graphical lasso. *Biostatistics 9*, 3 (2008), 432–441.

[37] Frydenberg, M. The chain graph markov property. *Scandanavian Journal of Statistics 17* (1990), 333–353.

[38] Futreal, P. A., Coin, L., Marshall, M., Down, T., Hubbard, T., Wooster, R., Rahman, N., and Stratton, M. R. A census of human cancer genes. *Nat. Rev. Cancer 4* (Mar 2004), 177–183. [PubMed Central:PMC2665285] [DOI:10.1038/nrc1299] [PubMed:14993899].

[39] Gehlenborg, N., O'Donoghue, S., Baliga, N., Goesmann, A., Hibbs, M., Kitano, H., Kohlbacher, O., Neuweger, H., Schneider, R., Tenenbaum, D., et al. Visualization of omics data for systems biology. *Nature methods 7* (2010), S56–S68.

[40] Gentleman, R. C., Carey, V. J., Bates, D. M., and others. Bioconductor: Open software development for computational biology and bioinformatics. *Genome Biology 5* (2004), R80.

[41] Gerlowski, L., and Jain, R. Physiologically based pharmacokinetic modeling: principles and applications. *Journal of Pharmaceutical Sciences 72*, 10 (1983), 1103–1127.

[42] Gerstung, M., Baudis, M., Moch, H., and Beerenwinkel, N. Quantifying cancer progression with conjunctive bayesian networks. *Bioinformatics 25*, 21 (2009), 2809–2815.

[43] Geva-Zatorsky, N., Rosenfeld, N., Itzkovitz, S., Milo, R., Sigal, A., Dekel, E., Yarnitzky, T., Liron, Y., Polak, P., Lahav, G., et al. Oscillations and variability in the p53 system. *Molecular Systems Biology 2*, 1 (2006).

[44] Gevaert, O., De Smet, F., Timmerman, D., Moreau, Y., and De Moor, B. Predicting the prognosis of breast cancer by integrating clinical and microarray data with bayesian networks. *Bioinformatics 22*, 14 (2006), e184–e190.

[45] Goodman, L., Wintrobe, M., Dameshek, W., Goodman, M., Gilman, A., and McLennan, M. Nitrogen mustard therapy; use of methyl-bis (beta-chloroethyl) amine hydrochloride and tris (beta-chloroethyl) amine hydrochloride for hodgkin's disease, lymphosarcoma, leukemia and certain allied and miscellaneous disorders. *JAMA* (1946), 125–132.

[46] Gordon, A. D. *Classification*. Chapman & Hall /CRC, 1999.

[47] Hageman, R. S., Leduc, M. S., Korstanje, R., Paigen, B., and Churchill, G. A. A Bayesian framework for inference of the genotype-phenotype map for segregating populations. *Genetics 187* (2011), 1163–1170.

[48] Hammerman, P. S., Hayes, D. N., Wilkerson, M. D., ..., and Thomson, E. Comprehensive genomic characterization of squamous cell lung cancers. *Nature 489*, 7417 (Sep 2012), 519–525. [PubMed Central:PMC3466113] [DOI:10.1038/nature11404] [PubMed:22960745].

[49] Han Liu, Kathryn Roeder, L. W. Stability approach to regularization selection (stars) for high dimensional graphical models. *arXiv:1006.3316 [stat.ML]* (2010).

[50] Heiden, M. V. Targeting cancer metabolism: a therapeutic window opens. *Nature Reviews Drug Discovery 10* (2011), 671–684.

[51] Hoffmann, A., Levchenko, A., Scott, M., and Baltimore, D. The iκb-nf-κb signaling module: temporal control and selective gene activation. *Science 298*, 5596 (2002), 1241.

[52] Horvath, S. *Weighted Network Analysis*. Spinger, 2011.

[53] I. Tsamardinos, C. Aliferis, A. S. Algorithms for large scale markov blanket discovery. *In: Proceedings of the sixteenth international Florida artificial intelligence research society conference* (2003).

[54] Imoto, S., Higuchi, T., Goto, T., Tashiro, K., Kuhara, S., and Miyano, S. Combining microarrays and biological knowledge for estimating gene networks via Bayesian networks. *Journal of Bioinformatics and Computational Biology 2* (2004), 77–98.

[55] Jerby, L., and Ruppin, E. Predicting drug targets and biomarkers of cancer via genome-scale metabolic modeling. *Clinical Cancer Research 18*, 20 (2012), 5572–5584.

[56] Jerby, L., Wolf, L., Denkert, C., Stein, G., Hilvo, M., Oresic, M., Geiger, T., and Ruppin, E. Metabolic associations of reduced proliferation and oxidative stress in advanced breast cancer. *Cancer Research* (2012).

[57] Jiang, D. Cluster analysis for gene expression data: a survey. *IEEE Transactions on Knowledge and Data Engineering 16* (2004), 1370–86.

[58] Johnstone, R., Ruefli, A., and Lowe, S. Apoptosis:: A link between cancer genetics and chemotherapy. *Cell 108*, 2 (2002), 153–164.

[59] Kahn, C., Laur, J., and Carrera, G. A bayesian network for diagnosis of primary bone tumors. *Journal of Digital Imaging 14* (2001), 56–57.

[60] Kahn Jr, C., Roberts, L., Wang, K., Jenks, D., and Haddawy, P. Preliminary investigation of a bayesian network for mammographic diagnosis of breast cancer. In *Proceedings of the Annual Symposium on Computer Application in Medical Care* (1995), American Medical Informatics Association, p. 208.

[61] Kanehisa, M., Goto, S., Furumichi, M., Tanabe, M., and Hirakawa, M. Kegg for representation and analysis of molecular networks involving diseases and drugs. *Nucleic acids research 38*, suppl 1 (2010), D355–D360.

[62] Kanehisa, M., Goto, S., Sato, Y., Furumichi, M., and Tanabe, M. KEGG for integration and interpretation of large-scale molecular data sets. *Nucleic Acids Res. 40*, Database issue (Jan 2012), D109–114.

[63] Kitano, H., et al. Computational systems biology. *Nature 420*, 6912 (2002), 206–210.

[64] Koboldt, D. C., Fulton, R. S., McLellan, M. D., ..., and Palchik, J. D. Comprehensive molecular portraits of human breast tumours. *Nature 490*, 7418 (Oct 2012), 61–70. [PubMed Central:PMC3465532] [DOI:10.1038/nature11412] [PubMed:23000897].

[65] Lähdesmäki, H., Shmulevich, I., and Yli-Harja, O. On learning gene regulatory networks under the boolean network model. *Machine Learning 52*, 1 (2003), 147–167.

[66] Lauritzen, S. *Graphical Models*. Clarendon Press, 1996.

[67] Lautitzen, S. L. Mixed graphical association models. *Scandanavian Journal of Statistics 16* (1989), 273–306.

[68] Lee, J., Gianchandani, E., and Papin, J. Flux balance analysis in the era of metabolomics. *Briefings in Bioinformatics 7*, 2 (2006), 140–150.

[69] Li, Y., Tesson, B. M., Churchill, G. A., and Jansen, R. C. Critical reasoning on causal inference in genome-wide linkage and association studies. *Trends in Genet 26*, 12 (2010), 493–498.

[70] Livnat Jerby, T., and Ruppin, E. Computational reconstruction of tissue-specific metabolic models: application to human liver metabolism. *Molecular systems biology 6*, 1 (2010).

[71] Lord, C., and Ashworth, A. Biology-driven cancer drug development. *BMC Biology 8*, 38 (2010).

[72] Lu, W., Pelicano, H., and Huang, P. Cancer metabolism: Is glutamine sweeter than glucose? *Cancer cell 18*, 3 (2010), 199–200.

[73] MacLachlan, M., Sundnes, J., and Spiteri, R. A comparison of non-standard solvers for odes describing cellular reactions in the heart. *Computer Methods in Biomechanics and Biomedical Engineering 10*, 5 (2007), 317–326.

[74] Martin, G. Cell signaling and cancer. *Cancer Cell 4*, 3 (2003), 167–174.

[75] Matthews, L., Gopinath, G., Gillespie, M., Caudy, M., Croft, D., De Bono, B., Garapati, P., Hemish, J., Hermjakob, H., Jassal, B., et al. Reactome knowledgebase of human biological pathways and processes. *Nucleic acids research 37*, suppl 1 (2009), D619–D622.

[76] McLendon, R., Friedman, A., Bigner, D., Van Meir, E. G. ., Guyer, M., Ozenberger, B., Peterson, J., and Thomson, E. Comprehensive genomic characterization defines human glioblastoma genes and core pathways. *Nature 455* (Oct 2008), 1061–1068. [PubMed Central:PMC2671642] [DOI:10.1038/nature07385] [PubMed:18772890].

[77] Medes, G., Thomas, A., and Weinhouse, S. Metabolism of neoplastic tissue. iv. a study of lipid synthesis in neoplastic tissue slices in vitro. *Cancer research 13*, 1 (1953), 27.

[78] Miller, L. D., Smeds, J., George, J., Vega, V. B., Vergara, L., Ploner, A., Pawitan, Y., Hall, P., Klaar, S., Liu, E. T., and Bergh, J. An expression signature for p53 status in human breast cancer predicts mutation status, transcriptional effects, and patient survival. *Proc. Natl. Acad. Sci. U.S.A. 102*, 38 (Sep 2005), 13550–13555.

[79] Morris, M., Saez-Rodriguez, J., Sorger, P., and Lauffenburger, D. Logic-based models for the analysis of cell signaling networks. *Biochemistry 49*, 15 (2010), 3216–3224.

[80] Mukherjee, S., and Speed, T. Network inference using informative priors. *Proceedings of the National Academy of Sciences 105*, 38 (2008), 14313.

[81] N. Wermuth, S. L. L. On substantive research hypotheses, conditional ijdependence graphs and graphical chain models (with discussion). *Journal of the Royal Statistical Society B 52* (1990), 21–72.

[82] Neto, E., Ferrara, C., Attie, A., and Yandell, B. Inferring causal phenotype networks from segregating populations. *Genetics 179* (2008), 1089–1100.

[83] Neto, E., Keller, M., Attie, A., and Yandell, B. Causal graphical models in systems genetics: a unified framework for joint inference of causal network and genetic architecture for correlated phenotypes. *Annals of Applied Statistics 4*, 1 (2010), 320–339.

[84] Nigro, J., Misra, A., Zhang, L., Smirnov, I., Colman, H., Griffin, C., Ozburn, N., Chen, M., Pan, E., Koul, D., et al. Integrated array-comparative genomic hybridization and expression array profiles identify clinically relevant molecular subtypes of glioblastoma. *Cancer research 65*, 5 (2005), 1678.

[85] Oberhardt, M., Palsson, B., and Papin, J. Applications of genome-scale metabolic reconstructions. *Molecular Systems Biology 5*, 320 (2009).

[86] Ogata, H., Goto, S., Sato, K., Fujibuchi, W., Bono, H., and Kanehisa, M. Kegg: Kyoto encyclopedia of genes and genomes. *Nucleic acids research 27*, 1 (1999), 29.

[87] Ohta, M., Tokuda, Y., Suzuki, Y., Kubota, M., Makuuchi, H., Tajima, T., Nasu, S., Yasuda, S., and Shohtsu, A. Whole body pet for the evaluation of bony metastases in patients with breast cancer: comparison with 99tcm-mdp bone scintigraphy. *Nuclear medicine communications 22*, 8 (2001), 875–879.

[88] Ookhtens, M., Kannan, R., Lyon, I., and Baker, N. Liver and adipose tissue contributions to newly formed fatty acids in an ascites tumor. *American Journal of Physiology-Regulatory, Integrative and Comparative Physiology 247*, 1 (1984), R146–R153.

[89] Orton, R., Sturm, O., Vyshemirsky, V., Calder, M., Gilbert, D., and Kolch, W. Computational modelling of the receptor-tyrosine-kinase-activated mapk pathway. *Biochemical Journal 392*, Pt 2 (2005), 249.

[90] P. Buhlmann, S. v. d. G. *Statistics for High-Dimensional Data*. Springer, 2011.

[91] P. Spirtes, C. G. An algorithm for fast recovery of sparse causal graphs. *Social Science Computing Review 9* (1991), 62–72.

[92] Pavlopoulos, G., Wegener, A., Schneider, R., et al. A survey of visualization tools for biological network analysis. *Bio-data mining 1* (2008), 12.

[93] Pearl, J. *Probabilistic Reasoning in Intelligent Systems: Networks of Plausible Inference*. Morgan Kaufmann, 1988.

[94] Pearl, J. *Probalistic Reasoning in Intelligent Systems: Networks of Plausible Inference.* San Francisco: Morgan Kaufmann, 1988.

[95] Piras, V., Hayashi, K., Tomita, M., and Selvarajoo, K. Enhancing apoptosis in trail-resistant cancer cells using fundamental response rules. *Scientific reports 1* (2011).

[96] R Development Core Team. *R: A Language and Environment for Statistical Computing.* R Foundation for Statistical Computing, Vienna, Austria, 2011. ISBN 3-900051-07-0.

[97] Reitz, R., Mendrala, A., Corley, R., Quast, J., Gargas, M., Andersen, M., Staats, D., and Conolly, R. Estimating the risk of liver cancer associated with human exposures to chloroform using physiologically based pharmacokinetic modeling. *Toxicology and applied pharmacology 105*, 3 (1990), 443–459.

[98] Remington, D. L. Effects of genetic and environmental factors on trait network predictions from quantitative trait locus data. *Genetics 181* (2009), 1087–1099.

[99] Ribba, B., Saut, O., Colin, T., Bresch, D., Grenier, E., and Boissel, J. A multiscale mathematical model of avascular tumor growth to investigate the therapeutic benefit of anti-invasive agents. *Journal of theoretical biology 243*, 4 (2006), 532–541.

[100] S. Hojsgaard, D. Edwards, S. L. *Graphical Models with R.* Springer, 2012.

[101] Sanga, S., Sinek, J., Frieboes, H., Ferrari, M., Fruehauf, J., and Cristini, V. Mathematical modeling of cancer progression and response to chemotherapy. *Expert review of anticancer therapy 6*, 10 (2006), 1361–1376.

[102] Sean, D., and Meltzer, P. S. GEOquery: a bridge between the Gene Expression Omnibus (GEO) and BioConductor. *Bioinformatics 23* (Jul 2007), 1846–1847. [DOI:10.1093/bioinformatics/btm254] [PubMed:17496320].

[103] Selvarajoo, K. Macroscopic law of conservation revealed in the population dynamics of toll-like receptor signaling. *Cell Commun. Signal 9*, 9 (2011).

[104] Selvarasu, S., Karimi, I., Ghim, G., and Lee, D. Genome-scale modeling and in silico analysis of mouse cell metabolic network. *Mol. BioSyst. 6*, 1 (2009), 152–161.

[105] Shampine, L., Gladwell, I., and Thompson, S. *Solving ODEs with MATLAB.* Cambridge Univ Pr, 2003.

[106] Shepherd, R., Forbes, S. A., Beare, D., Bamford, S., Cole, C. G., Ward, S., Bindal, N., Gunasekaran, P., Jia, M., Kok, C. Y., Leung, K., Menzies, A., Butler, A. P., Teague, J. W., Campbell, P. J., Stratton, M. R., and Futreal, P. A. Data mining using the Catalogue of Somatic Mutations in Cancer BioMart. *Database (Oxford) 2011* (2011), bar018. [PubMed Central:PMC3263736] [DOI:10.1093/database/bar018] [PubMed:21609966].

[107] Shlomi, T., Benyamini, T., Gottlieb, E., Sharan, R., and Ruppin, E. Genome-scale metabolic modeling elucidates the role of proliferative adaptation in causing the warburg effect. *PLoS Computational Biology 7*, 3 (2011), e1002018.

[108] Sleeman, B., and Levine, H. Partial differential equations of chemotaxis and angiogenesis. *Mathematical methods in the applied sciences 24*, 6 (2001), 405–426.

[109] Sullivan, D., Gabbard Jr, J., Shukla, M., and Sobral, B. Data integration for dynamic and sustainable systems biology resources: challenges and lessons learned. *Chemistry & biodiversity 7*, 5 (2010), 1124–1141.

[110] Thangue, N. L., and Kerr, D. Predictive biomarkers: a paradigm shift towards personalized cancer medicine. *Nature Reviews Clinical Oncology 8* (2011), 587–596.

[111] Varmus, H., et al. Science funding: Provocative questions in cancer research. *Nature 481*, 7382 (2012), 436–437.

[112] Wagner, G., Pavlicev, M., and Cheverud, J. The road to modularity. *Nature Reviews Genetics 8*, 12 (2007), 921–931.

[113] Warburg, O., et al. On the origin of cancer cells. *Science 123*, 3191 (1956), 309–314.

[114] Watts, D., and Strogatz, S. Collective dynamics of small-worldnetworks. *nature 393*, 6684 (1998), 440–442.

[115] Werhli, A. V., and Husmeier, D. Reconstructing gene regulatory networks with bayesian networks by combining expression data with multiple sources of prior knowledge. *Stat Appl Genet Mol Biol 6* (2007), Article15.

[116] Werner, S., Barken, D., and Hoffmann, A. Stimulus specificity of gene expression programs determined by temporal control of ikk activity. *Science 309*, 5742 (2005), 1857.

[117] Whittaker, J. *Graphical Models in Applied Multivariate Statistics*. Wiley, 1990.

[118] Wilson, T. J., and Ge, S. X. Arraysearch: A web-based genomic search engine.

[119] Zhang, J., and Wiemann, S. Kegggraph: a graph approach to kegg pathway in r and bioconductor. *Bioinformatics 25*, 11 (2009), 1470–1471.

[120] Zhukov, N., and Tjulandin, S. Targeted therapy in the treatment of solid tumors: practice contradicts theory. *Biochemistry 73*, 5 (2007), 605–618.

In: Recent Advances in Systems Biology Research
Editors: A. X. C. N. Valente, A. Sarkar and Y. Gao
ISBN: 978-1-62948-736-6

Chapter 11

Systems Biology Approach for Human Complexity and Personalized Cardiology

George E. Louridas*[*] *and Katerina G. Louridas
Thessaloniki, Greece

Abstract

Purpose of Review

Heart failure and atherosclerotic coronary artery disease constitute major health problems with significant morbidity and mortality. Heart diseases are complex and self-organized biological entities, which can be approached by the holistic concept of systems biology. The methodology of systems biology gives the opportunity to work on the emerging field of personalized cardiology that is defined by genomic and proteomic advances.

Recent Findings

There are significant advances in molecular biology and adequate data from the fields of genomics and proteomics to define biological pathways, networks and functional modules related to clinical entities. Systems biology approach to cardiac diseases, like heart failure and atherosclerosis, delineates pathology, diagnosis, progression and treatment. Such an approach is based in three emerging disciplines, omics, modeling and complex networks, and in two directions of systems biology that involves functional composition (bottom-up direction) and decomposition (top-down direction).

* Corresponding author at: Dimokratias 69, Thessaloniki, 552 36, Greece. Tel:+30 2310347049; fax:+30 2310346695; Email address: louridasg@gmail.com

Summary

The present review embraces the progress made in genomic and proteomic data integrating them with clinical cardiovascular information thus advancing the idea of a more personalized cardiology.

Keywords: Systems biology, personalized cardiology, heart failure, coronary artery disease

1. Introduction

Systems biology is a new discipline that emerged to explain the complexities of human physiological status and disease partly with functional delineation of genes and molecules but mainly by reconstructing biochemical supramolecular networks, modules and phenotypes. The concept of systems biology considers the biological networks as 'systems' of cooperating and interacting components with hierarchical structuring [1]. The integration of these 'systems' produces complex functional behaviors in all biological levels. Systems biology is the study of the structure and function of cells and whole organisms, rather than the analysis of the cellular components or individual parts of an organism [2]. Thus, systems biology approach describes, explains and correlates the integrated processes of various biological networks in all levels of biological structure from cells to the whole organism based in three emerging disciplines, omics, modeling, and complex networks. In recent years, two directions of systems biology are developed, the 'bottom-up' (functional composition) and the 'top-down' (functional decomposition) direction. The 'bottom-up' systems biology direction examines the mechanisms through which functional properties arise and leans on computational modeling based on the hierarchical network construction of the various biological processes. The 'top-down' systems biology direction depends on –omics (genomics, proteomics, transcriptomics, metabolomics) data in order to decompose biological processes, identify molecular interaction networks and explain complex biological networks or modules involved in the genesis of clinical phenotypes, or to define a more individualized therapy [3,4]. (Figure 1).

Heart failure (HF) and atherosclerotic coronary artery disease (CAD) constitute major human health problems with significant morbidity and mortality. Heart diseases are complex and self-organized biological entities with dynamic interactions in the level of molecular pathways, and between different biological networks and modular functional units. Therefore, heart diseases as complex systems are dynamic phenomena with top-down and bottom-up hierarchical interactions across different spatial levels [3]. The clinical phenotypes of heart diseases represent the interacting emergent properties of the complex and distorted biological systems that are involved in the disease process. In systems biology, the term 'emergence' is addressed to the appearance of holistic properties in complex biological systems from simpler interacting subsystems. The new properties of the complex biological systems are not additive or equal to the sum of the properties of the biological subsystems. The new unpredicted properties that emerge from the combination of the lower-level units are called emergent properties [5].

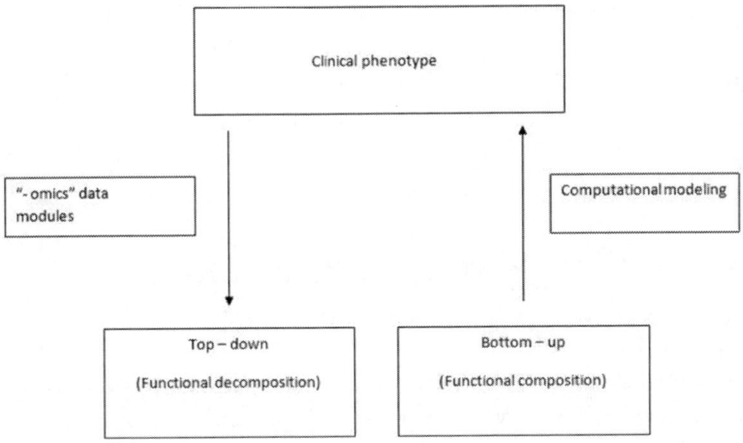

Figure 1. The top-down direction is based on "-omics" data and tries to explain complex networks or modules. The bottom-up direction examines the mechanisms for functional properties with the help of computational modeling.

Personalized cardiology is a new challenge for systems biology because of the increased complexity and interactions between abnormally regulated biological networks. This approach, intends to modify the practice of cardiology producing a more predictive, preventive and individualized cardiology [6]. The integrative methodology of systems biology applied to the complex syndrome of heart failure is able to incorporate dynamic data into networks in order to explain the adaptive mechanisms in heart failure [7]. Also, the understanding of the complex atherosclerotic cardiovascular disease advances with the application of genomics, proteomics and metabolomics and allows personalized diagnostic and therapeutic approaches [8]. Developments in individual genetic profiling and predisposition to cardiac disease open new opportunities to medical practitioners and promise additional possibilities to the patients with heart diseases. Personalized cardiology refers to the individualized diagnosis and treatment according to special characteristics of each patient. The categorization of the patients in small subgroups is rested on the concept of patients' susceptibility to the disease and the response to the standard therapy. Scientists from different fields are cooperating to transfer knowledge from biology to clinical practice while recent research in the genetic and cellular function of cardiac pathology put together researchers with medical practitioners.

Over the last 20 years the advances in the fields of genomics, proteomics and metabolomics produced a new reality in explaining human complexity and variability. In addition to the improvement in medical management, the new developments in genetics, proteomics and metabolomics introduce changes in medical ethics and cause ethical and legal dilemmas for society and the medical profession [9]. Personal genetic testing and identification of the genetic code is based on the individual's will for genetic information on possible inherited genes that are specific to a heart disease. The genetic code identification provokes ethical questions about who has access and owns the personal genetic information, and if the genetic information is a private matter or is a social responsibility. Also, medical professionals and the public should be prepared to function with the new scientific reality of personalized cardiology and genetic information. Personalized medical treatment will change pharmaceutical manufacturing as the new tailored drugs would be addressed to specific groups of people with a high manufacturing drug cost. Probably, in the first application, a personalized healthcare system would be addressed to affluent societies, but later on will be extended to the whole global population. In personalized medicine, the philosophy and the ethical guidelines for initiating a clinical trial are affected as each patient is converted into a clinical trial of one with a shared responsibility between the patient and the researchers [10].

2. Human Complexity

Complex system is considered a structure with many degrees of freedom, or a system with many interactions between included diverse components [11]. In human physiology, complexity and variability are poorly understood parameters with significant implications for a patient's management. The medical practitioner is called to give answers to complex human diseases caused by a combination of genetic and environmental parameters. Systems biology approach is entangled with the pathophysiological research of various human disorders including metabolic and cardiovascular diseases [12]. Systems biology analysis constructs phenotypes of cardiovascular diseases predicting genesis, progression and management. Human heart diseases are complex processes involving many interconnected elements with unpredictable behavior. To understand the effect of systems biology on human heart diseases and personalized cardiology, it is vital to define the importance of human complexity and variability. Human complexity is determined by the 35 thousand genes of the human genome, the 1-10 million proteins of the human proteome and the 2-3 thousand small molecules named metabolites that reside in the human cells or tissues [13]. The impact of an allele varies according to the macromolecular level that its effect is exercised: an allele is encoded as DNA, transcribed into RNA and translated into protein. Therefore, the effect of an allele is studied with DNA sequence, RNA expression and protein expression [14]. In human variability are involved various genetic differences, like gene mutation, genetic drift, and translational or post-translational modifications, environmental parameters and an unpredictable diversity of molecular processes. Coronary artery disease is a multifactorial disease characterized by both genetics and environmental factors [15]. Environmental factors like diet and lifestyle contribute to CAD onset and progression. Systems and network biology has the capacity to increase our knowledge of nutrition's influence on metabolic pathways and homeostasis, and to disclose the relationship between genes, dietary habits and human health [16]. The complexity of foods, culture and lifestyle, and the variety of metabolic processes involved in a disease, are critical factors for personalizing dietary advice of people with chronic diseases [17]. In the USA, there has been a significant decline in the cardiovascular mortality from 632 to 223 deaths per 100,000 people during the last 40 years. This decline has been primary credited to changes in lifestyle and diet and less to new therapeutic management [13]. Therefore, the development of individualized cardiology with specific drugs addressed to subgroups of people probably will increase the number of the patients reacting favorably to innovative therapeutic drug management. The systems biology proposes a method with significant impact in defining biochemical pathways and networks implicated in health and disease.

Recently, the Encyclopedia of DNA Elements (ENCODE) project has systematically mapped in the human genome, regions of transcription, transcription factor association, chromatin structure and histone modification [18]. ENCODE is designed to collect data from the uncharted non-coding RNA molecule area of the human genome which was excluded by the Human Genome Project. Thus, the ENCODE project will enquire about 80% of the genome, including 70,000 'promoter' regions that control gene expression and 400,000 'enhancer' regions that regulate expression of distant genes [19]. The data collected from this uncharted area of the human genome will reveal the biological complexity and the way that a few protein coding genes are regulating the human body. This project will identify elements related to sequence variants linked to human disease and provide information for further biomedical research [18].

3. Personalized Cardiology

Systems biology emerged from classical molecular biology and is characterized by a more holistic approach to modern biology than molecular biology [20]. The classical diagnostic and therapeutic approach to human disease is based on reductionist principles of experimentation and clinical analysis after the onset of a disease. This reductionist approach requires replacement by the more holistic approach of the systems biology [21]. The Personalized Medicine Coalition describes personalized medicine as "Personalized medicine is the ability to determine an individual's unique molecular characteristics...to diagnose...select treatments...reduce possible adverse reactions....is the ability to predict an individual's susceptibility to diseases" [22]. Therefore, personalized cardiology is a predictive, preventive, diagnostic and therapeutic application of the holistic methodology of systems biology, based on large data stores from molecular medicine, bioinformatics and complex biological networks (Figure 2). To prevent a disease, personalized cardiology needs prior knowledge at the molecular level before the onset of the disease. The success of modern biotechnology and biological research to clarify cellular complexities, to model various biological entities and to produce experimental validations, motivated the research in the field of human cardiovascular disease. The general name 'omics' encompasses the new disciplines approaching genes, proteins, RNA, and metabolites. Thus, the present article examines the progress made in the fields of genomics (genes), epigenetics, proteomics (proteins), transcriptomics (RNA) and metabolomics (metabolites), in advancing a more individualized approach to clinical cardiovascular diseases. The clinical cardiovascular practice is reconsidered due to the complicated nature of genomic and environmental risks, while the personalized approach is refined particularly in complex entities, like CHD and HF. In HF the various clinical phenotypes with complicated hemodynamics, fluid retention and neurohormonal adaptations, needs a more sophisticated methodology in clinical screening and novel technologies to predict a tailored therapeutic approach. Cardiac biomarkers are correlated with disease states and have an important role in the application of a personalized healthcare system.

Recently the advances in genome programming produced a new scientific field of stem cell technologies which can create, in "petri dish", models of patients' diseases. This technique of "disease in a dish" permits to take patients' own cells and with quantitative microscopy to measure the cellular defect if a disease that affects those cells is present [23].

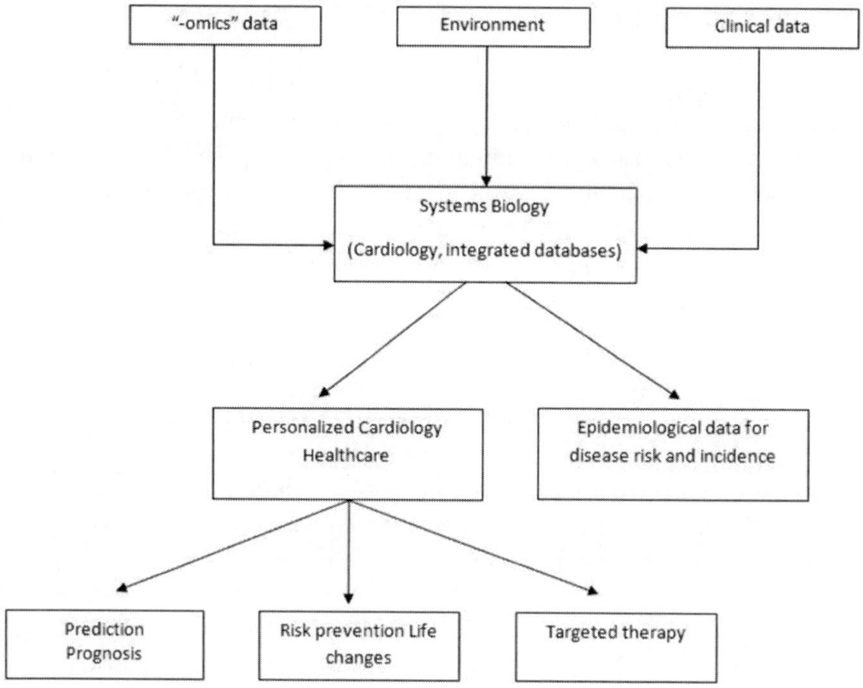

Figure 2. Interdependence between systems biology, epidemiology and personalized cardiology.

3.1. Cardiac Biomarkers

Cardiac biomarkers are indicators of normal biological or pathogenic processes with an increased role in disease management, drug discovery and pharmacological responses to therapeutic intervention. A cardiac biomarker is a protein molecule that is related to the underlying disease and provides important information for diagnosis, prognosis and follow up in a patient with progressive clinical condition. The first cardiac specific biomarkers used in acute myocardial infarction (AMI) were the creatine kinase-MB (CK-MB), first introduced in 1972, and the cardiac troponin (cTn), first introduced in 1989. The CK-MB together with aspartate transaminase and lactate dehydrogenase were the original cardiac biomarkers for the diagnosis of AMI. The original application of cardiac biomarkers to the diagnosis of acute cardiac events (AMI) has been expanded to include further applications for staging of cardiac disease, timing of cardiac events and prognostication [24]. The brain natriuretic peptide (BNP) and NT-proBNP first application in clinical practice was addressed primary to the diagnosis of HF in the emergency department and to the prognosis of HF in an outpatients program. Nevertheless, the role of the natriuretic peptides (NPs) was proved minor in the management of critically ill patients under intensive care [25]. The cardiac biomarkers play an important role in personalized medicine as they improve the therapeutic and diagnostic processes. Moreover, the development of personalized cardiology is based on molecular diagnostic technology used for biomarkers discovery, and is related to pharmacogenomics and pharmacoproteomics. The cardiac biomarkers are distinguished in three groups: Inflammatory biomarkers, biomarkers from myocardial injury and biomarkers from myocardial stress (Figure 3).

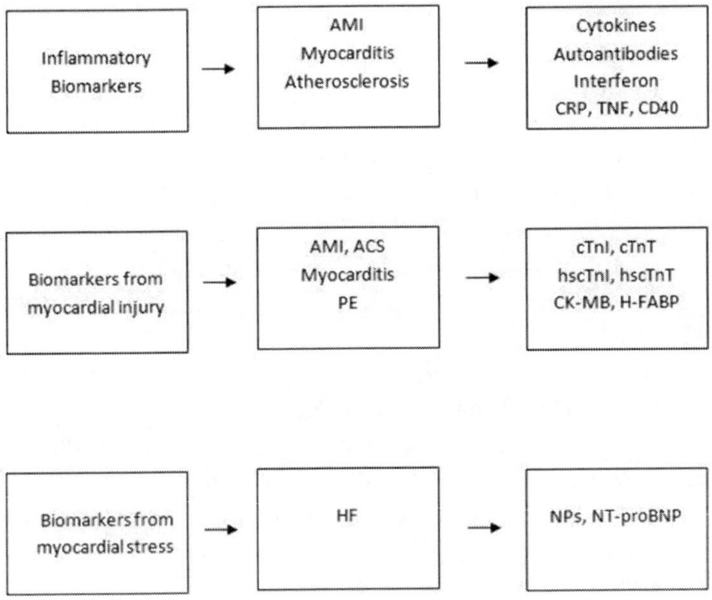

Figure 3. Current cardiac biomarkers used in clinical practice. ACS= acute coronary syndromes, AMI= acute myocardial infarction, PE= pulmonary embolism, HF= heart failure.

Cardiac Inflammatory Biomarkers

Patients surviving of an AMI enter the process of infarct healing which is a highly complex cascade of events [26]. In AMI, the inflammatory response phase starts 12-16 hours after the ischemic onset and triggers off self-repair and protection, while pathological autoimmune response leads to ventricular remodeling and HF [27]. During the immune response a number of immune mediators like cytokines, autoantibodies to myosin and tropomyosin, and interferon (INFγ) are mobilized. Despite that, clinically useful circulating inflammatory biomarkers for diagnosis and prognosis of AMI are not yet available [27].

The present knowledge of the cellular and molecular pathogenesis of post-viral and autoimmune myocarditis is based on animal models. In the animal model, the acute injury (myocyte damage) exposes the cardiac myosin and activates the innate immune system, resulting in a few weeks time in a robust inflammation mediated by T lymphocytes and antibodies. In humans, the progression of a viral myocardial infection to a dilated cardiomyopathy reflects the above inflammatory processes [28]. Despite the above experimental findings in post-viral myocarditis clinically useful inflammatory biomarkers are at the moment not available.

In atherosclerosis the contribution of the inflammation to the atherothrombotic process gives us a new insight into the mechanisms of the disease. In the atherosclerotic vascular inflammation are involved inflammatory cytokines that activate the generation of endothelial adhesion molecules, proteases and other mediators. In the inflammatory biomarkers discovered for cardiovascular risk prediction and stratification are included: CRP, tumor necrosis factor (TNF)-a, CD40 ligand, adiponectin, interleukin 18, and matrix metalloproteinase 9 [29]. There is a potential of the above inflammatory biomarkers to predict primary cardiovascular events and to play a significant role in the stratification and follow-up of patients with progressive atherothrombosis.

Biomarkers from Myocardial Injury

An acute myocardial injury involves an increase of cTn emerging 2-4 hours after symptoms initiation and remaining elevated for the next 7-14 days. The recent redefined criteria for an AMI are based on the diagnostic importance of the cTn as the central focus of this adjustment [30]. The elevated cTn has become an integral part in the diagnosis of AMI and is associated with poor prognosis in patients with acute ischemia [31]. The fundamental role of cTn as a biomarker expanded to include patients with acute coronary syndrome (ASC) or patients with secondary myocardial damage from other clinical situations like post-viral myocarditis and pulmonary embolism. There is greater sensitivity and specificity of cTn relative to CK-MB values. Therefore, are recent recommendations that cTn should replace CK-MB as the diagnostic standard on the prevalence of acute non-ST-elevation myocardial infarction (NSTEMI) [32]. This change to a cTn standard will increase the number of patients diagnosed with myocardial infarction with important clinical and health care implications. Consequently, the criteria to diagnose an AMI are the change of cTn blood levels and the clinical picture together with ECG or imaging demonstration of myocardial loss [33].

The troponin is a complex of proteins that are involved in the contraction of skeletal and cardiac muscle. The troponin consists of three subunits, TnT (binds to tropomyosin), TnI (binds to actin), and TnC (binds to calcium ions). Only the TnI (produced exclusively in the heart) and TnT (some found in the skeletal muscle) are clinically related to the heart. Commonly, in clinical practice, the increased levels of cardiac troponin I (cTnI) and cardiac troponin T (cTnT) in peripheral blood indicate necrotic myocardium. Despite this prevailing opinion that the detected increase in cardiac troponins is abnormal and caused by cardiac necrosis, it has been demonstrated that measurable amounts of troponin are present in the blood of healthy individuals. Thus, the cTnI and cTnT isoforms are cardiac structural proteins present in the blood of healthy adults and in patients with cardiac damage. The cTns can be released from cardiomyocytes of healthy adults due to the process of the 'physiological renewal' of the human myocardium and probably intensified by physical exercise or old age [34]. The application of the high-sensitivity cTn (hscTn), succeeding the fourth generation cTn assays, increases the diagnostic potential. The new generation of high sensitive (hs) cTnI and cTnT immunoassays are characterized by detection limits that have a profound effect on clinical practice. High-sensitive cTn assays are differentiated from common cTn assays by detecting of cTn in healthy persons and defining exactly of what is 'normal' (the 99th percentile). Therefore, 'detectable' troponin levels are considered to be normal and should be differentiated from 'elevated' levels [35]. Recent multicentre studies demonstrate that hs-cTn assays increase the early diagnosis of AMI [36]. The admission hs-cTnT assay, compared with the standard fourth generation cTnT, enlarged the NSTEMI detection from 61.5% on presentation to 100% within 6 hours, while in the whole myocardial infarction diagnoses increased by 34.6% [37].

The CK-MB levels are increased within 2-4 hours after an acute myocardial injury and their values are restored to normal in 36-72 hours [38]. To detect myocardial damage, the cTnI is a more specific and sensitive cardiac biomarker than CK-MB, because of low CK-MB specificity. In an analysis of 1719 consecutive patients with suspected ACS, was found that the 10% of those had an elevated CK-MB but normal cTn [39]. Thus, the CK-MB, previously the most important cardiac biomarker, has been replaced in clinical practice by the cTn which is more effective for myocardial damage detection.

The heart-type fatty acid binding protein (H-FABP) is located in the cytoplasm and it is a sensitive biomarker of myocardial damage released by both ischemia and necrosis [24]. In the first three hours of AMI, the diagnostic sensitivity of H-FABP is 93.1%, higher than CK-MB and cTn [40]. This advantage due to the rapid H-FABP release from the myocardial cell probably will be overwhelmed by the new methodology of hs-cTn.

Biomarkers from Myocardial Stress

The heart is a multifunctional and interactive organ with a cardiac endocrine function being parallel to the contractile function. The cardiac endocrine function is an essential component of the integrated systems of the body with a significant role for cardiovascular and renal systems regulation and homeostasis [41]. In the early stages of heart failure, the increased plasma levels of NP_s are important to maintain renal sodium excretion and sodium balance, and consequently for volume and pressure homeostasis [42].

Some of the natriuretic peptides (NPs) are true cardiac natriuretic hormones (CNHs) that are produced and secreted by atrial or ventricular cardiomyocytes. The natriuretic peptide family includes many peptides with their receptors, the atrial natriuretic peptide (ANP), the brain natriuretic peptide (BNP), the C-type natriuretic peptide (CNP), dendroaspis natriuretic peptide (DNP), and urodilatin. The biological processes of NPs and endothelins are closely linked up, with ANP and BNP inhibiting endothelin-1 (ET-1) production and ET-1 activating NPs synthesis [43]. The NPs are synthesized from polypeptide precursors with most prominent the pre-prohormone BNP. The pre-prohormone BNP is released by ventricular myocytes and cleaved into the biologically active BNP and the inactive NT-proBNP with wide clinical application. The earlier research involved ANP and later on the CNP, DNP and urodilatin, but recently the interest is shifted to the clinical application of BNP and NT-proBNP.

There are data supporting the hypothesis that the production and secretion of NPs is the consequence of complex integration of mechanical, chemical, hemodynamic, humoral, ischemic, and inflammatory effects [44]. The functional status of the NPs depends not only on the production and secretion by the cardiomyocytes, but also on the peripheral activation of circulating inactive precursors of NPs and the activation of guanylyl cyclase membrane receptors (signal transducers) for the transduction of the NPs signal (signal molecules) into intracellular signals [44]. The NPs induce natriuresis, diuresis, vasorelaxation, aldosterone antagonism, decrease of the sympathetic outflow, inhibition of vasopressin release and reduction of cardiac preload [45,46]. The NPs are assumed to counteract hypertension and blood volume expansion with consecutive hyperaldosteronism, in patients with heart failure, hypervolemia, renal failure, and hypertension.

The ANP is a cardiac hormone implicated in the homeostasis of blood volume and arterial blood pressure. The ANP is produced primarily in the cardiac atria, and the dominant stimulus for its release is the increased atrial wall tension produced by the increased intravascular volume [47]. De Bold et al., [48], demonstrated that intravenous injection of rat atrial myocardial extract in anesthetized bioassay rats produced a rapid and potent natriuretic response with hypotension and diuresis. The natriuresis was explained with the presence of an atrial substance that inhibited the sodium chloride transport in the renal tubule. Also, in an experimental study in isolated rat atria preparation, atrial stretch resulted in an immediate and significant increase in ANP release [49]. Further experiments strengthened the atrial stretch speculation for ANP release. Lang et al., have described in isolated rat heart a specific radioimmunoassay for ANP, and its stimulation and secretion on volume loading [50]. In children with chronic renal failure the plasma concentrations of ANP were correlated positively with volume status [51]. In patients with heart failure the ANP levels were raised compared to normal people while the highest plasma ANP concentrations were found in those with more severe heart failure [52]. Therefore, the raised levels of ANP are connected with the volume status and atrial distension.

The BNP originally identified in extracts of porcine brain. In humans, the BNP is present in the brain and it is secreted by cardiac ventricles after increased ventricular wall stress [53]. The BNP exhibits natriuretic and hypotensive effects similar to ANP. In patients with dilated cardiomyopathy, the secretions of ANP and BNP from the left ventricle are regulated mainly by the wall tension of the left ventricle, and the peripheral plasma level of these hormones may be used as a measure of left ventricular dysfunction [54]. It is unclear to what extent the cellular stretch or wall stress contribute to increased serum BNP concentration, but it appears that cellular stretch is a major trigger for BNP release [55]. In symptomatic patients with aortic valve stenosis, the BNP release is controlled by systolic and diastolic load, justifying the possibility that the myocardial stretch modulates BNP release in human pressure overload hypertrophy/failure [56]. In the Breathing Not Properly Multinational Study was demonstrated that a serum BNP value of 100pg/ml had a sensitivity of 86% to detect abnormal diastolic dysfunction and to discriminate patients with HF from those without HF [57]. In a prospective study of 600 patients with dyspnea, the increased levels of NT-proBNP (>450pg/ml for patients <50 years of age and >900pg/ml for patients >or=50 years of age) were highly sensitive and specific for the diagnosis of acute HF [58].

The CNP is found in endothelial and vascular smooth muscle cells, and in central nervous system [59]. The CNP has hypotensive properties similar to ANP and BNP, but is deprived of significant natriuretic activity [60].

Novel Cardiac Biomarkers

The strategy to discover and develop novel biomarkers in cardiovascular diseases is addressed in both experimental and clinical field. Extended clinical studies and proteomics technologies are valuable tools to identify new candidate biomarkers specific for coronary artery disease and heart failure. Already, the biomarker development has been enlarged and diversified to include cardiac biomarkers for inflammatory processes (C-reactive protein-CRP), acute muscle injury (cTn, CK-MB), and myocardial wall stress/stretch in HF patients - due to pressure or volume overload (BNP/NT-proBNP, ANP) [24]. Genetic and acquired animal models of HF have been used and their data were integrated with other information contained in tissue analysis from patients or explanted hearts with advanced form of HF [61]. In the new realm of personalized cardiology, the association of genetic haplotype information with proteomic biomarkers and clinical phenotypes leads to a more precise and individualized diagnosis and therapy [62]. The field of clinical proteomics can identify new disease biomarkers in body fluids, cells and tissues. New protein array formats and spectrometry technologies have a potential for use in clinical diagnostics [63]. Novel cardiovascular biomarkers are evaluated by the pharmaceutical industry for their screening, predictive, diagnostic, prognostic and therapeutic value. Recently, a large number of circulating plasma substances is under consideration for their potential utility as cardiac biomarkers. In the near future will be brought in clinical practice novel biomarkers like high sensitive cardiac troponin (hscTn), new natriuretic peptides (NPs), copeptin, choline and lipoprotein-associated phospholipase A2 (LP-PLA2) [6,64]. In patients after percutaneous coronary interventions (PCI) or ACS, have been identified new biomarkers with high clinical sensitivity and specificity. The minor allele of rs5065 ANP gene variant is associated with increased susceptibility to ACS with unfavorable prognostic value in patients with coronary disease [65]. The systematic screening for ANP gene variants probably will improve the individual cardiovascular risk prediction and outline a specific tailored pharmacological treatment for the patients carrying rs5065 minor allele. New predictive biomarkers for type 2 diabetes and cardiovascular disease have been explored with little success. This research was addressed to the subclinical disease and to the biochemical pathways of insulin resistance and atherosclerosis, but a further assessment requires a more sophisticated and integrated approach using novel "-omics" technologies [66].

A novel class of biomarkers are the circulating microRNAs (miRNAs) which are involved in many biological processes and have an important role to play in the regulation of the cardiovascular system (see transcriptomics session). In AMI, novel miRNAs biomarkers of cardiac damage are examined that up-regulate the miR-1, -133a, -133b, and -499-5p plasma levels, both in humans and mice [67]. The elevated cardiac-specific miR-208a in plasma is another novel biomarker for early detection of myocardial injury in humans[68].

3.2. Genomics and Proteomics

Genomic and proteomic advances challenge the conventional cardiology and provide insights to genesis and progression of a variety of heart diseases. Monogenic transmission has been reported from a miscellany of cardiovascular diseases, such as hypertrophic cardiomyopathy [69], long QT syndrome [70], Marfan syndrome [71], and arrhythmogenic right ventricular cardiomyopathy [72]. In complex cardiovascular diseases, without monogenic transmission, there is a genetic rationale with interconnection and feedback mechanisms between various genes, and reciprocal interplay between epigenetic superstructure and environmental changes. In those complex heterogeneous cardiovascular conditions are included HF, myocardial infarction and atherosclerosis. The genetic relationship between genes and clinical phenotype is not direct, but involves proteomic and transcriptomic data for a more complete picture in order to explain the epigenetic and environmental influences [61]. The risk for cardiovascular diseases is increased with genetic variability at the 9p21 chromosome locus which is related to non-coding RNA gene. About 21% of general population is homozygous for the 9p21 variant and the estimated risk of myocardial infarction is 1.64 times greater than that of noncarriers. Genetic variability on 9p21 and its associated loci could increase the risk for cardiovascular diseases and myocardial infarction about 30% [73, 74]. Furthermore, the 9p21 and its associated loci are linked up with stroke. The genetic variation at the CDKN2A/CDKN2B locus on chromosome 9p21 predicts stroke in hypertensive patients, and is independent of classical cardiovascular risk factors [75]. Also, the CDKN2A locus is a new candidate gene potentially contributing to vascular dementia and Alzheimer's disease [76].

Pharmacotherapy is the main scientific field of clinical implementation of new genomic and proteomic technologies. Pharmacological targeted therapy is addressed to the modern genomic and proteomic methodology for the identification of particular molecular targets in deranged biochemical pathways. These molecular/pharmacological targets are multiple protein nodes (molecules) located in biological networks and/or the involved interconnected biological pathways. Major causes of interindividual variation of drug efficiency are genetic polymorphisms of drug-metabolizing enzymes, like cytochrome P450 CYP2D6 gene, generating distinct human subgroups with different competence to perform drug biotransformation reactions [77]. The study of genetic polymorphisms in drug metabolizing enzymes in order to elucidate different or unwanted drug effects is termed pharmacogenetics or pharmacogenomics [78]. Thus, the elucidation of the existent genetic polymorphisms facilitates the identification of the right drug for the right patient, an important healthcare paradigm for individualized medicine.

The oral anticoagulants are important agents for the treatment and prophylaxis of thromboembolic disorders. There is a large interindividual variability in dosage titration with unpredictable anticoagulation response and with serious complications due to insufficient initial control [79]. The use of anticoagulant drugs is an example of personalized cardiology as it requires a more individualized dose for loading and maintenance. The bleeding risk renders an individualized approach to antithrombotic therapy more important. Warfarin is a coumarin derivative and a common anticoagulant drug for maintenance therapy and prevention of thromboembolic episodes. It displays a narrow therapeutic window with indeterminate dosing and therapeutic effect. The response to the warfarin is determined by polymorphisms in genes related to metabolism or to pharmacodynamic response. Patients with CYP2C9 (CYP450 family of genes) variant alleles need lower warfarin dose [80]. Variants in the gene VKORC1 encoding vitamin K epoxide reductase complex 1 are connected with reduced expression of VKORC1 and diminished response to warfarin [81]. Patients with various VKORC1 haplotypes, are separated into low-, intermediate-, and high-dose warfarin groups, with the responsible molecular mechanism located at the transcriptional level [82]. Therefore, the warfarin's therapeutic dose is linked to polymorphisms in genes encoding for CYP and VKORC1.

Recently, a novel group of antiplatelet drugs have been developed with the potential to improve the safety and efficiency of the antiplatelet therapy in ACS patients treated invasively. Dual antiplatelet therapy (DAPT) in patients with ACS or after PCI and stent implantation is the cornerstone of optimal therapy. In these patients, the dual antiplatelet therapy with aspirin and clopidogrel, a P2Y12 ADP receptor antagonist, is used in everyday clinical practice and associated with substantial protective effect of stent thrombosis with an incremental bleeding risk. Many new P2Y12 ADP receptor antagonists have been approved for clinical application with considerable debate over their potential bleeding risk and efficacy. In the published trials, it is hard to compare and differentiate the therapeutic function and safety of P2Y12 antagonists because of the variations in the study designs.

The American College of Cardiology Foundation/American Heart Association (ACCF/AHA) published new guidelines for the use of clopidogrel, prasugrel, and ticagrelor [83]. These guidelines provide no direction on the drug selection or the optimal dual drug combination or the possible risk from adverse bleeding events or the contribution of genotyping in drug preference. This position is supported by the fact that no large prospective trial implied therapeutic improvement after a change to new antiplatelet therapy grounded on different platelet function or genetic testing. There is significant interindividual variability in antiplatelet response to clopidogrel. Recent studies intend to evaluate the affiliation between genetic variation of cytochrome P450 (CYP) isoenzymes and the pharmacokinetic response of the P2Y12 receptor antagonists in patients with ACS. There is variability in the antiplatelet effect due to the presence of CYP2C19 alleles that inhibit the metabolism of clopidogrel to the active form [84]. There is significant increase in stent thrombosis in patients having the CYP2C19 polymorphism and treated with clopidogrel [85]. Thus, in patients with ACS genotyping for CYP2C19 loss of function variant should be examined [86]. Nevertheless, these genetic polymorphisms are expected in fewer than 20% of patients with a low response to clopidogrel, as other causes also provide variability in the reduced antiplatelet effect [87]. In clinical practice, many clinical and demographic variables and multiple genetic factors are responsible for the variability in antiplatelet response to clopidogrel [88]. In patients with ACS undergoing PCI, clopidogrel therapy is inefficient in 10-30% of them, and the resistance and suboptimal response to aspirin is 5.5-9.5% and 23.8% accordingly [89,90]. Application of functional testing methods for measuring the antiplatelet effect of clopidogrel on the P2Y12 receptor would guide to a more individualized cardiology [91]. In three hundred patients treated with PCI, the clopidogrel platelet reactivity decreased over time while the genotype (CYP2C19*2, *17, CYP3A5*3, and ABCB1) affected 18% of this tendency [92,93].

In comparison with clopidogrel, prasugrel therapy reduced significantly the rate of stent thrombosis and ischemic events (9.9% vs 12.1%, p<0.001) in ACS patients with coronary stenting [94]. Nevertheless, prasugrel was associated with an increased risk of major (2.4 % vs 1.8%, p<0.03) and fatal (0.4% vs 0.1%) bleeding, while mortality was not significantly different between the two treatment groups. Ticagrelor, a new antiplatelet agent with greater antagonism of the P2Y12 than clopidogrel, was compared with clopidogrel in the PLATO trial and demonstrated a considerable decrease (1.9%) in the value of the primary composite endpoint (cardiovascular death, myocardial infarction and stroke) [95]. Ticagrelor's success was accomplished regardless of the genetic CYP2C19 polymorphism, a fact that explains the greater tendency for bleeding with ticagrelor compared with clopidogrel [96].

In contrast to the ACCF/AHA guidelines, the European Society of Cardiology (ESC) guidelines recommend the ticagrelor or prasugrel instead of clopidogrel in patients with ACS [97]. The ticagrelor is recommended (Class I, level B) for all patients with ACS despite if they were treated invasively or noninvasively, using proton pump inhibitors (PPIs) or having genetic CYP2C19 polymorphism that characterize low responders to clopidogrel. The prasugrel is recommended (Class I, level B) particularly for diabetic patients treated with PCI provided that are not high risk patients for bleeding. The clopidogrel is favoured (Class I, level A) only if ticagrelor or prasugrel are not recommended.

The beta-blocking drugs are critical therapeutic agents for heart failure patients, because they prolong the life, decrease the mortality rate that is related to heart failure, and reduce the incidence of arrhythmic sudden cardiac death [98]. The Multicenter Automatic Defibrillator Implantation Trial-II (MADIT-II) demonstrated that patients receiving the higher doses of beta-blockers had significant reduction in the risk for VT (ventricular tachycardia) or VF (ventricular fibrillation) requiring ICD (implantable cardioverter defibrillator) therapy compared with patients not receiving beta-blockers [99].

The beta1 and beta 2 adrenergic receptors (AR) contribute to the regulation of the cardiovascular system. The two beta-AR subtypes are polymorphic and their polymorphisms are examined for their possible relation to the cardiovascular diseases. The beta-AR polymorphisms are considered therapeutic targets and can affect drug responses, but are not inducing a cardiovascular disease like CAD, hypertension or HF. In the beta1 gene are recorded two major single nucleotide polymorphisms (SNPs), the Ser49Gly and Arg389Gly, and in the beta 2 gene are documented three major SNPs, the Arg16Gly, GIn27Glu and

Thr164lle [100]. Patients with heart failure been homozygous for the Arg389 beta1 polymorphism are considered good responders to beta-blockers, and when were treated with carvedilol had left ventricular functional improvement [101]. Also, patients homozygous for the Arg389 beta1 polymorphism treated with bucindolol, showed a 38% mortality reduction and 34% reduction in mortality or hospitalization compared with patients receiving placebo [102]. Patients homozygous for the Gly389 beta1 polymorphism should be poor responders or non-responders to beta-blockers [100]. In transgenic mice that expressed either human Arg389 beta1 or Gly389 beta1 receptors, Arg389 beta1 hearts exhibited upregulated expression of genes associated with inflammation, programmed cell death, and extracellular matrix [103].

There are three classes of therapeutic agents acting on the renin-angiotensin-aldosterone system (RAAS): the β-adrenergic antagonists (beta-blockers) (reduce levels of renin), the angiotensin-converting enzyme inhibitors (ACE) (reduce levels of angiotensin II) and the AT_1 receptor blockers (ARBs) (antagonize the binding of angiotensin II to AT_1 receptors). The ACE inhibitors to a significant extent decrease cardiovascular mortality, myocardial infarction, and hospitalization for HF recurrence in patients with asymptomatic or symptomatic left ventricular systolic dysfunction. The ACE deletion allele (ACE-D) is connected with increased RAAS activation and in patients with left ventricular systolic dysfunction demonstrated reduced transplant-free survival [104]. In patients with ACE-D allele polymorphism and left ventricular systolic dysfunction the use of ACE inhibitors reduced the effect of the ACE-D allele on the increased activation of RAAS, while treatment with beta-blockers and ACE inhibitors had greater beneficial effect in the patients with DD polymorphism [105].

3.3. Transcriptomics

Transcriptomics is the scientific field concerned with the study of the transcriptome, that embraces the complete set of the mRNA transcripts (generated by the genes) of a cell, tissue, or organism. The genome is fixed for a given cell line, while the transcriptome which is the pattern of gene expression, changes with the type of tissue, hormones, drugs, disease, stage of development, and external environmental conditions. The DNA microarray (DNA chip) is a technology that allows the quantification of many thousands of mRNA transcripts in a single reading, identifies new molecular abnormalities and contributes to the discovery of novel clinical biomarkers [106]. A microarray contains numerous small DNA molecules and can be used to detect DNA or RNA, converting the mRNAs to single-stranded complementary DNAs (cDNAs) after reverse transcription.

Systolic HF is a complex syndrome that involves multiple cellular and hormonal mechanisms and is the outcome of a variety of clinical conditions and gene mutations. The clinical phenotypes of HF are characterized by reduced ventricular contraction and ventricular dilatation. The phenotypically similar models of HF are generated by the involvement of a variety of genes with different expression changes and heterogeneous transcriptional response. Gene expression and transcription analysis in HF has been undertaken in many experimental and human HF microarray studies. Barth et al., using a meta-analysis of 28 experimental and human HF microarray studies, demonstrated that gene expression in failing myocardium is characterized by downregulation of metabolic transcripts and concomitant upregulation of cell signaling pathways [107]. Microarray analyses of the failing myocardium demonstrate the importance of cardiac myosin light chain kinase (MLCK) in HF. Moreover, the microarray analyses establish the significant role of many genes involved in mitochondrial dysfunction and oxidative phosphorylation, and the importance of three extracellular molecules-periostin, pliotrophin and SERPINA3- that could be evolved to new diagnostic and therapeutic targets in patients with HF [108]. The cardiac regulatory network is controlled by transcription factors and transcriptional coactivators which are the primary targets of both physiological and pathological signals. Recently, was explored the myocardin- (MYOCD) related family of transcriptional cofactors (MRTF-A and MRTF-B), which is involved in cardiac dysfunction and in the expression of target genes associated to cardiac growth and adaptation to stress [109].

MicroRNAs (miRNAs or miRs) are short (22-nucleotide long) noncoding RNAs that are binding to messenger RNAs transcripts and thus regulate gene expression at the post-transcriptional level through post-transcriptional mRNA silencing [110]. The miRNAs can negatively regulate or inhibit gene expression and mediate mRNA degradation or post-transcriptional repression (translational inhibition) [111]. The miRNAs are implicated in every biological process and several of those are identified in the blood of humans. The misexpression of miRNAs has been exposed as a pathogenic mechanism in several diseases including cardiovascular pathologies [112]. Numerous studies have established that miRNAs expression is changing in cardiac hypertrophy and HF progression, by targeting genes that regulate functions of cardiac remodeling process like myocyte hypertrophy or myocyte loss, and myocardial fibrosis [113]. Furthermore, transgenic mouse studies showed that individual miRNAs have an impact on stem cell differentiation, progression of myocardial hypertrophy and failure, fibroblast proliferation, angiogenesis, ion channel function, and apoptosis [114]. The serum miRNAs levels are being evaluated as disease biomarkers with important diagnostic and therapeutic repercussions in the field of cardiovascular diseases. The development of small miRNA-targeting molecules has the ability to act in many biological targets in a single biological response pathway [115]. The miRNAs expression seems to be more sensitive than mRNA to changes in the clinical status of HF. Thus, miRNAs levels in myocardium or plasma could enhance clinical diagnostics in myocardial dysfunction [115]. Further experimental and clinical research on miRNAs probably will increase our ability to prevent sudden cardiac death and improve myocardial failure prospects through modulation of cardiac miRNAs [116].

3.4. Metabolomics

The term 'metabolome' characterizes the whole number of small molecules (metabolites) that reside in the human cells, tissues, organs or organism, and represent the end-products of cellular processes. The metabolites are low molecular weight biological substances with an important role on health and disease. The field of metabolomics with its own computational infrastructure combines chemical data with biological outcomes linking elements of bioinformatics with cheminformatics [117]. The discipline of metabolomics refers to the systematic study of metabolite profiles that are related to a variety of cellular metabolic processes in a cell, organ, or organism and has an impact on the discovery of new drugs and biomarkers [118]. Therefore, the metabolomics translation to cardiovascular biomarker discovery improves the identification and confirmation of a cardiovascular disease. Metabolic profiling exposes and amplifies changes that are emerged in the proteome and the genome, and delineates a more precise picture of the human phenotype in health and disease [119]. Thus, in accordance with systems biology approach the information deducted from metabolomics is used in order to decode disease mechanisms, diagnose and treat cardiovascular diseases. In transgenic mouse model, the integration of gene transcript, protein and metabolite profiling are involved in the characterization of the three processes of atherosclerosis: lipid metabolism, inflammation, and tissue changes [120]. Barderas et al. [121], reported three different metabololomic approaches for clinical course assessment and patient recovery: metabolomic fingerprinting (metabolites changed during the course of a disease), metabolic profiling (metabolites involved in a targeted biological pathway), and metabolomic footprinting (monitoring secreted metabolites). The chemical diversity of the metabolome requires appropriate methodology to characterize metabolite data. The nuclear magnetic resonance spectroscopy (NMR) and mass spectrometry (MS) have been employed in experimental and epidemiological studies to elucidate metabolite profiling and metabolic disturbances of human diseases [122]. Senn et al. [118], believe that "with technological advances in metabolomic tools and more sophisticated bioinformatics and analytical techniques, the ability to measure low-molecular-weight metabolites in biospecimens provides a unique insight into established and novel metabolic pathways."

3.5. Tailoring Clinical Care

The classical disease models are characterized by their limited comprehension of disease onset and therapy. In contrast, the systems biology approach, based on a large number of data, is characterized by the concept of "the most useful therapy to be tailored to the precise individual biological status". The conventional definitions in the predictive, diagnostic and therapeutic domains are surpassed by the accumulation of 'omics' data and new technologies. The increased knowledge and understanding of the complexity in the molecular processes, biochemical pathways and biological networks directed us to a profound cognitive change and a reappraisal of human diseases. The systems biology concept contributed to the evolution of novel technologies with new explanation to the human complexity and prompted the emergence of individualized medicine. A new concept of patient stratification is needed in which individuals are biologically subclassified for personalized therapy and for drug safety and efficacy [123]. Disease risk assessment is predicted from large epidemiological studies relating demographic and clinical features with eventual end-results. The current epidemiological studies represent multivariable models and in the future should incorporate in their design genomic and proteomic features [61].

Human diseases are complex biological syndromes produced by a combination of genetic and environmental factors. The personal biological reaction to pharmaceutical therapy and the molecular basis of drug efficacy and toxicity are the outcome of the interactions between genetic profiles and environmental factors [124]. A 'conditional metabolic phenotype' is considered the starting point of an individual in a multivariate metabolic space that contains many physical, chemical, genetic and environmental components that determines the outcome of a drug or dietary intervention and is the basis of the personalized healthcare paradigm [125]. The traditional drug manufacturing industry disregards the complexity of biological systems, and approaches human diseases with an oversimplified strategy. It is addressed preferably to the 'disease' as an entity than to underlying biomolecular networks. The pharmaceutical industry should be designed to discover biological assays with multiple biological targets as the treated diseases are complex entities and require therapies that display many pharmacological effects.

Systems biology approach in the attempt to elucidate the cause of HF progression to a more advanced myocardial dysfunction stage, integrates genes and molecules, interprets molecular networks and modular functional units, and defines the interaction of mechanical dysfunction and cardiac remodeling [126]. Moreover, the human HF from the point of view of systems biology is a complex clinical syndrome that has been caused by a variety of diseases. In the realm of the clinical syndrome are involved many clinical appearances or clinical phenotypes with symptoms that include retention of fluid in the lungs and periphery [126]. Genetically based interindividual differences are important determinants of left ventricular remodeling. A personalized assessment and follow-up in patients with HF arise with the application of molecular imaging techniques for the elucidation of the cellular and molecular mechanisms that are involved in left ventricular remodeling [127]. Recently, Neeland et al. [128], attempted to establish if biomarkers of subclinical myocardial injury and hemodynamic stress are able to detect asymptomatic patients with left ventricular hypertrophy (LVH) who are in danger to develop HF or increased cardiovascular (CV) mortality [128]. They found that minimal elevations of hscTn or NT-proBNP "identify a malignant subphenotype of LVH with high risk for progression to HF and CV death".

The symptomatic HF is a syndrome with increased prevalence, morbidity and mortality. In our aged population the presence of co-morbidities like diabetes mellitus, CAD and hypertension is expected to increase the prevalence of HF in the near future. The present mode of therapy retards the progression of the HF syndrome but new therapies are needed for prevention and functional or structural repair of the diseased myocardium. The development of novel HF therapies needs testing the reputed therapeutic strategies in suitable HF animal models [129]. The pathological features of human HF should be present in an animal model in order to detect significant causes of HF or assess conjectural preventive or repairing therapies. Properly evaluated animal models give the opportunity to improve clinical care, look into molecular pathways involved in early or late periods of HF, and identify novel therapeutic targets or biomarkers for HF progression [129].

The intrinsic complexity of the cardiac signalling networks puts a limit on the interpretation of animal experimental findings and their significance on cardiac physiology and disease. Thus, sophisticated computational models have been developed that can identify biological mechanisms and reduce the complexity of large data sets [130]. Thus, in a normal human left coronary artery bifurcation computational model, the wall shear stress oscillation was investigated by applying non-Newtonian blood properties and phasic flow. It was found that the lateral walls of the bifurcation, where low and oscillating wall shear stress is observed, are more susceptible to atherosclerosis [131].

In end-stage renal disease, the prevalence of atrial fibrillation (AF) is 13%-27%, and is considered to be a significant factor of increased morbidity and mortality [132]. An individualized therapeutic approach is required when referring for anticoagulation patients with chronic kidney disease (CKD) and AF. In these patients with complex pathology, considering anticoagulation is a difficult decision, balancing the chance of arterial thrombosis from the untreated atrial fibrillation with the possibility of bleeding from the use of anticoagulants. The CKD patients are in a high risk for both thromboembolic events and hemorrhagic complications. A high number of pathophysiologic mechanisms are involved that increase the coagulation (tissue plasminogen abnormalities) or the risk of bleeding (platelet irregularities). An individualized stratification algorithm is proposed to the dilemma of treating patients with CKD and AF [133]. In hospitalized chronic dialysis patients with AF, the use of coumadin was associated with improved long-term survival compared to those that were not anticoagulated [134]. Hypertrophic cardiomyopathy (HCM) is a common myocardial disorder characterized by cardiac hypertrophy due to de novo mutations or to inheritance in an autosomal dominant Mendelian pattern. Human genetic studies demonstrated that dominant mutations in the proteins that make up the contractile apparatus of sarcomere are credited for the disease. Mutations in more than 11 genes encoding proteins of the cardiac sarcomere (>1,400 variants) are responsible for the clinical manifestations of HCM [69]. Genotyping in HCM is a powerful tool for family screening and diagnosis. Thus, genetic testing expands the HCM disease spectrum and increases the diagnostic potentiality to trace HCM patients with different natural history and varied clinical manifestations of the myocardial disorder. The recent progress in the genetic screening and the application of the molecular information in the clinical level is accomplished by the collaboration between multidisciplinary teams trained in clinical diagnostics and molecular genetics [135]. In HCM patients, the association of gene-based diagnosis with various clinical phenotypes increases the opportunity for a more individualized medicine. Novel therapeutic targets are under clinical assessment and new strategies are changing the course of the disease progression, and the natural history of cardiac symptoms, arrhythmias and heart failure [136].

Conclusion

Human cardiac diseases are complex and self-organized biological entities approached with the holistic concept of systems biology. The acceptance that human disease complexity can be explored by using systems biology approach has increased the prospects for the development of personalized cardiology. The validation of cardiac biomarkers in large-scale clinical trials is an important step for personalized cardiology to emerge as a reality. The tailoring of medical therapy in identified subpopulations intends to disclose patients' susceptibility to a disease and detect their response to a particular treatment. In clinical research, the integration of clinical and epidemiological information with genomic and proteomic data is a precondition for a personalized cardiology approach.

References

[1] Louridas GE, Kanonidis IE, Lourida KG. Systems biology in heart diseases.*Hippokratia*.2010; 14:10-16.

[2] Kitano H. Systems biology: a brief overview. *Science*.2002; 295:1662-1664.

[3] Louridas GE, Lourida KG. A conceptual paradigm of heart failure and systems biology approach. *Intern J Cardiol*. 2012; 159:5-13.

[4] Bruggeman FJ, Westerhoff HV. The nature of systems biology. *Trends Microbiol.* 2007; 15:45-50.

[5] Novikoff AB. The concept of integrative levels and biology. *Science*. 1945; 101:209-215.

[6] Louridas GE, Lourida KG. The new biology: a bridge to clinical cardiology. *Hippokratia*.2012; 16:106-112.

[7] Dewey FE, Wheeler MT, Ashley EA. Systems biology of heart failure, challenges and hopes. Current Opinion in *Cardiology*.2011; 26:314-321.

[8] Rosenson RS. New technologies personalize diagnostics and therapeutics. *Curr Atheroscler Rep*.2010; 12:184-186.

[9] Torrey T. What ethical, legal and moral questions are raised by personalized medicine? 2012. About.com Inc, a part of The New Times Company.

[10] Sharrer GT. Personalized medicine: ethics for clinical trials. *Methods Mol Biol*. 2012; 823:35-48.

[11] Nunez PL. Brain, Mind, and the Structure of Reality. New York: Oxford University Press Inc; 2010.

[12] Drake TA, Ping P. Thematic review series: systems biology approaches to metabolic and cardiovascular disorders. Proteomics approaches to the systems biology of cardiovascular diseases. *J Lipid Res*.2007; 48:1-8.

[13] Naylor S, Chen JY. Unraveling human complexity and disease with systems biology and personalized medicine. *Per Med*.2010; 7:275-289.

[14] Lobo I. Biological complexity and integrative levels of organization. *Nature Education*. 2008; 1(1).

[15] Watkins H, Farrall M. Genetic susceptibility to coronary artery disease: From promise to progress. *Nature Reviews Genetics*. 2006; 7:163-173.

[16] Panagiotou G, Nielsen J. Nutritional systems biology: definitions and approaches. *Annu Rev Nutr*.2009; 29:329-339.

[17] Kaput J. Nutrigenomics research for personalized nutrition and medicine. *Curr Opin Biotecnol*.2008; 19:110-120.

[18] ENCODE Project Consortium, Bernstein BE, Birney E, Dunham I, Green ED, Gunter C, Snydes M. An integrated encyclopedia of DNA elements in the human genome. *Nature*. 2012; 489:57-74.

[19] Maher B. ENCODE: The human encyclopaedia. *Nature*. 2012; 489:46-48.

[20] Gatherer D. So what do we really mean when we say that systems biology is holistic? *BMC Systems Biology*.2010; 4:22 doi:10.1186/1752-0509-4-22.

[21] Loscalzo J, Barabasi AL. Systems biology and the future of medicine. WIREs Syst Biol Med.2011 doi:10.1002/ws-bm.144.

[22] Personalized Medicine Coalition (PMC). What is Personalized Medicine? 2012. www.partners.org

[23] Reed JC, Jackson M. Disease in a dish: The ultimate personalized medicine. *Medscape*, Dec 04, 2012.

[24] McLean AS, Huang SJ. Cardiac biomarkers in the intensive care unit. *Ann of Intensive Care*. 2012;2:8.doi:10.1186/2110-5820-2-8.

[25] Christenson R. What is the value of B-type natriuretic peptide testing for diagnosis, prognosis, or monitoring of the critically ill adult patients in intensive care? *Clin Chem Lad Med*. 28; 46:1524-1532.

[26] Blankesteijn WM, Creemers E, Lutgens E, Cleutjens JP, Daemen MJ, Smits JF. Dynamics of cardiac wound healing following myocardial infarction: observations in genetically altered mice. *Acta Physiol Scand.* 2001; 173:75-82.

[27] Liao YH, Cheng X. Autoimmunity in myocardial infarction. *Int J Cardiol.* 2006; 12:21-26.

[28] Cooper L. The heat is off: immunosupression for myocarditis revisited. *Eur Heart J.* 2009; 30:1936-1939.

[29] Packard RR, Libby P. Inflammation in atherosclerosis: from vascular biology to biomarker discovery and risk prediction. *Clin Chem.* 2008; 54:24-38.

[30] Thygesen K, Alpert J, White H. Joint ESC/ACCF/AHA/WHF Task Force for the Redefinition of Myocardial Infarction. Universal definition of myocardial infarction. *Eur Heart J.*2007; 28:2525-2538.

[31] Heidenreich P, Alloggiamento T, Melsop K, McDonald K, Go A, Hlatky M. The prognostic value of troponin in patients with non-ST elevation acute coronary syndromes: a meta-analysis. *J Am Coll Cardiol.* 2001; 38:478-485.

[32] Kontos MC, Fritz LM, Anderson FP, Tatum JL, Ornato JP, Jesse RL. Impact of the troponin standard on the prevalence of acute myocardial infarction. *Am Heart J.* 2003; 146:446-452.

[33] European Heart Journal. Universal definition of myocardial infarction. *Eur Heart J.* 2007; 28:2525-2538.

[34] Giannoni A, Giovannini S, Clerico A. Measurement of circulating concentrations of cardiac troponin I and T in healthy subjects: a tool for monitoring myocardial tissue renewal? *Clin Chem Lab Med.* 2009; 47:1167-1177.

[35] Twerenbold R, Reichlin T, Reiter M, Muller C. High-sensitive cardiac troponin: friend or foe? *Swiss Med Wkly.* 2011; 10; 141:w13202. doi: 10.4414/smw.2011.13202.

[36] Reiter M, Twerenbold R, Reichlin T, Haaf P, Peter P, Meissner J, et al. Early diagnosis of acute myocardial infarction in the elderly using more sensitive cardiac troponin assays. *Eur Heart J.* 2011; 32:1379-1389.

[37] Giannitsis E, Becker M, Kurz K, Hess G, Zdunek D, Katus H. High-sensitivity cardiac troponin T for early prediction of evolving non-ST-segment elevation myocardial infarction in patients with suspected acute coronary syndrome and negative troponin results on admission. *Clin Chem.* 2010; 56:642-650.

[38] Sobel B, Shell W. Serum enzyme determination in the diagnosis and assessment of myocardial infarction. *Circulation.* 1972; 45:471-482.

[39] Lin JC, Apple FS, Murakami MM, Luepker RV. Rates of positive cardiac troponin I and creatine kinase MB mass among patients hospitalized for suspected acute coronary syndromes. *Clin Chem.* 2004; 50:333-338.

[40] Tanaka T, Sohmiya K, Kitaura Y, Takeshita H, Morita H, Ohkaru Y, et al. Clinical evaluation of point-of-care-testing of heart-type fatty acid-binding protein (H-FABP) for the diagnosis of acute myocardial infarction. *J Immunoassay Immunochem.* 2006; 27:225-238.

[41] Clerico A, Recchia FA, Passino C, Emdin M. Cardiac endocrine function is an essential component of the homeostatic regulation network: physiological and clinical implications. *Am J Physiol Heart Circ Physiol.* 2006; 290: H17-H29.

[42] Dietz JR. Mechanisms of atrial natriuretic peptide secretion from the atrium. *Cardiovascular Research.*2005; 68:8-17.

[43] Piechota M, Banach M, Jacon A, Rysz J. Natriuretic peptides in cardiovascular diseases. *Cell Mol Biol Lett*. 2008; 13:155-181.

[44] Clerico A, Giannoni A, Vittorini S, Passino C. Thirty years of the heart as an endocrine organ: physiological role and clinical utility of cardiac natriuretic hormones. *Am J Physiol Heart Circ Physiol* 2011; 301:H12-20.

[45] Berendes E, Van Aken H, Raufhake C, Schmidt C, Assmann G, Walter M. Differential secretion of atrial and brain natriuretic peptide in critically ill patients. *Anesth Analg*.2001; 93:676-682.

[46] Steele MK, Gardner DG, Xie PL, Schultz HD. Interactions between ANP and ANG II in regulating blood pressure and sympathetic outflow. *Am J Physiol*.1991; 260:R1145-1151.

[47] Levin ER, Gardner DG, Samson WK. Natriuretic peptides. *N Engl J Med*.1998; 339:321-328.

[48] De Bold AJ, Borenstein HB, Veress AT, Sonnenberg H. A rapid and potent natriuretic response to intravenous injection of atrial myocardial extract in rats. *Life Sci*.1981; 28:89-94.

[49] Mangat H, de Bold AJ. Stretch-induced atrial natriuretic factor release utilizes a rapidly depleting pool of newly synthesized hormone.*Endocrinology*.1993; 133:1398-1403.

[50] Lang RE, Tholken H, Ganten D, Luft FC, Ruskoaho H, Unger T. Atrial natriuretic factor-a circulating hormone stimulated by volume loading. *Nature*.1985; 314:264-266.

[51] Rascher W, Tulassay T, Lang RE. Atrial natriuretic peptide in plasma of volume-overloaded children with chronic renal failure. *Lancet*.1985; 2:303-305.

[52] Tikkanen I, Fyhrquist F, Metsarinne K, Leidenius R. Plasma atrial natriuretic peptide in cardiac disease and during infusion in healthy volunteers.*Lancet*.1985; 2:66-69.

[53] Sudoh T, Kangawa K, Minamino N, Matsuo H. A new natriuretic peptide in porcine brain. *Nature*.1988; 332:78-81.

[54] Yasue H, Yoshimura M, Sumida H, Kikuta K, Kugiyama K, Jougasaki M et al. Localization and mechanism of secretion of B-type natriuretic peptide in comparison with those of A-type natriuretic peptide in normal subjects and patients with heart failure.*Circulation*.1994; 90:195-203.

[55] Alter P, Rupp H, Rominger MB, Vollrath A, Czemy F, Figel JH, et al. B-type natriuretic peptide and wall stress in dilated human heart. *Mol Cell Biochem*. 2008; 314:179-191.

[56] Vanderheyden M, Goethals M, Verstreken S, De Bruyne B, Muller K, Van Schuerbeeck E, Bartunek J. Wall stress modulates brain natriuretic peptide production in pressure overload cardiomyopathy. *J Am Coll Cardiol*. 2004; 44:2349-2354.

[57] Maisel AS, McCord J, Nowak RM, Hollander JE, Wu AH, Duc P, Omland T, et al. Bedside B-TYPE natriuretic peptide in the emergency diagnosis of heart failure with reduced or preserved ejection fraction. Results from the Breathing Not Properly Multinational Study. *J Am Coll Cardiol*. 2003; 41:2010-2017.

[58] Januzzi JL Jr, Camargo CA, Anwaruddin S, Baggish AL, Chen AA, Krauser DG, et al. The N-terminal Pro-BNP investigation of dyspnea in the emergency department (PRIDE) study. *Am J Cardiol*. 2005; 95:948-954.

[59] Takahashi K, Totsune K, Sone M, et al. Human brain natriuretic peptide-like immunoreactivity in human brain. *Peptides*. 1991; 13:121-123.

[60] Stingo AJ, Clavell AL, Aarhus LL, Burnett JC. Cardiovascular and renal actions of C-type natriuretic peptide. *Am J Physiol.* 1992; 262(1 Pt 2):308-312.

[61] Ouzounian M, Lee DS, Gramolini AO, Emili A, Fukuoka M, Liu PP. Predict, prevent and personalize: Genomic and proteomic approaches to cardiovascular medicine.*Can J Cardiol.* 2007; 23(Suppl A):28A-33A.

[62] Arab S, Gramolini AO, Ping P, Kislinger T, Stanley B, van Eyk J, et al. Cardiovascular proteomics: tools to develop novel biomarkers and potential applications. *J Am Coll Cardiol.* 2006; 48:1733-1741.

[63] Apweiler R, Aslanidis C, Deufel T, Gerstner A, Hansen J, Hocsstrasser D, et al. Approaching clinical proteomics: current state and future fields of application in fluid proteomics. *Clin Chem Lab Med.* 2009; 47:724-744.

[64] Searle J, Danne O, Muller C, Mockel M. Biomarkers in acute coronary syndrome and percutaneous coronary intervention. *Minerva Cardioangiol*. 2011; 59:203-224.

[65] Barbato E, Bartunek J, Mangiacapra F, Sciarretta S, Stanzione R, Delrue L, et al. Influence of rs5065 atrial natriuretic peptide gene variant on coronary artery disease. *J Am Coll Cardiol*. 2012; 59:1763-1770.

[66] Herder C, Karakas M, Koenig W. Biomarkers for the prediction of type 2 diabetes and cardiovascular disease. *Clin Pharmacol Ther*. 2011; 90:52-66.

[67] D' Alessandra Y, Devanna P, Limana F, Straino S, Di Carlo A, Brambilla PG, et al. Circulating microRNAs are new and sensitive biomarkers of myocardial infarction. *Eur Heart J.* 2010; 31:2765-2773.

[68] Wang GK, Zhu JQ, Zhang JT, Li Q, Li Y, He J, et al. Circulating microRNA: a novel potential biomarker for early diagnosis of acute myocardial infarction in humans. *Eur Heart J.* 2010; 31:659-666.

[69] Maron BJ, Maron MS, Semsarian C. Genetics of hypertrophic cardiomyopathy after 20 years, clinical perspectives. *J Am Coll Cardiol.* 2012; 60:705-715.

[70] Modell SM, Lehmann MH. The long QT syndrome family of cardiac ion channelopathies: A HuGE review. *Genet Med.* 2006; 8:143-155.

[71] Kaartinen V, Warburton D. Fibrillin controls TGF-beta activation. *Nat Genet*. 2003; 33:331-332.

[72] Calkins H. Arrhythmogenic right ventricular dysplasia/cardiomyopathy. *Curr Opin Cardiol*. 2006; 21:55-63.

[73] Helgadottir A, Thorleifsson G, Manolescu A, Gretarsdottir S, Blondal T, Jonasdottir A, et al. A common variant on chromosome 9p21 affects the risk of myocardial infarction. *Science*. 2007; 316:1491-1493.

[74] Kovacic JC, Castellano JM, Fuster V. The links between complex coronary disease, cerebrovascular disease, and degenerative brain disease. *Ann. N. Y. Acad Sci.* 2012; 1254:99-105.

[75] Wahlstrand B, Orho-Melander M, Delling L, Kjeldsen S, Narkiewicz K, et al. The myocardial infarction associated CDKN2A/CDKN2B locus on chromosome 9p21 is associated with stroke independently of coronary events in patients with hypertension. *J Hypertens*. 2009; 27:769-773.

[76] Zuchner S, Gilbert JR, Martin ER, Leon-Guerrero CR, Xu PT, Browning C, et al. Linkage and association study of late-onset Alzheimer disease families linked to 9p21.3. *Ann Hum Genet*. 2008; 72:725-731.

[77] Meyer UA, Zanger UM. Molecular mechanisms of genetic polymorphisms of drug metabolism. Annu Rev *Pharmacol Toxicol.* 1997; 37:269-296.
[78] Srivastava P. Drug metabolism and individualized medicine. *Curr Drug Metab*. 2003; 4:33-44.
[79] Kamali F, Pirmohamed M. The future prospects of pharmacogenetics in oral anticoagulation therapy. *Br J Clin Pharmacol*. 2006; 61:746-751.
[80] Aithal GP, Day CP, Kesteven PJ, Daly AK. Association of polymorphisms in the cytochrome P450CYP2C9 with warfarin dose requirement and risk of bleeding complications. *Lancet*. 1999; 353:717-719.
[81] Rost S, Fregin A, Ivaskevicius V, Conzelmann E, Hortnagel K, Pelz HJ. Mutations in VKORC1 cause warfarin resistance and multiple coagulation factor deficiency type 2. *Nature*. 2004; 427:537-541.
[82] Rieder MJ, Reiner AP, Cage BF, Nickerson DA, Eby CS, McLeod HL, et al. Effect of VKORC1 haplotypes on transcriptional regulation and warfarin dose. *N Engl J Med*. 2005; 352:2285-2293.
[83] 2011 ACCF/AHA/SCAI Guideline for percutaneous coronary intervention. Levine GN and committee members. *Circulation*. 2011; 124:2574-2609.
[84] Brandt T, Close SL, Iturria SJ, Payne CD, Farid NA, Ernest CS, et al. Common polymorphisms of CYP2C19 and CYP2C9 affect the pharmacokinetic and pharmacodynamic response to clopidogrel but not prasugrel. *J Thromb Haemost*. 2007; 5:2429-2436.
[85] Montalescot G, Hulot JS, Collet JP. Stent thrombosis: who's guilty? *Eur Heart J.*2009; 30:2685-2688.
[86] Wright RS, Anderson JL, Adams CD, Bridges CR, Casey DE Jr, Ettinger SM, et al., 2011. ACCF/AHA focused update incorporated into the ACC/AHA 2007 Guidelines for the management of patients with unstable angina/non-ST-elevation myocardial infarction. *J Am Coll Cardiol*. 2011; 57:e215-367.
[87] Hochholzer W, Trenk D, Fromm MF, Valina CM, Stratz C, Bestehorn HP, et al. Impact of cytochrome P450 2C19 loss-of-function polymorphism and of major demographic characteristics on residual platelet function after loading and maintenance treatment with clopidogrel in patients undergoing elective coronary stent placement. *J Am Coll Cardiol.* 2010; 55:2427-2434.
[88] Trenk D, Zolk O, Fromm MF, Neumann FJ, Hochholzer W. Personalizing antiplatelet therapy with clopidogrel. Clin Pharmacol Ther. 2012. Sep 5, doi: 10.1038 /clpt.2012.133. [Epub ahead of print].
[89] Podolec J, Gajos G, Budziaszek L, Czaniecka M, Kleczkoska A, Zmudka K. New approach to interventional cardiology treatment with personalized medicine. *Przegl Lek*. 2008; 65:850-857.
[90] Geisler T, Gawaz M. Individualized antiplatelet therapy:what can a clinical score contribute? *Hamostaseologie*. 2009; 29:360-367).
[91] Price MJ, Barker CM. Functional testing methods for the antiplatelet effect of P2Y12 receptor antagonists. *Biomark Med*. 2011; 5:43-51.
[92] Campo G, Parrinello G, Ferraresi P, Lunghi B, Tebaldi M, Miccoli M, et al. Prospective evaluation of on-clopidogrel platelet reactivity over time in patients treated with percutaneous coronary intervention. *J Am Coll Cardiol*. 2011; 57:2474-2483.

[93] Angiolillo DJ. Editorial: Unraveling myths of platelet function and genetic testing. *J Am Coll Cardiol*. 2011; 57:2484-2486.

[94] Wiviott SD, Braunwald E, McCabe CH, Montalescot G, Ruzyllo W, Gottlieb S, et al; TRITON-TIMI 38 Investigators. Prasugrel versus clopidogrel in patients with acute coronary syndromes. *N Engl J Med*. 2007; 357:2001-2015.

[95] Wallentin L, Becker RC, Budaj A, Cannon CP, Emanuelsson H, Held C, et al. Ticagrelor versus clopidogrel in patients with acute coronary syndromes. *N Engl J Med*. 2009; 361:1045-1057.

[96] Tantry US, Bliden KP, Wei C, Storey RF, Armstrong M, Butler K., et al. First analysis of the relation between CYP2C19 genotype and pharmacodynamics in patients treated with ticagrelor versus clopidogrel: the ONSET/OFFSET and RESPOND genotype studies. *Circ Cardiovasc Genet*. 2010; 3:556-566.

[97] ESC Guidelines for the management of acute coronary syndromes in patients presenting without persistent ST-segment elevation. Hamm CW and committee members. *Eur Heart J*.2011; 32:2999-3054.

[98] Effect of metoprolol CR/XL in chronic heart failure: Metoprolol CR/XL Randomised Intervention Trial in Congestive Heart Failure (MERIT-HF). *Lancet*. 1999; 353:2001-2007.

[99] Brodine W, Tung R, Lee J, Hockstad ES, Moss AJ, Zareba W, et al. Effects of beta-blockers on implantable cardioverter defibrillator therapy and survival in the patients with ischemic cardiomyopathy (from the Multicenter Automatic Defibrillator Implantation Trial-II). *Am J Cardiol*. 2005; 96:691-695.

[100] Brodde OE. Beta1-and beta 2-adrenoceptor polymorphisms and cardiovascular diseases. *Fundam Clin Pharmacol*. 2008; 22:107-125.

[101] Mialet PJ, Rathz DA, Petrashevskaya NN, Hahn HS, Wagoner LE, Schwartz A, et al. Beta 1-adrenergic receptor polymorphisms confer differential function and predisposition to heart failure. *Nat Med*. 2003; 9:1300-1305.

[102] Liggett SB, Mialet-Perez J, Thaneemit-Chen S, Weber SA, Greene SM, Hodne D, et al. A polymorphism within a conserved beta (1)-adrenergic receptor motif alters cardiac function and beta-blocker response in human heart failure. *Proc Acad Sci USA*. 2006; 103:11288-11293.

[103] Swift SM, Gaume BR, Small KM, Aronow BJ, Liggett SB. Differential coupling of Arg- and Gly389 polymorphic forms of the beta1-adrenergic receptor leads to pathogenic cardiac gene regulatory programs. *Physiol Genomics*. 2008; 35:123-131.

[104] McNamara DM, Holubkov R, Janosko K, Palmer A, Wang JJ, MacGowan GA, et al. Pharmacogenetic interactions between beta-blocker therapy and the angiotensin-converting enzyme deletion polymorphism in patients with congestive heart failure. *Circulation*. 2001; 103:1644-1648.

[105] Huang WY, Sun M, Zhou HY, Pei ZF, Yang TL. Effect of the polymorphism of the angiotensin-converting enzyme gene on the drug treatment in patients with chronic heart failure. Zhong Nan Da Xue Xue Bao Yi Xue Ban. 2004; 29:686-689.

[106] Heidecker B, Hare JM. The use of transcriptomic biomarkers for personalized medicine. *Heart Fail Rev*. 2007; 12:1-11.

[107] Barth AS, Kumordzie A, Frangakis C, Margulies KB, Cappola TP, Tomaselli GF. Reciprocal transcriptional regulation of metabolic and signalling pathways correlates with disease severity in heart failure. *Circ Cardiovasc Genet*. 2011; 4:475-483.

[108] Asakura M, Kitakaze M. Global gene expression profiling in the failing myocardium. *Circ J.* 2009; 73:1568-1576.
[109] Mikhailov AT, Torrado M. In search of novel targets for heart disease: myocardin and myocardin-related transcriptional cofactors. *Biochem Res.* 2012; 2012:973723. Epub 2012, May 17.
[110] Dorn GW2nd. MicroRNAs in cardiac disease. *Transl Res.* 2011; 157:226-235.
[111] Zorio E, Medina P, Rueda J, Millan JM, Arnau MA, Beneyto M, et al. Insights into the role of microRNAs in cardiac diseases: from biological signaling to therapeutic targets. *Cardiovasc Hematol Agent Med Chem.* 2009; 7:82-90.
[112] Condorelli G, Latronico MV, Dorn GW 2nd. microRNAs in heart disease: putative novel therapeutic targets? *Eur Heart J.* 2010; 31:649-658.
[113] Topkara VK, Mann DL. Role of microRNAs in cardiac remodeling and heart failure. *Cardiovasc Drugs Ther.* 2011; 25:171-182.
[114] Port JD, Sucharov C. Role of microRNAs in cardiovascular disease: therapeutic challenges and potentials. *J Cardiovasc Pharmacol.* 2010; 56:444-453.
[115] Dorn GW 2nd. Therapeutic potential of microRNAs in heart failure. *Curr Cardiol Rep.* 2010; 12:209-215.
[116] Wang N, Zhou Z, Liao X, Zhang T. Role of microRNAs in cardiac hypertrophy and heart failure. *IUBMB Life.* 209; 61:566-571.
[117] Wishart DS. Current progress in computational metabolomics. *Brief Bioinform.* 2007; 8:279-293.
[118] Senn T, Hazen SL, Tang WH. Translating metabolomics to cardiovascular biomarkers. *Prog Cardiovasc Dis.* 2012; 55:70-76.
[119] Mercuro G, Bassareo PP, Deidda M, Cadeddu C, Barberini L, Atzori L. Metabolomics: a new era in cardiology? J Cardiovasc Med (Hagerstown). 2011; 12:800-805.
[120] Oresic M, Clish CB, Davidov EJ, Verheij E, Vogels J, Havekes LM, et al. Phenotype characterization using integrated gene transcript, protein and metabolite profiling. *Appl Bioinformatics.* 2004; 3:205-217.
[121] Barderas MG, Laborde CM, Posada M, de la Cuesta F, Zubiri I, Vivanco F, et al. Metabolomic profiling for identification of novel potential biomarkers in cardiovascular diseases. J Biomed Biotechnol. 2011; 2011:790132. Epub 2011 Jan 2.
[122] Rhee EP, Gerszten RE. Metabolomics and cardiovascular biomarker discovery. *Clin Chem.* 2012; 58:139-147.
[123] Nicholson JK. Global systems biology, personalized medicine and molecular epidemiology. Molecular Systems Biology. Editorial. 2006; doi: 10.1038/msb4100095.
[124] Nicholson JK, Holmes E, Lindon JC, Wilson ID. The challenges of modelling mammalian biocomplexity. *Nat Biothechnol.* 2004; 22:1268-1274.
[125] Nicholson JK, Holmes E, Wilson ID. Gut microbes, mammalian metabolism and personalized healthcare. *Nat Rev Microbiol.* 2005; 3:2-8.
[126] Louridas GE, Lourida KG. Systems biology and biomechanical model of heart failure. *Curr Cardiol Rev.* 2012; 8:220-230.
[127] Shirani J, Dilsizian V. Molecular imaging targets of cardiac remodeling. *Curr Cardiol Rep.* 2009; 11:148-154.
[128] Neeland IJ, Drazner MH, Berry JD, Ayers CR, deFilipp C, Seliger SI, et al. Biomarkers of chronic cardiac injury and hemodynamic stress identify a malignant phenotype of

left ventricular hypertrophy in the general population ONLINE FIRST. J Am Coll Cardiol. 2012. doi: 10.1016/jack.2012.10.012.

[129] Houser SR, Margulies KB, Murphy AM, Spinale FG, Francis GS, Prabhu SD, et al. AHA Scientific Statement.Animal Models of Heart Failure. *Circulation Research.* 2012; 111:131-150.

[130] Yang JH, Saucerman JJ. Computational models reduce complexity and accelerate insight into cardiac signalling networks. *Circulation Research.* 2011; 108:85-97.

[131] Soulis JV, Giannoglou GD, Chatzizisis YS, Farmakis TM, Giannakoulas GA, Parcharidis GE, Louridas GE. Spatial and phasic oscillation of non-Newtonian wall shear stress in human left coronary artery bifurcation: an insight to atherogenesis. *Coron Artery Dis*. 2006; 17:351-358.

[132] Genovesi S, Pogliani D, Faini A, Valsecchi MG, Riva A, Stefani F, et al. Prevalence of atrial fibrillation and associated factors in a population of long-term hemodialysis patients. *Am J Kidney Dis*. 2005; 46:897-902.

[133] Reinecke H, Brand E, Mesters R, Schabitz WR, Fisher M, Pavenstadt H et al. Dilemmas in the management of atrial fibrillation in chronic kidney disease. J Am Soc Nephrol. 2009; 20:705-711.

[134] Abbott KC, Trespalacios FC, Taylor AJ, Agodoa LY. Atrial fibrillation in chronic dialysis patients in the United States: Risk factors for hospitalization and mortality. *BMC Nephrol*. 2003; 4:1-10.

[135] Seidman CE, Seidman JG. Identifying sarcomere gene mutations in hypertrophic cardiomyopathy: a personal history. *Circ Res*. 2011; 108:743-750.

[136] Wang L, Seidman JG, Seidman CE. Narrative review: harnessing molecular genetics for the diagnosis and management of hypertrophic cardiomyopathy. *Ann Intern Med.* 2010; 152:513-520.

In: Recent Advances in Systems Biology Research
Editors: A. X. C. N. Valente, A. Sarkar and Y. Gao
ISBN: 978-1-62948-736-6

Chapter 12

Football, Airplanes, Diabetes and Systems Biology

***Alexandra R. Petrasek*[1] *and Danny Petrasek*[2]**
[1] Tulane University, New Orleans, LA
[2] California Institute of Technology, Pasadena ,Ca

The Blind Man and the Elephant (modified from the original Kipling poem)

It was six men of Indostan, (Though all of them were blind),
who went to see an elephant, to satisfy his mind.

The first approached the elephant, and, happening to fall,
against his broad and sturdy side, 'Its nothing but a wall!'

The second feeling of the tusk, cried: 'Ho! what have we here,
this wonder of an elephant, is very like a spear!'

The third approached the animal, and with his hands did take
the squirming trunk ... 'I see,' quoth he, the elephants a snake!'

The fourth reached out his eager hand, and felt about the knee:
'Tis clear enough the elephant is very like a tree.'

The fifth, who chanced to touch the ear, Said; 'E'en the blindest man
This marvel of an elephant, is very like a fan!'

The sixth seized on the swinging tail, that fell within his scope,
'I see,' quothe he, 'the elephant is very like a rope!'

And so these men of Indostan, disputed loud and long,
Though each was partly in the right, and all were in the wrong!

Abstract

The spectacular success of the reductionist approach in Biology and Medicine has yielded an unprecedented treasure of detailed information characterizing individual components of cellular processes. A growing realization that a complete understanding of an individual isolated cell cannot predict the physiology of a more complex organism creates a new challenge. Systems biology as defined here represents the idea that new integrating principles are needed to understand the physiology at the organism level. This evolution in Biological science recapitulates the path that physics experienced many times when attempting to amalgamate disparate phenomena into a unified principle or system. Physics is our most successful predictive science and the lessons gleaned from its approach are likely of great value in advancing Biology and Medicine. In this chapter we survey several physiological and clinical phenomena with the idea that solutions for them may be motivated by principles adapted from physical science.

I. Introduction: Football, Planes, Automobiles, and Systems Biology

Imagine you are trying to understand the rules of American Football. The first approach could be to study photographs of the various objects involved in the game: the players on each team, the ball, the cheerleaders and fans. What are the chances that you could even guess a single rule? An alternative approach might be to attend and view an entire game of football (even without commentary) and try to make sense of the play. I believe we would all agree that the probability of deciphering the rules of football is increased with the second method. The latter represents a systems approach. The view held by Systems Science is that there are behaviors, properties and dynamics that can only be appreciated in an integrated context as opposed to studying individual components (reductionist approach). One has little hope of discerning the rules of football by studying the shape of the ball or staring at photographs of the players. To Illustrate this view further , suppose you are presented with a box containing every single part belonging to a Boeing 777 commercial airline (approximately 150,000 components). By studying these parts, what are the chances that you could figure out how airplanes fly? If the parts were from a car, would the problem be any easier? In other words, the reductionist view often implies that a system is merely a sum of its parts, as opposed to the systems view, which holds that certain properties are discernible only by observing and studying the integrated system.

Defining Systems Biology

The term systems biology has unfortunately already taken on multiple definitions. The one definition that is consistent with my view is described by Leroy Hood [1, 2], founder for the Institute of Systems Biology in Washington: "We call these properties and functions that arise from the interacting parts in a system 'emergent properties.' The concept of emergent properties is central to the study of systems. Any function performed by a system that is not the result of a single part in the system, but rather is the result of interacting parts in the system, is an emergent property."

An Information Treasure

The reductionist strategy has yielded tremendous information about the fundamental structure of biology. Yet, we have not been able to optimally leverage this information treasure into a deeper understanding of macro or whole-body physiology. In this chapter, we will describe selected examples that display emergent properties of a system of cells. Later we discuss new systems principles as applied to the formation of endocrine glands and the analysis and management of clinical Diabetes.

II. The Reductionist Approach has Dominated Biological Science for the Past Half Century

Reductionism is generally defined as an approach to understanding the nature of complex things by reducing them to their component parts (or to simpler or more fundamental things), implying that a complex system is nothing but the sum of its parts. The pursuit to understand and eradicate human disease has always been the central motivation for medical science. In that regard, the reductionist approach, beginning with the identification of the cell as a unit of living matter and the consequential observation of sub-cellular structures, launched an extremely fruitful course which ultimately achieved landmark discoveries. The most notable among these include the double helical DNA molecule and the genetic code. Moreover, the recent decade has yielded an accelerating and unprecedented amount of molecular and cell biological data which symbolically culminated with the complete sequencing of the human genome.

The challenges in analyzing and integrating these results are ongoing and have inspired whole new fields of biological computation, genomics, and informatics. For Clinicians however, the need has always been to understand and exploit the relationship between molecular mechanisms and system (whole body) physiology. The gulf between our knowledge of the cell and molecular physiology and our understanding of the macro physiology underscores the need to establish systems biology/physiology principles.

III. The Limits of the Reductionist Approach: Can a "Complete" Knowledge of a Single Cell Explain All of Human Physiology?

Until recently many if not most biologists held the belief that a full understanding of a single cell in isolation will yield a complete understanding of all biology. This idea has been eroded if not flat out rejected [3]. Countless examples in developing cancer drug therapy demonstrate on the one hand, efficacy of candidate pharmaceuticals against in vitro cancer cell lines and on the other hand, their spectacular failure when scaling up to treat an animal or human being. The following examples from developmental biology and endocrine physiology demonstrate the divergent behavior between single cell biology and systems biology.

The Holy Grail of Biology: Cell Differentiation

We all begin life as a single self replicating cell, creating exact copies and forming a growing continous sheet without any end in sight. At one miraculous moment, this sheet of identical cells differentiate to become the nervous system, the musculo-skeletal system, or the circulatory system! Given that every one of our cells has the identical genome, how do the cells know to determine their fate? Moreover, how do terminally differentiated cells maintain their identity? Escape from this identity defines neoplasia and in its most evil manifestation, carcinoma. What is the cell-to-cell communication that permits this coordinated plan? Included in the answer to this question will be the solution to why we age!

C. elegans and Replaceable Parts

Developmental biologists are gaining insights into how the system or the collection of cells may influence the behavior of individual cells. The formation of organs in *Caenorhabditis elegans* (*C. elegans* is a worm commonly used as a biological model) represents a nicely documented example of the patterns that occur repeatedly while a system is intact. It should be appreciated that *C. elegans,* as a biological model, is transparent. Every cell in the living animal can be seen under the microscope from the fertilized egg to the 556 cells of the newly-hatched worm. Later, the 959 somatic cells, and a variable number germ cells of the adult worm can also be studied via light microscopy. Its entire genome has been sequenced and the fate of every single one of its cells can be tracked from conception. Remarkably, during *C. elegans* vulvar development, individual cells are replaceable parts. Paul Sternberg [4, 5] has shown that the individual cells making up the worm's vulva take on different roles depending on their spatial positioning. The anchor cell is the central cell whose function determines the fate of its neighbors. If the cell that is in position to be the anchor cell is removed from its position and exchanged with a neighboring cell, their respective roles will be reversed! Here is a demonstration that there is a system pattern or rule that is dependent on geometric or spatial factors, rather than an individual cell independently governing its own fate.

The Brilliance of Biological Architecture and the Role of Calcium

The role of calcium in cell signaling processes is nearly ubiquitous. Every one of our cells is continuously processing messages using calcium ions. The incredible fact that our fundamental physical structure, bone, is almost entirely composed of calcium--the very same essential signaling molecule--should cause us to pause and reflect at this marvelous biological engineering solution. It would be similar to an airplane whose wings and fuselage were made up of a material that was also the energy source (or more precisely, the source of communication technology) needed to run the airplane; perhaps we should design wings out of photovoltaic solar panels.

In mammals, serum calcium is tightly regulated by the parathyroid gland. The parathyroids secrete a hormone, PTH (parathyroid hormone), that mobilizes calcium from bone. PTH is then transported to distal cells and utilized in cellular signaling events. The specialized parathyroid cells possess unique calcium sensing receptors that when bound with calcium, inhibits PTH secretion, thus creating a feedback mechanism.

IV. Gestalt Physiology, the Whole Is Not a Sum of its Parts

An important illustration of a systems biology phenomenon occurs in normal parathyroid physiology. Fitzpatrick and Leong [6] were interested in studying and comparing the difference in secretory function between single isolated parathyroid cells and the secretion of individual parathyroid cells residing in a connected group of cells or a gland. They demonstrated that the parathyroid hormone secretion of a single cell within a group is significantly higher than the individual isolated cell. They further showed this to be a function of its connections to neighboring cells. Here is a simple but fundamentally important demonstration of an emergent property that is only visible in the physiology of a group/system of cells. This seemingly trivial observation has dramatic consequences when scaling up into a tumor!

Parathyroid adenomas are benign tumors composed of the parathyroid cells. These tumors no longer respond to calcium inhibition (calcium resistance) and in patients they create hypercalcemia, which causes serious pathology including severe bone demineralization and cardiac arrhythmias. Andre Van Herle and colleagues demonstrated that the property of "calcium resistance" (analogous to insulin resistance) of parathyroid tumors is a systems feature and not a feature of individual cells. Goodman [7] et al. studied parathyroid adenomas *in vitro* and showed that the production and secretion of PTH by the cells within the intact adenoma was nearly impossible to inhibit with increasing concentrations of calcium. When the very same cells were dispersed into small fragments or individual cells, the calcium resistance vanished and PTH was completely inhibited. The calcium resistance, occurring only in a group or connected system, is the basis for the disease and would not have been observed when studying the individual cell. The parallels to insulin resistance may be noteworthy. In a subsequent section of this chapter, I will offer a principle that could be the basis for how this phenomenon occurs.

V. Is Evolution a Smart Designer? Why Should We Have Endocrine Glands or Organs?

When contemplating the design and structure of a mammal, does it make sense to have specialized cells or glands all localized to a specific region in the body? The liver is localized to the right upper quadrant of our abdomen. It contains several specialized cells and functions that affect multiple physiological systems: hematological, endocrine and metabolism, gastrointestinal and more. The liver must export its products (coagulation factors, albumin, bile) to other areas of the body. One could argue against the design of the liver as an organ. The liver is physically vulnerable and could be injured in an accident. The pancreas is a gland that also contains multiple specialized regions with cells that export proteins (endocrine) to cells far away. The beta cells of the pancreas are constantly monitoring the blood glucose levels and adjusting their secretory amplitude and frequency as the need arises. Would it not make as much or even more sense if instead of cells relying on insulin from the pancreas or proteins from the liver, each local area of cells had their friendly neighborhood insulin and liver cells. This would satisfy all the local needs for glucose metabolism without having to invoke the entire pancreatic pathway. Every region would autonomously work independently of every other region. In fact, if there was an injury to one area of the body, it wouldn't affect the local function in other areas.

VI. Geometry and a Physiological Mechanism That Supports the Reason for Organs

Evolution seems to have provided us with the opposite solution in the form of endocrine glands. The features of the parathyroid gland, referenced above, demonstrate the efficiency of having cells in a group (i.e., an individual cell residing in a group of cells will produce significantly higher PTH for the same calcium concentration than an isolated cell). This extra efficiency is permitted by the spatial arrangement of the cells in groups that resemble a football huddle. These cells appear to interdigitate and this cellular handholding takes place in the precise location of the calcium sensing receptor. Nature "cleverly" places the calcium sensing receptors precisely in the same tiny cellular crypt--or location--where the PTH exocytotic secretion takes place. *Thus, any cell that discharges PTH obstructs the path of the inhibiting calcium molecules for the entire group,* which initiates a conflagration of PTH secretion that only subsides when the PTH is washed away. This mechanism [8] allows a group of cells (and in fact an entire gland) to respond to tiny fluctuations in serum calcium, in a robust manner, similar to how the central water department in a city responds to small changes in the cities water pressure. The energy savings using this mechanism, or one similar to it, may be at play in the beta cells of the pancreas and other glands. This again is a solution to an aspect of physiology that makes possible the required efficiency needed to support an increasingly bigger organism.

VII. Is There Any Hope to Understanding Systems Without Complete Details of Individual Parts? Lessons from Physics

A key philosophical and scientific question for a systems approach is the following: if there is an **incomplete understanding** of the details of the components of the system, is it still possible to form legitimate macro-physiological or systems biology principles? Clearly, if it ever happens, it will take some time to decipher all the molecular and cell biological rules that govern the complete function of a cell. Is it then folly to consider forming a systems biology approach given that we do not have all the detailed facts? A good heuristic for evaluating this question is to look at other scientific disciplines that have faced similar questions and examine the ideas and techniques that have worked there.

A notable success in physics is the field of thermodynamics. The laws of thermodynamics [9] are a fundamental part of how we understand nature. They, in fact, are accurate descriptions of ensembles of countless "invisible" components that, interacting under prescribed conditions, behave in predictable patterns. It would be futile, for example, to model the trajectory of individual molecules and their interactions with billions of other molecules. Here then is an example of a systems solution that does not need specific details of the components.

Another example is the theory of elasticity (Hooke's law), where the governing formula does not in fact need the detailed quantum mechanical description of each individual atom to accurately describe or predict the system properties that emerge. As physicians and biological scientists, we could ask ourselves whether there are systems approaches in other fields that could be applied to medical science and specifically to the understanding of endocrine physiology and diabetes.

VIII. Newton's Law and Diabetes: *F=ma*

We have all been passengers in cars cruising at 65 mph (100 km/hr) or in jets at 500 mph (800 km/hr) and yet, we don't feel like we are moving at all. However, we absolutely *do* feel the jolt of a slam on the brakes or a sudden jerk of an elevator moving upwards. That is, *what we really sense or feel are the accelerations or decelerations!*

This is a fundamental law of nature embodied in the most famous equation of classical physics: Newton's second law of motion. It states that ***force equals the product of mass and acceleration (F=ma)***. The law implies that forces are accelerations, and accelerations are forces.

If all we really sense are accelerations, then symptomology is really a reflection of our body's forces or accelerations. We are reminded of the "slow boil" of a frog experiment attributed to a 19th century physiologist. The claim is that as long as the temperature is very slowly raised a frog placed in a pot of water on a lit stove won't sense the increase in temperature even when the water reaches the boiling point. The clinical example that follows is a specific case of how ***F=ma*** may be applied as a simple systems physiology principle.

Endocrine Rounds and the Mysterious Case of Pseudo-Hypoglycemia

During my years as an endocrine fellow at UCLA, we were often summoned urgently by the nurses to examine our diabetic patients that were purported to be experiencing severe hypoglycemic episodes. The usual setting involved an in-hospital patient taking either oral agents or routine insulin therapy and being monitored by the "diabetes team". The patient's blood sugar was routinely sampled every 4 hours and the therapy administered accordingly. When such a patient began complaining of sudden and severe sweating, nausea, and light headiness, the nursing staff followed an algorithm that required measuring the glucose at once and then giving orange juice or administering a dextrose injection for the presumed hypoglycemia. Commonly we experienced that the hypoglycemic attacks were not accompanied with actual low blood sugar levels. With one specific patient, the blood sugar was either normal or more often significantly elevated. This presented a clinical conundrum that we struggled to explain. Why were patients who had normal or very elevated blood sugars experiencing hypoglycemic attacks? The morning blood sugar was, for example, 300 mg/dl, a dose of insulin was given, less than an hour would pass, a hypoglycemic attack would occur, and the measured blood sugar during the "hypoglycemic attack" was 200mg/dl. Moreover, our continuous over treatment of the blood glucose created a chaotic and unsustainable treatment cycle. We re-checked the glucometers, we questioned the patient about the symptoms, and we often began investigating alternative diagnoses. We also increased the frequency of monitoring the glucose. It became apparent that what seemed to be inducing these hypoglycemic attacks was not the glucose level but the steepness or deceleration of the glucose concentration! We concluded that the glucose plunged so steeply with insulin treatment that the deceleration in blood glucose elicited a force leading to "hypoglycemic type" reaction despite being euglycemic! We anecdotally proved this by changing the regimen to a slower acting insulin therapy--changing the slope of the glucose curve and subsequently avoiding the hypoglycemic responses completely. I submit that the application of the macro physiological idea of $F=ma$ provided a practical guide for our therapy and helped create a reasonable clinical treatment.

IX. Rocket Science and Clinical Diabetes: The Conservation of Uncertainty

Control theory is a branch of engineering used for systems analysis and design. It contains an extremely useful conservation principle, or law, known by several names, one of which is Bode's integral formula [10] or the "conservation of uncertainty." This concept has found its way into biological science through the work of leading control theorist John Doyle and others. Csete and Doyle [11] examined this principle in the setting of glycolysis and refer to it as the "conservation of fragility." Different names address the application, but what is an invariant is that there exists a mathematical expression that precisely describes this principle and predicts certain outcomes with the same validity as other physical laws--something that is generally lacking in the field of medical science. I will now describe the concept (without the math) and then give some potential examples as to how it may have a role in diabetes, metabolism, and other biological systems.

Jets with Sturdy Wings and Flimsy Toilet Seats

The level of complexity in engineering design today is approaching that of biological systems. Control engineers have to contend with designing features and functions that must be robust with respect to many potential conditions. The conservation of uncertainty, simply stated, is the notion that any system has a fixed quantity or measure of uncertainty that is always present. This fixed amount of uncertainty may be shifted around or distributed among the component parts, but it can never be done away with. The flip side is that there is a conservation of the robustness as well. In an engineered system, such as our Boeing 777 commercial jet, where there are 150,000 connected parts, the design principle is aimed at maintaining "certainty" or stability in certain critical features, such as making sure the wings or engines are secure by bolting them redundantly. The conservation principle thus implies that while the certainty is increased in some components, such as the wings, uncertainty is increased automatically in others. The design strategy would therefore be to cause the uncertainty to be distributed in components whose function is less critical for the safety of the passengers, such as toilet covers or reading light switches. The intuition here becomes more apparent in simple systems. For example, a water pipe, where the first half is strengthened with extra materials, can withstand greater pressures and thus increase certainty or stability there and automatically increase uncertainty in the second half of the pipe. Therefore, as the water pressure increases, the likelihood of a fracture and collapse is largely in the second half.

Physics, Clinical practice and the Metabolic Syndrome

One can reasonably posit that when applying similar constraints on a physiological system or more practically the clinical management of patients ,we are also manipulating a system in a similar way. Imagine a patient who has the common cluster of high cholesterol, hypertension, resistant diabetes, and atrial fibrillation. We could impose a typical regimen of pharmacological therapies so that the certainty of some physiological and biochemical responses is increased. This can be done by restricting the cholesterol from rising above a certain level, maintaining the diastolic blood pressure under 90 mm Hg, attempting to restrict the glucose levels from rising beyond a prescribed level, treating with digoxin to restrict the pulse rate, and elongating the pro time to a certain "therapeutic" domain. All these measures impose increased physiological certainty, but from a systems understanding, we are clearly distributing uncertainties to unknown components and physiological functions. Although anecdotal, I am certain that many physicians have witnessed patients who are difficult to control when they try to corral multiple physiological functions. The resistance to controlling may be the body's way of informing us about deeper systems properties, such as adaptation and robustness, which are being studied intensely at the cellular level today. The ACCORD trial experience warrants a careful examination of our clinical intuition and the underlying principles.

X. Diabetes, DCCT, ACCORD Trial and Sudden Death: The Failure of Intuition

Past studies that aimed for "tight" control of blood glucose, such as the Diabetes Control and Complications Trial (DCCT) [12]. It concluded that lower hemoglobin A1c levels were associated with improved cardiovascular health. In Type II Diabetes, we often see the triad (metabolic syndrome) of hypertension and hyperlipidemia along with insulin resistance. The Action to Control Cardiovascular Risk in Diabetes (ACCORD) trial seemed like a natural extension of the DCCT. In some sense the ACCORD trial was simply confirming what was held to be good medical practice. That is, with "tight" control of glucose, hyperlipidemia and hypertension would result in improved mortality and morbidity. When the announcement to suddenly halt the intensive glucose-lowering arm of the ACCORD trial was made, it received tremendous press. The basis for stopping the trial was that there was a rising number of unexpected deaths from cardiovascular events in the treatment group as compared to the control group. No specific cause could be identified. This confounded our clinical intuition! One important difference in the ACCORD trial is the study population. In the DCCT trial, exclusion criteria eliminated concurrent hypertension and hyperlipidemia. Whereas in the ACCORD study, one of the objectives was to examine the effect of intensive glycemic lowering in subjects who had concurrent hyperlipidemia and hypertension. The group that we expected to perform better clinically was in fact the group that was experiencing statistically more deaths! If this unexpected tragedy had a tiny silver lining, it may be that we now have a legitimate controlled study that demonstrates the predicted outcome of the uncertainty principle described.

The conservation of uncertainty states that attempts to rein in more certainty in one area automatically creates uncertainty in another area. When applying this principle to physiology, and in particular the ACCORD study, it is conceivable that attempts to constrain or increase certainty in glycemic control, blood pressure, and hyperlipidemia all at the same time, may have induced uncertainty or vulnerability in other physiological functions, potentially resulting in a catastrophic event.

In light of the recent ACCORD findings, it will be important to consider a range of potential mechanisms and explanations that may be at play. The conservation of uncertainty predicts the consequences resulting from restricting the system's range of physiological responses. These might include increased fragility in the microvasculature leading to myocardial infarction, stroke, hypoglycemia, or other neuroendocrine problems. A precise event cannot be specified without a detailed model of the system's complete physiology. Moving forward, when dealing with the complexity of several interconnected physiological functions, careful consideration should be given to studying the system biology along with the specific molecular and cellular mechanisms.

Summary

A Complete Understanding of Biology and Medicine Will Likely Require Both the Reductionist and the Systems Approach

Clinicians have always been faced with navigating the "system biology" of their patients. A large experience has accumulated that guides the practicing physician in manipulating the physiology so that the healthy state is restored and maintained. These practices are often based on a phenomenological understanding of the physiology. For medical science, the goal has always been to understand and exploit the relationship between molecular mechanisms and system (whole body) physiology. The gap between our knowledge at the cell level and our understanding of the macrophysiology underscores this need.

For Diabetology, a systems approach seems uniquely appropriate. Diabetes has all the features of a control feedback system where many of the key parameters are measurable and capable of being manipulated.

The closed loop artificial pancreas, where a reservoir of insulin and a man-made sensor already exist, can be an important opportunity to test systems theory and create a new approach that may be physiologically more accurate than what we are able to do currently. The debate concerning "tight" glycemic control will naturally extend into the domain of continuous glucose monitoring and continuous insulin infusion.

From Photographs to Movies

Our collective understanding of glycemic health comes from data sets that are fundamentally different in nature: epidemiological, molecular, transgenic models, and clinical trials. These data are usually static in nature, meaning that they are snapshots in time; interpreting them becomes much like trying to understand the rules of football by studying a few photographs from distinct, isolated moments in the game. Once continuous monitoring becomes more prevalent, we will have a better chance at guessing the rules because we will be, in essence, watching the entire movie. The importance of features such as glucose oscillations and overshoots/undershoots, will become more amenable to study and analysis. There have already been strong suggestions in the diabetes literature that pulsatile insulin release is fundamental to better glycemic control.

A favorite slogan among biologists is "*our ignorance is almost complete!*" Just as physics had in the past, biology is currently undergoing a transition from being a dominantly observational science, to a predictive and quantitative science. The study of biology requires knowing both the forest and the trees, and as such, the hope is that a systems study of biology will be regarded as a strong complement to our traditional biological methods.

References

[1] Leroy Hood Expounds the Principles, Practice and Future of Systems Biology, *Drug Discovery Today* Vol. 8, No.10 May 2003

[2] Naylor S, Cavanagh J. Status of systems biology--does it have a future? *Drug Disc Today Biosilica*. 2004;2(5):171-4.

[3] Yang J, Mani SA, Donaher JL, Ramaswamy S, Itzykson RA, Come C, Savagner P, Gitelman I, Richardson A, Weinberg RA. Twist, a master regulator of morphogenesis, plays an essential role in tumor metastasis. *Cell*. 2004;117(7):927-39.

[4] Karbowski J, Schindelman G, Cronin CJ, Seah A, Sternberg PW. Systems level circuit model of C. elegans undulatory locomotion: mathematical modeling and molecular genetics. J Comput Neurosci. 2007 Sept.

[5] Sherwood DR, Sternberg PW (July 2003). "Anchor cell invasion into the vulval epithelium in C. elegans". *Dev. Cell* 5 (1): 21–31.

[6] Fitzpatrick LA, Leong DA. Individual parathyroid cells are more sensitive to calcium than a parathyroid cell population. *Endocrinology*. 1990;126(3):1720

[7] Goodman WG, Veldhuis JD, Belin TR, Van Herle AJ, Juppner H, and Salusky IB. Calcium-sensing by parathyroid glands in secondary hyperparathyroidism. *J Clin Endocrinol Metab* 83:

[8] Petrasek, Danny, Ravi Samtaney, and Donald S. Cohen. Glandular regulation of interstitial diffusion: a model and simulation of a novel physiological mechanism. *Am J Physiol Endocrinol Metab* 283: E195–E206, 2002; 10.1152/ajpendo.00306.2001.

[9] Di Cera E, editor. Thermodynamics in biology. Oxford University Press; 2001

[10] Bode H (1945) Network Analysis and Feedback Amplifier Design. Melbourne, FL: Krieger

[11] Csete ME, Doyle JC (2002) Reverse engineering of biological complexity. *Science* 295: 1664–1669

[12] The Diabetes Control and Complications Trial Research Group. *N Engl J Med* 1993; 329:977-98

[13] Effects of Intensive Glucose Lowering in Type 2 Diabetes. The Action to Control Cardiovascular Risk in Diabetes Study Group. *N Engl J Med* 2008; 358:2545-2559.

In: Recent Advances in Systems Biology Research
Editors: A. X. C. N. Valente, A. Sarkar and Y. Gao
ISBN: 978-1-62948-736-6

Chapter 13

Cardiac Systems Biology

Seong-Eui Hong[1], Hong Ki Song[1], Taeyong Kim[1], Pradeep Kumar Sreenivasaiah[1] and Do Han Kim[1,*]
[1]School of Life Sciences and Systems Biology Research Center,
Gwangju Institute of Science and Technology (GIST),
Gwangju, Republic of Korea

Abstract

Cardiac systems biology has also been greatly advanced due to the recent development of high-throughput (HT) technologies such as the next-generation sequencing (NGS) and liquid chromatography-tandem mass spectrometry (LC-MS/MS). In this chapter, the main focus is on the introduction of cardiac systems biology which is advanced through the accumulation of data generated from the HT techniques.

Keywords: Heart, Systems Biology

1. Genomics

1.1. Genome-Wide Association Study for Cardiac Diseases

Over the last decade, genome-wide association study (GWAS) has contributed to the accumulation of data for genetic variants associated with a trait across individuals. After the completion of the human genome project in 2003, "the 1,000 Genomes Project" has accelerated the detailed catalog of genomic variants such as single nucleotide polymorphisms (SNPs), deletions, insertions and copy number variations (CNVs). According to the National

* E-mail: dhkim@gist.ac.kr.

Human Genome Research Institute (NHGRI), the current catalog includes >8,000 SNPs from ~1,500 publications (Hindorff et al., 2009).

More than 90 cases of GWAS on the heart have identified ~500 SNPs for 34 cardiac phenotypes and diseases. For example, Ellinor et al. recently reported the genetic variants in 6,707 individuals (European ancestry) with atrial fibrillation and 52,426 individuals (European ancestry) without atrial fibrillation (Ellinor et al., 2012). The authors had identified atrial fibrillation susceptibility loci: PRRX1, WNT8A, CAV1, C9orf3, KCNN3-PMVK, SYNE2, HCN4, ZFHX3, PITX2 and SYNPO2L. Among them, four loci (ZFHX3, PRRX1, CAV1 and KCNN3-PMVK) were overlapped with GWAS on individuals having Japanese ancestry. Among the test individuals, 843 subjects were with and 3,350 subjects were without atrial fibrillation. Lu et al. have contributed to identification of genetic variants on various cardiac phenotypes which were conducted by GWAS on the individuals having Chinese Han ancestry. Among the test individuals, 1,515 were with and 5,019 were without coronary artery disease (CAD) (Lu et al., 2012). They identified four novel loci (TTC32-WDR35, GUCY1A3, C6orf10-BTNL2 and ATP2B1) and four susceptibility loci which were already identified in the European population. Among the total SNPs detected to date, 111 SNPs related to CAD are illustrated in Figure 1.

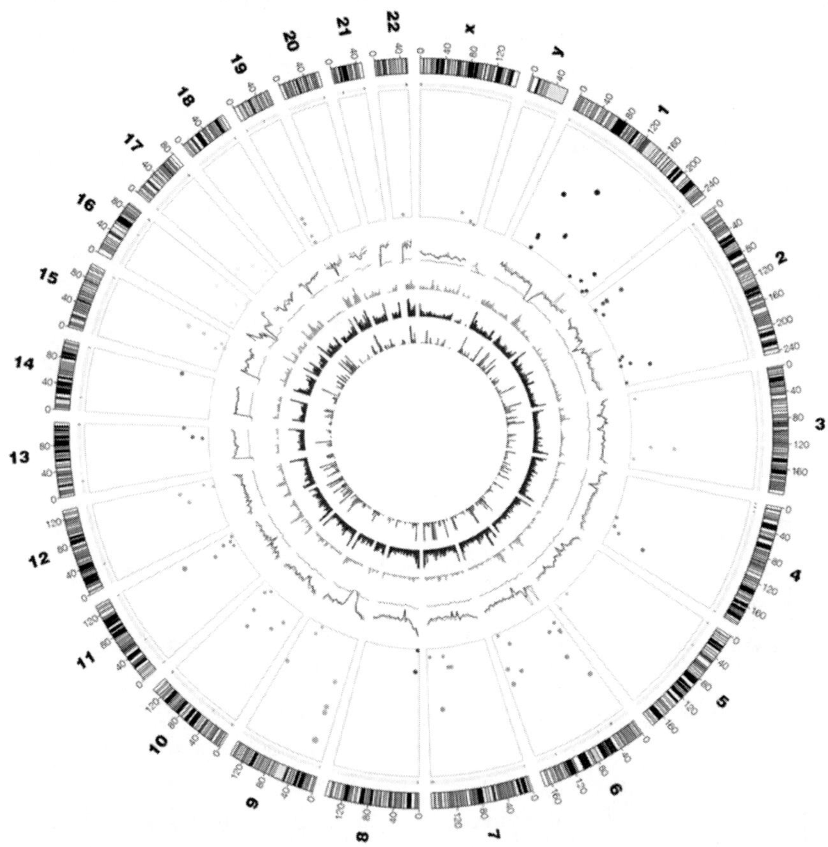

Figure 1. 111 SNPs related to coronary artery diseases were obtained from GWAS catalog of NHGRI (Hindorff et al., 2009) and are presented in colored circles using GWASrap (Li et al., 2012). Each SNPs were investigated by GWAS approach with a significance of $p < 1.0 \times 10^{-5}$. Detailed information for each SNPs can be obtained from http://www.genome.gov/gwastudies.

Copy number variation (CNV) is a parameter known to be associated with various diseases. Approximately, 12% of human genome has CNV and this structural variation appears to be more frequently occurring than SNPs (Stankiewicz et al., 2010). Redon et al., identified 285 OMIM genes having CNVs which were determined by global mapping of CNVs in 270 human genomes. Among these CNVs, ~20 genes were closely related to cardiac diseases (e.g., CASQ2 and RYR2) (Redon et al., 2006). In case of sporadic congenital heart disease, Soemedi et al. found rare genetic deletions which were observed in ~5% of sporadic congenital heart disease. They further characterized paternally transmitted properties for 10 CNVs (Soemedi et al., 2012). For familial dilated cardiomyopathy (DCM), rare CNV occurred in the exon region of BAG3 gene. Norton et al., found a 8,733 bp deletion in exon 4 of BAG3 leading to the symptoms in DCM patients (Norton et al., 2011). The association of rare CNVs to cardiac diseases was also observed in left-sided congenital heart disease (LS-CHD). Hitz et al., newly identified 25 rare CNVs observed in ~10% of LS-CHD patients (Hitz et al., 2012).

Although the approaches to identify the genetic variants such as SNPs and CNVs for cardiac diseases have been emerging, in-depth studies for causing the effects of the genetic variants on cardiac diseases have not fully conducted. Especially, genetic susceptibility variants in noncoding regions (e.g., variants on 9p21 and 4q25) make it difficult to understand the detailed pathophysiological mechanisms (Gudbjartsson et al., 2007; McPherson et al., 2007; Helgadottir et al., 2008; Lemmens et al., 2009; Smith et al., 2009). However, continually accumulating knowledge for genetic variants and rapidly developing techniques would help understand the inheritable pathological mechanism of genetic variants for various cardiac diseases in the future.

1.2. Exome Sequencing

Owing to NGS technologies which have been developed recently, the sequencing speed and its cost have been dramatically decreased. However, the detailed catalogs for genetic variants such as SNPs and CNVs in large population will take a time. To further reduce the sequencing cost and to efficiently identify the potential variants, Sarah et al., designed the targeted capture method, so called exome sequencing for protein-coding regions comprising 180,000 exons in human genome (Ng et al., 2009). This powerful technique has been widely adopted to extend the current knowledge for genetic variants, especially disease-causing variants. NHLBI Grand Opportunity Exome Sequencing Project (ESP) was established to discover novel genes and mechanisms contributing to heart, lung and blood disorders.

Recent exome sequencing studies have great success in identification of cardiac disease-causing variants or mutations. Regalado et al., conducted exome sequencing for patients who suffered from thoracic aortic aneurysms leading to acute aortic dissections (TAAD) (Regalado et al., 2011).

The authors newly identified four mutations in SMAD3 genes and subsequent sequencing of families with TAAD revealed that these mutations were responsible for 2% of familial TAAD. Pathological mutations causing TAAD were also identified by Boileau et al., They revealed that whole-exome sequencing of two unrelated families with TAAD showed the frame-shift mutation in exon 6 and a nonsense mutation in exon 4 in TGFB2 gene (Boileau et al., 2012). Besides TAAD, the precise knowledge for the variants and mutations responsible for the cardiac disease such as dilated cardiomyopathy (DCM) and congenital heart disease (CHD) have been recently reported (Arrington et al., 2012; Norton et al., 2012).

Exome sequencing is limited to detection of genetic variants occurred in only protein coding regions, however, this time and cost-effective approach is still attractive alternative to obtain high coverage in sequence depth. In near future, exome sequencing would be considered as robust diagnostic solution for various cardiac diseases.

2. Transcriptomics

2.1. Genes for Cardiac Diseases

Microarray technology has been highly successful in obtaining large-scale transcriptomic data. However, the technology has potential problems associated with poor specification, reproducibility and interplatform inconsistency. An alternative method to solve the problems is 'tag sequencing'. It has been reported that quantification of transcription by tag counting—via e.g., Solexa Genome Analyzer—is more reliable than by using various microarray platforms ('t Hoen et al., 2008). Moreover, recent advances in HT cDNA sequencing (RNA-Seq) have enabled to identify novel genes and splice variants, and to quantify genome-wide expression profiles through a single assay.

In this section, we will describe a HT method of NGS for detecting mRNA. Contrary to conventional microarray techniques, sequence-based approaches can directly determine cDNA sequences.

The first step of RNA-Seq is generation of a cDNA library with adapter molecules attached to either one or both ends to a given sequence. This cDNA library is then sequenced in HT manner to obtain the insert sequences (Wang et al., 2009).

The first report for heart transcriptome analysis using RNA-Seq was published in 2011. Matkovich et al. used RNA-Seq to investigate the transcriptomic profile of pathological cardiac hypertrophy induced by Gα transgenic mice. In the study, they obtained 18.2 million tag reads from 8 heart cDNA libraries using Illumina's genome analyzer. Subsequent analysis also revealed that RNA-Seq resulted in more reproducibility and higher reliability compared to that of microarray (Matkovich et al., 2010).

In our laboratory, we conducted a separate study to clarify the mechanisms for physiological cardiac hypertrophy (PHH) and pathological cardiac hypertrophy (PAH) by transcriptomic profiling with the use of RNA-Seq. Among ~13 million reads for both hypertrophy models, a dramatic transcriptomic remodeling was found in PAH with 8.3 times more abundant differentially expressed genes (DEGs) than in PHH. The unique transcriptomic signatures were also captured through the identification of exon variants occurring during hypertrophy.

Numerous genes undergo alternative splicing and are involved in transcription and phosphorylation related to gain or loss of protein domains, changes of activity and localization (Song et al., 2012).

The major obstacle of RNA-Seq is still the cost and the limited bioinformatics tool for biologists. However in the near future, the cost of sequencing will fall continuously and user friendly web-based analysis tools (Galaxy; https://main.g2.bx.psu.edu) will become available.

2.2. Transcription Factors Associated with Cardiac Diseases

Transcription factors (TF) are proteins that bind to specific DNA sequence, thereby regulating the expression of transcripts. This function of TF may act alone or by interacting with other proteins. Therefore, the identification of TF regulating gene is crucial for understanding the mechanism behind the entire gene network. Chromatin immunoprecipitation followed by NGS (ChIP-Seq) is one method to identify TF binding regions. ChIP-Seq is used for analyzing genome-wide profiling of DNA-binding proteins. Identification of genomic regions that control specific TF could serve as an important tool to understand the regulatory mechanisms for gene expression (Park 2009).

One recent study evaluated the genomic region that is controlled by multiple TFs in the atria originated HL1 cell line. Among many TFs, NKX2-5, GATA4, TBX, SRF and MEF2A were focused, because these factors are essential for initiation and maintenance of cardiac gene expression. After overexpression of biotinylated TFs in HL-1 cells, they performed biotin-mediated ChIP-Seq. As a result, they identified thousands of prospective cardiac regulatory sequences and suggested a possibility of multiple TF co-occupancy in the same genomic region (He et al., 2011). Another study by Matthew et al. was focused on the enhancer associated protein p300. With ChIP-Seq with p300, 130 candidate regions were found in mouse embryonic heart tissue (Blow et al., 2010). Since ChIP-Seq is a powerful tool for understanding the regulatory effect of TFs, the accumulation of ChIP-seq data for cardiac TFs will facilitate the understating of gene regulatory network in the heart.

2.3. miRNA Involving Post-transcriptional Regulation

MicroRNAs (miRNAs) are endogenous ~22 nucleotide RNAs mediating the gene expression as a post-translational regulator. These small RNAs are involved in complementary base-pair interactions with 3'-non-coding region of specific messenger RNAs, thereby negatively regulating target gene mRNA stability or translation. Growing evidence supports the notion that miRNAs have crucial roles in thc heart, especially in the biological processes such as embryonic development, apoptosis, and hypertrophy. In 2012, Yuanxin et al. identified 715 miRNAs in the heart using miRNA-Seq. Among them, they found 62 cardiac hypertrophy-associated microRNAs. In their study, after comparison between miRNA and mRNA profiles, 66 miRNA-targeted cardiac transcripts were identified (Hu et al., 2012). Hsu et al., identified the mRNA and miRNA profiles in human left and right atria using RNA-Seq and miRNA-Seq methods (Hsu et al., 2012).

The identification of target mRNA for specific miRNA is important to understand the biological function of miRNA. Recently, computational programs using sequence alignment algorithm (e.g., TargetScan and miRBase) have been established to predict specific miRNA target (Lewis et al., 2003; Kozomara and Griffiths-Jones 2011). However, miRNA and its target mRNA are usually less complementary and short seed sequence of miRNAs make false-positive results. The advance of miRNA profiling techniques with NGS makes it possible to identify the miRNA-targeted mRNA. RNA-induced silencing complexes (RISC)-Sequencing (RISC-Seq) is one of the important methods to identify the functional relationship between mRNAs and miRNAs. In RISC-Seq, isolating mRNAs sequestered by microRNAs into RISCs has been used to profile miR-targeted mRNAs. RISC-Seq is an useful method to identify the in vivo target of specific miRNA in heart (Matkovich et al., 2011).

3. Proteomics

3.1. Cardiac Proteomics

Recently, the knowledge on cardiac proteomes has rapidly growing due to various approaches. For instance, in-depth subproteomes for nucleus, mitochondria and extracellular matrix (ECM) have been investigated. Franklin et al. reported cardiac nuclear proteome of 1,048 proteins, using liquid chromatography-tandem mass spectrometry (LC-MS/MS) (Franklin et al., 2011). They applied powerful analytical techniques to identify distinct protein isoforms including 54 histone variants. Mitochondrial proteome is also essential to understand energetics in the constantly beating heart. Label-free MS performed by Goudarzi et al. identified four important biomarker candidates (four and a half LIM, destrin, heat shock protein 2 and chaperonin-containing TCP1) for atrial fibrillation (AF) with the use of mitochondrial proteomes of patients (Goudarzi et al., 2011). Barallobre-Barreiro et al. identified several novel ECM molecules contributing to cardiac remodeling in ischemia/reperfusion model using LC/MS-MS (Barallobre-Barreiro et al., 2012). One of the obstacles in cardiac proteomics is associated with the complicated protein interactions. To understand the cardiac-specific proteasome complexes, the interacting partners of 20S proteasome were identified using multi-dimensional chromatography and blue native polyacrylamide gel electrophoresis (BN-PAGE) together (Gomes et al., 2009).

Rapid advance in proteomic techniques makes it possible to generate large-scale proteome with significantly enhanced resolution. With enormous amount of data, the quantification technique is critical to identify most powerful candidates of biomarker. Recently, multiple reaction monitoring (MRM)/selective reaction monitoring (SRM) has emerged as a powerful quantification technique for proteome (Picotti et al., 2012). Keshishian et al. measured one to two signature peptides derived from six proteins including cardiac troponins for hypertrophic obstructive cardiomyopathy patients (Keshishian et al., 2009).

3.2. Cardiac Phosphoproteome

Alteration of post-translational modification (PTM) (e.g., phosphorylation, glycosylation, ubiquitination, nitrosylation, methylation, acetylation, lipidation and sumoylation) is involved in the pathogenesis of various heart diseases. So far, phosphoproteome for heart was most widely reported among the various PTM. As phosphorylation of protein is best described mechanism to regulate the various signaling pathways, the phosphoproteome response to signaling agonists or to certain cardiac diseases have been reported. For example, Chu et al., identified the phosphoproteome associated with cardiac β-adrenergic signaling using 2D gel electrophoresis in combination with ^{32}P autoradiography (Chu et al., 2004). Christensen et al., identified 1,183 phosphorylation sites regulated by angiotensin II using stable isotope labeling with amino acids in a cell culture (SILAC) method and further found the importance of protein kinase D in Gαq-pathways (Christensen et al., 2010). In addition, few efforts on quantitation of cardiac mitochondrial phosphoproteome have been reported (Lam et al., 2012).

Although the major obstacle for identifying phosphoproteome is associated with the extremely low coverage of phosphoproteins, the improved detection sensitivities of the new methods (e.g., immobilized metal affinity chromatography (IMAC), metal oxide affinity chromatography (MOAC, SRM/MRM)) are promising.

4. Metabolomics

4.1. Metabolomics for Cardiac Diseases

To date, over 40,000 metabolites have been identified and deposited in Human Metabolome Database (HMDB). Metabolomics is especially important for the constantly beating heart. Several lines of evidence have shown that metabolic transition is associated with various cardiac diseases (Depre et al., 1999; Lopaschuk et al., 2010; Rowe et al., 2010).

The metabolic changes have been recently observed in the various cardiac diseases. Shah et al., conducted mass spectrometry-based profiling of 69 metabolite in patients with coronary artery disease and suggested the importance of metabolites involving in the branched-chain amino acid pathway and urea cycle pathway (Shah et al., 2010). Turer et al., profiled 63 metabolites in patients with either coronary artery disease or left ventricular dysfunction during surgical ischemia/reperfusion (I/R) (Turer et al., 2009). The authors found the global suppression of metabolic fuel uptake in the dysfunctional ventricle due to I/R and further suggested perioperative metabolic monitoring in cardiac surgical patients. Detailed metabolome for I/R is well documented in the recent review (Rasmiena et al., 2013).

4.2. Combined -Omics Strategies to Study Heart Diseases

Several studies have been performed to examine the heart diseases using the combined strategies of metabolomics and other '-omics. For example, Gieger et al. interrogated the association of genetic variants and the 363 metabolites in 284 male participants (Gieger et al., 2008). In this study, the authors identified four genetic variants in the genes (e.g., FADS1, LIPC, SCAD and MCAD) involving in lipid metabolism via GWAS. These variants were significantly concordant with the metabolic capacities regarding to the fatty acid synthesis, beta-oxidation and breakdown of triglycerides. Another example of integrative analysis between metabolomics and proteomics was shown in the study investigating the effect of perhexiline on cardiac system. Yin et al., examined the cardioprotective effect of perhexiline, known as a potent anti-anginal drug, using combined proteomics and metabolomics in mice (Yin et al., 2012). The authors suggested that perhexiline might be involved in rebalancing mechanism between carbon and nucleotide phosphate fluxes to maintain cardiac output. Comparative study between transcriptome and metabolome was also performed. Lu et al., investigated the effect of fenofrate and fish oil targeting plasma trigelycerides in mice using combined transcriptomic and metabolomics approach (Lu et al., 2011).

Since metabolites are the metabolic end-product, metabolomics would contribute to the most intuitive diagnostic biomarkers. However, for better understanding of the pathological mechanism for cardiac disease, in-depth study for the transcriptomic- or proteomic signatures should be accompanied.

5. Structural Systems Biology

5.1. The Advent of Structural Systems Biology

The advent of "Structural systems biology" is due to the importance of the relationships between the structures of the parts (e.g., proteins) and the mechanistic interpretations how the parts function as modules and as a whole (Aloy et al., 2006; Telesco et al., 2012). Although a number of protein interaction networks have been published from various species using both experimental and computational methods, precise molecular details for the interactions are available for only limited number of cases (Aloy and Russell 2006; Telesco and Radhakrishnan 2012). Precise 3-dimensional (3D) structures solved by X-ray crystallography and NMR spectroscopy provide the crucial atomic details of protein bindings. However, due to the difficulties in obtaining large amounts of structural information for protein assemblies, the computational prediction methods are frequently used for structural systems biology. As of December, 2012, total 80,000 protein structures have been deposited in Protein Data Bank (RCSB; http://www.rcsb.org) and they are available for public use.

"Structural genomics" consortiums have been organized for determining the 3D structures of all proteins in a given organism by X-ray crystallography, NMR spectroscopy and computational methods. Structural genomics projects have attempted to do HT determination of protein structures. This has been performed in dedicated centers worldwide (e.g., SGC, PSI, CESG, RIKEN SGPI, PSF, OPPF, SPINE). Protein structures are closely linked to protein functions. Therefore performing structural genomics has to provide the knowledge of protein functions. However, in a number of cases the functional information still lacks and now some of the consortiums are putting a lot of emphases on understanding the structure-function relationships.

5.2. Structures of Protein Assemblies Can Be Determined by the Combined Methods of X-Ray Crystallography, NMR, Cryo-Electron Microscopy and Computational Methods

The individual protein structure can be determined now in a relatively short time with sufficient materials prepared using modern techniques. However, setting-ups for both purification and crystallization conditions studying the protein complexes are extraordinary tasks and they may require many years of efforts. Therefore, there is an enormous time gap between the progresses in developing protein networks from HT data and the progresses in solving the complex 3D protein structures. If the 3D structures of the protein assemblies are solved, they may provide information such as the strength of interactions. For instance, domain-domain interaction is supposed to be stronger than domain-peptide interaction.

Computational approaches have become important to determine the protein interactions in a confined biological system (Lu et al., 2003). A typical computational method is used to predict the structures of binding motifs and binding partners by such methods as homology modeling (Gray 2006). Some groups have developed methods predicting atomic details for pairs of interacting proteins (e.g., HADDOCK) (Lu et al., 2003). In the cases that the similar structural information is not available, the domain structures are analyzed bioinformatically and the domain-domain interactions are investigated on the basis of already reported information (InterPReTS) (Sprinzak et al., 2001). Peptides having a particular consensus sequence pattern or linear motif can be used for predicting binding partners (NetPhos; PhosphoELM; iSPOT) (Neduva et al., 2005).

5.3. Constructing Protein Assembly Models with Hybrids of High and Low Resolution Techniques

Putative structural models of protein assemblies in the mammalian systems have been obtained in combination of high and low resolution techniques (Aloy and Russell 2006; Vastrik et al., 2007; Telesco and Radhakrishnan 2012). For instance, the hybrid methods for determining the structure of small nuclear ribonucleoprotein particles (snRNPs) which bind pre-messenger RNA to form spliceosome were reported (Aloy and Russell 2006). To do their study, various methods such as X-ray crystallography, cryo-electron microscopy (EM) and chemical crosslinking were used. The hybrid approaches have been used to propose molecular structures of various molecular complexes (Aloy and Russell 2006; Telesco and Radhakrishnan 2012). Cryo-EM can even catch protein complexes in different conformational states (Muller et al., 2008).

The ryanodine receptor (RyR) (monomeric form: 550 kDa) is an intracellular Ca^{2+} release channel located in sarcoplasmic reticulum which is involved in the key processes in excitation-contraction coupling in striated muscles. It is an enormously large tetrameric hub protein in the muscles and could bind more than 20 intracellular and luminal proteins such as calmodulin, FKBP, triadin, junctin, and calumenin (Song et al., 2011). We have attempted to build up an assembly model using various methods such as X-ray crystallography, NMR, cryo-EM and the computational docking methods (Song et al., 2011). We could get insights in the mechanisms regarding to how the hub proteins interact with other proteins in signal transduction.

6. Modeling

Mathematical modeling and simulation are powerful approaches for characterizing the biological complexity. The application of modeling and simulation is especially important to investigate the cardiac systems, because various functional signaling, proliferative signaling and biophysical feedback mechanisms are involved in the cardio-vascular systems.

Mathematical modeling for cardiac system started with the cardiac action potential model based on Hodgkin-Huxley equation (Noble 1960). The cardiac model was further evolved to include the force generation of cardiac muscle (Rice et al., 1998; Rice et al., 2000; Niederer et al., 2006), calcium-induced calcium release process (Jafri et al., 1998; Hinch et al., 2006; Winslow et al., 2006) and mitochondrial ATP production (Cortassa et al., 2003; Vo et al., 2007). Recently cardiac modeling and simulation have been oriented to the signaling pathways.

Cardiac contractile function is enhanced by sympathetic stimulation. Catecholamine released from sympathetic nerve ending stimulates beta adrenergic receptor of cardiomyocyte and activates cAMP-dependent protein kinase, thereby activity of calcium handling machinery is positively regulated and cardiac contractility is enhanced. This process has been frequently targeted by modeling approach, because both cardiac inotropy and the development of cardiac disease are affected by the regulatory role of this pathway. Saucerman et al., developed a mechanistic systems model of beta adrenergic signaling pathway to examine the functional role of phospholamban, L-type calcium channel, ryanodine receptor and troponin I phosphorylation (Saucerman et al., 2004). Since the compartmentation of cAMP and PKA has been known to be important determinants of cellular response, spatiotemporal regulation of cAMP and PKA was modeled intensively by Saucerman et al. (Saucerman et al., 2006). With live-cell imaging, the modeling approach provides quantitative understanding of spatiotemporal information of the signaling pathways. Based on the models, Yang et al. extended the beta adrenergic signaling model including phospholeman to clarify its role in the signaling system (Yang et al., 2012). By simulating Na^+ and sarcoplasmic reticulum Ca^{2+} clamping, they identified the role of phospholeman and phospholamban: SR unloading and SR loading, respectively. Furthermore, phospholamban forms a negative feed-forward loop in beta adrenergic signaling cascade and dominates the beta adrenergic desensitization, thereby accelerating beta adrenergic receptor-induced inotropy. This approach is a prime example of modeling and simulation to unravel the intrinsic property of signaling system.

Calcineurin-NFAT signaling pathway is one of the most important signaling pathways for the development of cardiac hypertrophy. RCAN1 is a well-known key regulator of this signaling system. However, there has been a big debate on the functional role of RCAN1 on the regulation of calcinuerin-NFAT signaling between an inhibitory role and a facilitative role. Shin et al. developed a mathematical model in which the functional role of RCAN1 changes in dose-dependent manner (Shin et al., 2011).

7. Data Integration

Cardiac systems biology deals with massive amount of disparate multi-dimensional data now mostly being generated by HT technologies from cardiac tissue. The challenge is connecting these disparate data from multiple biological levels into one sophisticated useful resource, yet keeping it as a manageable simple for it to be used in computational modeling and analysis studies resulting in useful biological insights (Villoslada and Baranzini 2012; Powelland Acharya 2012; Bellasi et al., 2012; Patel et al., 2012; Svensson et al., 2012).

There are many methodologies of data integration put forth (Schneider and Jimenez 2012; Wu et al., 2012; Clarke et al., 2012; Nakayama et al., 2007; Carlson et al., 2010). The most common approaches to data integration including data warehousing (centralized databases), federated databases, hybrid databases (federated + centralized), multi-agent web service based systems, and the latest technologies such as semantic web technology and cloud computing paradigm (Sreenivasaiah and Kim 2010). Adoption of some of these are seamless process but some are very daunting task. Deciding which solutions to adopt depends on the task in hand and the availability of the resources (Hobbie et al., 2012; Lee and Smith 2012). Users are encouraged to refer to excellent reviews on these data integration technologies and their application (Seoane et al., 2012; Bernstein 2011; Gedela 2011; Kim et al., 2010; Hibbert et al., 2010).

The semantic technology is framed with common standard terminologies (ontologies), resource description language (RDF) and the Web Ontology Language (OWL) (Baitaluk and Ponomarenko 2010; Wang et al., 2012; Quan 2007). The core of semantic data integration is the concept of ontology (standard terminologies and relationship representing domain knowledge) (Assele et al., 2012). Here the ontology acts as a mediator for reconciliation of the heterogeneities between disparate data sources (Phillppi and Kohler 2004). Semantic technology is envisioned to revolutionize massive data integration over web and will enable machine to "understand" and respond to complex human request based on their meaning.

Adoption of globally accepted interoperability and information standards is the suitable path to achieve data integration (Sreenivasaiah et al., 2012). Significant work has been done in field of biological standards and ontologies. Users are encouraged to visit a consolidated resource for biostandards and ontologies (http://biostandards.info and http://www.bioontology.org) and also refer to literature pool reviewing these standards (Chervitz et al., 2011; Stromback et al., 2007; Musen et al., 2012).

Another set of useful resource other than collections of HT datasets are manual curated resources from literature (Sreenivasaiah et al., 2012; Crosft et al., 2011; Kanehisa et al., 2012). Generally these small-scale data sets are believed to have high data quality and have well defined context (Sreenivasaiah et al., 2012; Gong et al., 2008; Yurev 2012). Quality contextual data aids in accurate analysis of biological data and helps development of more meaningful hypothesis (Alam-Faruque et al., 2001; Grigoryev et al., 2008). However, it is not unlikely that errors get introduced into manually curated data (Soh et al., 2010; Cusick et al., 2009). Whereas, in future high quality data resources can be generated using new technologies of HT datasets and also innovative text mining approaches may aid in automatically extracting the required information precisely from the literatures (Kozhenkov and Baitaluk 2012).

The table provides highlights and links to databases and web-applications supporting the integration of ~omics data related to cardiovascular systems biology.

Databases and Tools	Description and Link
CADgene database	Coronary Artery Disease Gene Database (CADgene) which consists of information of the comprehensive CAD related genes manually curated from literature. http://www.bioguo.org/CADgene/
Cardio	A resource for molecular biology information on cardiovascular disease. http://cardio.bjmu.edu.cn/
Cardiovascular Comparative Genomics Database (CVCGD)	This database includes well-studied cardiovascular genes. http://pga.lbl.gov/cvcgd.html

Databases and Tools	Description and Link
Tox-database.net	Database aims to provide a scientifically useful, easily usable and clearly verifiable set of information describing not only IKr (hERG), but also other human cardiomyocyte specific ionic channels inhibition data (IKs, INa, ICa).
ER Stress induced pathway maps	This website provides an interactive diagram describing endoplasmic reticulum stress pathways with more emphasis towards its role in cardiac physiology and pathology. Available at : http://cidms.org/pathways/er_stress/index.html
IPAVS: Integrated Pathway Resources, Analysis and Visualization System.	IPAVS provides a collection of highly-structured manually curated human pathway data, it also integrates biological pathway information from several public databases and provides several tools to manipulate, filter, browse, search, analyze, visualize and compare the integrated pathway resources. Several pathways published by IPAVS are in of cardiovascular system. The key aspect of IPAVS is presentation of integrated data in biological contexts such as diseases, physiology, drug-action etc. Freely available at : http://ipavs.cidms.org
CardioGenomics	A legacy collection of data linking genes to function, dysfunction, and structural abnormalities of the cardiovascular system caused by clinically relevant genetic and environmental stimuli that was generated by investigators under the auspices of the CardioGenomics Program for Genomic Applications (PGA). http://cardiogenomics.med.harvard.edu/home
Human Cardiac Muscle Transcriptional Profile	The unigene-based human cardiac muscle transcriptional profile. The tissue specific function of the genes is calculated by counting number of ESTs corresponding to each UniGene entry. http://telethon.bio.unipd.it/GETProfiles/Cardiac_Muscle/index.html
HP-2DPAGE	The human myocardial two-dimensional electrophoresis protein database for human proteins. It contains data on proteins identified on the 2-D PAGE maps of ventricle and atrium of human heart.
Cardiac Integrated Database Management System(CIDMS)	CIDMS is an integrated web resource of genes expressed in heart and significantly coupled to cardiac patho-physiology. Currently CIDMS houses mammalian data (Human, Mouse, Rat).
Taverna workbench	Taverna workbench allows users to designing and executing complex workflows (created by the myGrid project) integrating many different tools and web services. Most importantly Taverna allows bench biologist to construct highly complex analyses over computational resources on all computational platforms (UNIX, PC or Mac). Freely available at: http://taverna.sourceforge.net
BioInv Index	Main aim of this project is to provide single entry point for researchers willing to deposit experimental datasets and /or download similar datasets. It supports multiple omic assay types. The data is stored in single format with standard metadata and terminologies. BioInvestigation Index is currently a prototype tool freely accessible at: http://www.ebi.ac.uk/net-project/projects.html

Databases and Tools	Description and Link
Biostandards	A consolidated resource for life sciences and biomedical standards. http://biostandards.info/
PharmaGKB	The Pharmacogenomics Knowledge Base which has some data annotated for cardiovascular system. Available at: http://www.pharmgkb.org
OMIM	**OMIM**, Online Mendelian Inheritance in Man, a **database** of human genes and genetic disorders includes several cardiovascular system disorders. Available at: www.ncbi.nlm.nih.gov/omim
GAD: Genetic association database	The **Genetic Association Database** is an archive of human genetic association studies of complex diseases and disorders. Includes annotations for several cardiovascular system disorders. Available at : *geneticassociationdb.nih.gov/*
KEGG pathway and disease database	KEGG pathway is a collection of manually drawn pathway maps representing known knowledge on the molecular interaction and reaction networks. There are pathways curated for human diseases which include cardiovascular diseases. Available at: www.genome.jp/kegg/pathway.html. KEGG disease database annotates different types of diseases, including single-gene (monogenic) diseases, multifactorial diseases. Few cardiovascular diseases are included. Available at : http://www.genome.jp/kegg/disease/

References

Alam-Faruque Y., Huntley R. P., Khodiyar V. K. et al., (2011). The impact of focused Gene Ontology curation of specific mammalian systems. *PLoS One* 6, e27541.

Aloy P., Russell R. B. (2006). Structural systems biology: modelling protein interactions. *Nat. Rev. Mol. Cell Biol.,* 7, 188-197.

Arrington C. B., Bleyl S. B., Matsunami N. et al., (2012). Exome analysis of a family with pleiotropic congenital heart disease. *Circ. Cardiovasc. Genet*., 5, 175-182.

Assele Kama A., Primadhanty A., Choquet R. et al., (2012). Data Definition Ontology for clinical data integration and querying. *Stud. Health Technol. Inform.,* 180, 38-42.

Baitaluk M., Ponomarenko J. (2010). Semantic integration of data on transcriptional regulation. *Bioinformatics,* 26, 1651-1661.

Barallobre-Barreiro J., Didangelos A., Schoendube F. A. et al., (2012). Proteomics analysis of cardiac extracellular matrix remodeling in a porcine model of ischemia/reperfusion injury. *Circulation,* 125, 789-802.

Bellasi A., Block G. A., Ferramosca E. et al., (2012). Integration of clinical and imaging data to predict death in hemodialysis patients. *Hemodial. Int.*

Bernstein P. A. (2011). Data integration for data-intensive science. *OMICS* 15, 241.

Blow M. J., McCulley D. J., Li Z. et al., (2010). ChIP-Seq identification of weakly conserved heart enhancers. *Nat. Genet.,* 42, 806-810.

Boileau C., Guo D. C., Hanna N. et al., (2012). TGFB2 mutations cause familial thoracic aortic aneurysms and dissections associated with mild systemic features of Marfan syndrome. *Nat. Genet.,* 44, 916-921.

Carlson D., Farkash A., Timm J. T. (2010). A model-driven approach for biomedical data integration. *Stud. Health Technol. Inform.,* 160, 1164-1168.

Chervitz S. A., Deutsch E. W., Field D. et al., (2011). Data standards for Omics data: the basis of data sharing and reuse. *Methods Mol. Biol.,* 719, 31-69.

Christensen G. L., Kelstrup C. D., Lyngso C. et al., (2010). Quantitative phosphoproteomics dissection of seven-transmembrane receptor signaling using full and biased agonists. *Mol. Cell Proteomics,* 9, 1540-1553.

Chu G., Egnaczyk G. F., Zhao W. et al., (2004). Phosphoproteome analysis of cardiomyocytes subjected to beta-adrenergic stimulation: identification and characterization of a cardiac heat shock protein p20. *Circ. Res.,* 94, 184-193.

Clarke D., Bhardwaj N., Gerstein M. B. (2012). Novel insights through the integration of structural and functional genomics data with protein networks. *J. Struct. Biol.,* 179, 320-326.

Cortassa S., Aon M. A., Marban E. et al., (2003). An integrated model of cardiac mitochondrial energy metabolism and calcium dynamics. *Biophys. J.,* 84, 2734-2755.

Croft D., O'Kelly G., Wu G. et al., (2011). Reactome: a database of reactions, pathways and biological processes. *Nucleic Acids Res.,* 39, D691-697.

Cusick M. E., Yu H., Smolyar A. et al., (2009). Literature-curated protein interaction datasets. *Nat. Methods,* 6, 39-46.

Depre C., Vanoverschelde J. L., Taegtmeyer H. (1999). Glucose for the heart. *Circulation,* 99, 578-588.

Ellinor P. T., Lunetta K. L., Albert C. M. et al., (2012). Meta-analysis identifies six new susceptibility loci for atrial fibrillation. *Nat. Genet.,* 44, 670-675.

Franklin S., Zhang M. J., Chen H. et al., (2011). Specialized compartments of cardiac nuclei exhibit distinct proteomic anatomy. *Mol. Cell Proteomics,* 10, M110 000703.

Gedela S. (2011). Integration, warehousing, and analysis strategies of Omics data. *Methods Mol. Biol.,* 719, 399-414.

Gieger C., Geistlinger L., Altmaier E. et al., (2008). Genetics meets metabolomics: a genome-wide association study of metabolite profiles in human serum. *PLoS Genet.,* 4, e1000282.

Gomes A. V., Young G. W., Wang Y. et al., (2009). Contrasting proteome biology and functional heterogeneity of the 20 S proteasome complexes in mammalian tissues. *Mol. Cell Proteomics,* 8, 302-315.

Gong L., Owen R. P., Gor W. et al., (2008). PharmGKB: an integrated resource of pharmacogenomic data and knowledge. *Curr. Protoc. Bioinformatics,* Chapter 14, Unit 14 17.

Goudarzi M., Ross M. M., Zhou W. et al., (2011). Development of a novel proteomic approach for mitochondrial proteomics from cardiac tissue from patients with atrial fibrillation. *J. Proteome Res.,* 10, 3484-3492.

Gray J. J. 2006. High-resolution protein-protein docking. *Curr. Opin. Struct. Biol.,* 16, 183-193.

Grigoryev D. N., Mathai S. C., Fisher M. R. et al., (2008). Identification of candidate genes in scleroderma-related pulmonary arterial hypertension. *Transl. Res.,* 151, 197-207.

Gudbjartsson D. F., Arnar D. O., Helgadottir A. et al., (2007). Variants conferring risk of atrial fibrillation on chromosome 4q25. *Nature,* 448, 353-357.

He A., Kong S. W., Ma Q., Pu W. T. (2011). Co-occupancy by multiple cardiac transcription factors identifies transcriptional enhancers active in heart. *Proc. Natl. Acad. Sci. U S A,* 108, 5632-5637.

Helgadottir A., Thorleifsson G., Magnusson K. P. et al., (2008). The same sequence variant on 9p21 associates with myocardial infarction, abdominal aortic aneurysm and intracranial aneurysm. *Nat. Genet.,* 40, 217-224.

Hibbert M., Lohrey J., Melnikoff S. (2010). Integration of data for research. *Stud. Health Technol. Inform.,* 151, 461-475.

Hinch R., Greenstein J. L., Winslow R. L. (2006). Multi-scale models of local control of calcium induced calcium release. *Prog. Biophys. Mol. Biol.,* 90, 136-150.

Hindorff L. A., Sethupathy P., Junkins H. A. et al., (2009). Potential etiologic and functional implications of genome-wide association loci for human diseases and traits. *Proc. Natl. Acad. Sci. U S A,* 106, 9362-9367.

Hitz M. P., Lemieux-Perreault L. P., Marshall C. et al., (2012). Rare copy number variants contribute to congenital left-sided heart disease. *PLoS Genet.,* 8, e1002903.

Hobbie K. A., Peterson E. S., Barton M. L. et al., (2012). Integration of data systems and technology improves research and collaboration for a superfund research center. *J. Lab. Autom.,* 17, 275-283.

Hsu J., Hanna P., Van Wagoner D. R. et al., (2012). Whole genome expression differences in human left and right atria ascertained by RNA sequencing. *Circ. Cardiovasc. Genet.,* 5, 327-335.

Hu Y., Matkovich S. J., Hecker P. A. et al., (2012). Epitranscriptional orchestration of genetic reprogramming is an emergent property of stress-regulated cardiac microRNAs. *Proc. Natl. Acad. Sci. U S A,* 109, 19864-19869.

Jafri M. S., Rice J. J., Winslow R. L. (1998). Cardiac Ca2+ dynamics: the roles of ryanodine receptor adaptation and sarcoplasmic reticulum load. *Biophys. J.,* 74, 1149-1168.

Kanehisa M., Goto S., Sato Y. et al., (2012). KEGG for integration and interpretation of large-scale molecular data sets. *Nucleic Acids Res.,* 40, D109-114.

Keshishian H., Addona T., Burgess M. et al., (2009). Quantification of cardiovascular biomarkers in patient plasma by targeted mass spectrometry and stable isotope dilution. *Mol. Cell. Proteomics,* 8, 2339-2349.

Kim T. Y., Kim H. U., Lee S. Y. (2010). Data integration and analysis of biological networks. *Curr. Opin. Biotechnol.,* 21, 78-84.

Kozhenkov S., Baitaluk M. (2012). Mining and integration of pathway diagrams from imaging data. *Bioinformatics,* 28, 739-742.

Kozomara A., Griffiths-Jones S. (2011). miRBase: integrating microRNA annotation and deep-sequencing data. *Nucleic Acids Res.,* 39, D152-157.

Lam M. P., Scruggs S. B., Kim T. Y. et al., (2012). An MRM-based workflow for quantifying cardiac mitochondrial protein phosphorylation in murine and human tissue. *J. Proteomics*, 75, 4602-4609.

Lee S., Smith C. A. (2012). Criteria for quantitative and qualitative data integration: mixed-methods research methodology. *Comput. Inform. Nurs.,* 30, 251-256.

Lewis B. P., Shih I. H., Jones-Rhoades M. W. et al., (2003). Prediction of mammalian microRNA targets. *Cell,* 115, 787-798.

Li M. J., Sham P. C., Wang J. (2012). Genetic variant representation, annotation and prioritization in the post-GWAS era. *Cell Res.,* 22, 1505-1508.

Lemmens R., Abboud S., Robberecht W. et al., (2009). Variant on 9p21 strongly associates with coronary heart disease, but lacks association with common stroke. *Eur. J. Hum. Genet.,* 17, 1287-1293.

Lopaschuk G. D., Ussher J. R., Folmes C. D. et al., (2010). Myocardial fatty acid metabolism in health and disease. *Physiol. Rev.,* 90, 207-258.

Lu L., Arakaki A. K., Lu H., Skolnick J. (2003). Multimeric threading-based prediction of protein-protein interactions on a genomic scale: application to the Saccharomyces cerevisiae proteome. *Genome Res.,* 13, 1146-1154.

Lu X., Wang L., Chen S. et al., (2012). Genome-wide association study in Han Chinese identifies four new susceptibility loci for coronary artery disease. *Nat. Genet.,* 44, 890-894.

Lu Y., Boekschoten M. V., Wopereis S. et al., (2011). Comparative transcriptomic and metabolomic analysis of fenofibrate and fish oil treatments in mice. *Physiol. Genomics,* 43, 1307-1318.

Matkovich S. J., Van Booven D. J., Eschenbacher W. H., Dorn G. W., 2nd. (2011). RISC RNA sequencing for context-specific identification of in vivo microRNA targets. *Circ. Res.,* 108, 18-26.

Matkovich S. J., Zhang Y., Van Booven D. J., Dorn G. W., 2nd. (2010). Deep mRNA sequencing for in vivo functional analysis of cardiac transcriptional regulators: application to Galphaq. *Circ. Res.,* 106, 1459-1467.

McPherson R., Pertsemlidis A., Kavaslar N. et al., (2007). A common allele on chromosome 9 associated with coronary heart disease. *Science,* 316, 1488-1491.

Muller S. A., Aebi U., Engel A. (2008). What transmission electron microscopes can visualize now and in the future. *J. Struct. Biol.,* 163, 235-245.

Musen M. A., Noy N. F., Shah N. H. et al., (2012). The National Center for Biomedical Ontology. *J. Am. Med. Inform. Assoc.,* 19, 190-195.

Nakayama M., Shibuya A., Inoue R. et al., (2007). An XML-based integration method for different types of patient data in cardiology. *AMIA Annu. Symp. Proc*: 1057.

Neduva V., Linding R., Su-Angrand I. et al., 2005. Systematic discovery of new recognition peptides mediating protein interaction networks. *PLoS Biol.,* 3, e405.

Ng S. B., Turner E. H., Robertson P. D. et al., (2009). Targeted capture and massively parallel sequencing of 12 human exomes. *Nature,* 461, 272-276.

Niederer S. A., Hunter P. J., Smith N. P. (2006). A quantitative analysis of cardiac myocyte relaxation: a simulation study. *Biophys. J.*, 90, 1697-1722.

Noble D. 1960. Cardiac action and pacemaker potentials based on the Hodgkin-Huxley equations. *Nature,* 188, 495-497.

Norton N., Li D., Rieder M. J. et al., (2011). Genome-wide studies of copy number variation and exome sequencing identify rare variants in BAG3 as a cause of dilated cardiomyopathy. *Am. J. Hum. Genet.,* 88, 273-282.

Norton N., Robertson P. D., Rieder M. J. et al., (2012). Evaluating pathogenicity of rare variants from dilated cardiomyopathy in the exome era. *Circ. Cardiovasc. Genet.,* 5, 167-174.

Park P. J. (2009). ChIP-seq: advantages and challenges of a maturing technology. *Nat. Rev. Genet.,* 10, 669-680.

Patel C. J., Chen R., Butte A. J. (2012). Data-driven integration of epidemiological and toxicological data to select candidate interacting genes and environmental factors in association with disease. *Bioinformatics,* 28, i121-126.

Philippi S., Kohler J. (2004). Using XML technology for the ontology-based semantic integration of life science databases. *IEEE Trans. Inf. Technol. Biomed.,* 8, 154-160.

Picotti P., Aebersold R. (2012). Selected reaction monitoring-based proteomics: workflows, potential, pitfalls and future directions. *Nat. Methods,* 9, 555-566.

Powell V. J., Acharya A. (2012). Disease prevention: data integration. *Science,* 338, 1285-1286.

Quan D. (2007). Improving life sciences information retrieval using semantic web technology. *Brief. Bioinform.,* 8, 172-182.

Rasmiena A. A., Ng T. W., Meikle P. J. (2013). Metabolomics and ischaemic heart disease. *Clin. Sci. (Lond),* 124, 289-306.

Redon R., Ishikawa S., Fitch K. R. et al., (2006). Global variation in copy number in the human genome. *Nature,* 444, 444-454.

Regalado E. S., Guo D. C., Villamizar C. et al., (2011). Exome sequencing identifies SMAD3 mutations as a cause of familial thoracic aortic aneurysm and dissection with intracranial and other arterial aneurysms. *Circ. Res.,* 109, 680-686.

Rice J. J., Jafri M. S., Winslow R. L. (1998). Modeling short-term interval-force relations in cardiac muscle. *Ann. N. Y. Acad. Sci.*, 853, 345-349.

Rowe G. C., Jiang A., Arany Z. (2010). PGC-1 coactivators in cardiac development and disease. *Circ. Res.,* 107, 825-838.

Saucerman J. J., McCulloch A. D. (2004). Mechanistic systems models of cell signaling networks: a case study of myocyte adrenergic regulation. *Prog. Biophys. Mol. Biol.*, 85, 261-278.

Saucerman J. J., Zhang J., Martin J. C. et al., (2006). Systems analysis of PKA-mediated phosphorylation gradients in live cardiac myocytes. *Proc. Natl. Acad. Sci. U S A,* 103, 12923-12928.

Schneider M. V., Jimenez R. C. (2012). Teaching the fundamentals of biological data integration using classroom games. *PLoS Comput. Biol.,* 8, e1002789.

Seoane J. A., Dorado J., Aguiar-Pulido V., Pazos A. (2012). Data Integration in Genomic Medicine: Trends and Applications. Contribution of the IMIA Working Group on Informatics in Genomic Medicine. *Yearb. Med. Inform.,* 7, 117-125.

Shah S. H., Bain J. R., Muehlbauer M. J. et al., (2010). Association of a peripheral blood metabolic profile with coronary artery disease and risk of subsequent cardiovascular events. *Circ. Cardiovasc. Genet.,* 3, 207-214.

Shin S. Y., Yang H. W., Kim J. R. et al., (2011). A hidden incoherent switch regulates RCAN1 in the calcineurin-NFAT signaling network. *J. Cell Sci.,* 124, 82-90.

Smith J. G., Melander O., Lovkvist H. et al., (2009). Common genetic variants on chromosome 9p21 confers risk of ischemic stroke: a large-scale genetic association study. *Circ. Cardiovasc. Genet.,* 2, 159-164.

Soemedi R., Wilson I. J., Bentham J. et al., (2012). Contribution of global rare copy-number variants to the risk of sporadic congenital heart disease. *Am. J. Hum. Genet.,* 91, 489-501.

Soh D., Dong D., Guo Y., Wong L. (2010). Consistency, comprehensiveness, and compatibility of pathway databases. *BMC Bioinformatics,* 11, 449.

Song D. W., Lee J. G., Youn H. S. et al., (2011). Ryanodine receptor assembly: a novel systems biology approach to 3D mapping. *Prog. Biophys. Mol. Biol.,* 105, 145-161.

Song H. K., Hong S. E., Kim T., Kim D. H. (2012). Deep RNA sequencing reveals novel cardiac transcriptomic signatures for physiological and pathological hypertrophy. *PLoS One,* 7, e35552.

Sprinzak E., Margalit H. (2001). Correlated sequence-signatures as markers of protein-protein interaction. *J. Mol. Biol.,* 311, 681-692.

Sreenivasaiah P. K., Kim D. H. (2010). Current trends and new challenges of databases and web applications for systems driven biological research. *Front. Physiol*., 1, 147.

Sreenivasaiah P. K., Rani S., Cayetano J. et al., (2012). IPAVS: Integrated Pathway Resources, Analysis and Visualization System. *Nucleic Acids Res.,* 40, D803-808.

Stankiewicz P., Lupski J. R. (2010). Structural variation in the human genome and its role in disease. *Annu. Rev. Med.,* 61, 437-455.

Stromback L., Hall D., Lambrix P. (2007). A review of standards for data exchange within systems biology. *Proteomics,* 7, 857-867.

Svensson E., van der Walt J. S., Barnes K. I. et al., (2012). Integration of data from multiple sources for simultaneous modelling analysis: experience from nevirapine population pharmacokinetics. *Br. J. Clin. Pharmacol.,* 74, 465-476.

‘t Hoen P. A., Ariyurek Y., Thygesen H. H. et al., (2008). Deep sequencing-based expression analysis shows major advances in robustness, resolution and inter-lab portability over five microarray platforms. *Nucleic Acids Res.,* 36, e141.

Telesco S. E., Radhakrishnan R. (2012). Structural systems biology and multiscale signaling models. *Ann. Biomed. Eng.,* 40, 2295-2306.

Turer A. T., Stevens R. D., Bain J. R. et al., (2009). Metabolomic profiling reveals distinct patterns of myocardial substrate use in humans with coronary artery disease or left ventricular dysfunction during surgical ischemia/reperfusion. *Circulation,* 119, 1736-1746.

Vastrik I., D'Eustachio P., Schmidt E. et al., (2007). Reactome: a knowledge base of biologic pathways and processes. *Genome Biol.,* 8, R39.

Villoslada P., Baranzini S. (2012). Data integration and systems biology approaches for biomarker discovery: challenges and opportunities for multiple sclerosis. *J. Neuroimmunol.,* 248, 58-65.

Vo T. D., Palsson B. O. (2007). Building the power house: recent advances in mitochondrial studies through proteomics and systems biology. *Am. J. Physiol. Cell. Physiol.,* 292, C164-177.

Wang Z., Gerstein M., Snyder M. (2009). RNA-Seq: a revolutionary tool for transcriptomics. *Nat. Rev. Genet.,* 10, 57-63.

Wang Z., Sagotsky J., Taylor T. et al., (2012). Accelerating cancer systems biology research through Semantic Web technology. *Wiley Interdiscip. Rev. Syst. Biol. Med.*

Winslow R. L., Tanskanen A., Chen M., Greenstein J. L. (2006). Multiscale modeling of calcium signaling in the cardiac dyad. *Ann. N. Y. Acad. Sci.,* 1080, 362-375.

Wu S., Xu Y., Feng Z. et al., (2012). Multiple-platform data integration method with application to combined analysis of microarray and proteomic data. *BMC Bioinformatics,* 13, 320.

Yang J. H., Saucerman J. J. (2012). Phospholemman is a negative feed-forward regulator of Ca2+ in beta-adrenergic signaling, accelerating beta-adrenergic inotropy. *J. Mol. Cell Cardiol.,* 52, 1048-1055.

Yin X., Dwyer J., Langley S. et al., (2012). Effects of perhexiline-induced fuel switch on the cardiac proteome and metabolome. *J. Mol. Cell Cardiol*.

Yuryev A. (2012). Contextual data integration in drug discovery. *Expert. Opin. Drug Discov.,* 7, 659-666.

In: Recent Advances in Systems Biology Research
Editors: A. X. C. N. Valente, A. Sarkar and Y. Gao
ISBN: 978-1-62948-736-6

Chapter 14

Systems Biology Approach to Transplant Immune Tolerance

Reza F. Saidi, M.D., F.I.C.S., F.A.C.S.[1,*]
and S. Kamran Hejazii Kenari, M.D.[1]
[1]Division of Organ Transplantation, Department of Surgery,
University of Massachusetts Medical School, Worcester, MA, US

Abstract

Organ transplantation is not only considered as the last resort therapy but also as a treatment of choice for many patients with end-stage organ damage. Recipient-mediated acute or chronic immune response is the main challenge after transplant surgery. Nonspecific suppression of host immune system is currently the only method used to prevent organ rejection. Lifelong immunosuppression will cause significant side effects such as infections, malignancies, chronic kidney disease, hypertension and diabetes. This is more relevant in children who have a longer life-expectancy so may receive longer period of immunosuppressive medications. Efforts to minimize or complete withdrawal of immunosuppression would improve the quality of life and long-term outcome of pediatric transplant recipients. System biology could potentially lead to studies to prevent rejection or induce tolerance by indentifying genes and their interaction which leads to allograft recognition and rejection.

Abbreviation

EBV:	Epstein-Barr virus
CNI:	calcineurin inhibitors
PTLD:	Post-transplant lymphoproliferative disorder
IS:	Immunosuppression (ive)

* E-mail: Reza.Saidi@umassmemorial.org; Tel: (508) 334-2023; Fax: (508) 856-1102.

CMV:	Cytomegalovirus
LTx:	Liver transplantation
MULTIPLE	
MYELOMA: MM	
OPTN:	Organ procurement and transplant network
PTSD:	Post traumatic stress disorders
HSV:	Herpes simplex virus
PSP:	
HLA:	Human leukocyte antigen
APC:	Antigen presenting cell
ESRD:	End-stage renal disease
MM:	Multiple myeloma
GVHD:	Graft versus host disease
LDLT:	Living donor liver transplantation
DDLT:	Deceased donor liver transplantation
Tx:	Transplantation
SCT:	Stem cell transplantation

Introduction

The concept of System Biology (SB) was first described by Hodgkin and Huxley. They showed the system of interactions between cell electrical potential and ion channel activity, which led to nerve conduction. [1] They received Nobel Prize in Physiology and Medicine in 1963. SB is a comprehensive and quantitative approach to study the interactions between different components of a biological system. Immune system is a complex apparatus consisted of cellular and humoral components. SB can perfrom a comprehensive analysis of complex interaction of different structures of immune system.

SB is trying to study the interactions between the different components of **biological systems**, and how these interactions give rise to the function and behavior of that system, such as allograft rejection. In the last decade, biology tried to explain biological phenomena in a reductionist approach, such as single gene or single factor. This approach has uncovered many components of the biological systems. However, to understand the dynamic properties of the system, the individual parts need to be studied in context with each other. SB integrates and combines data with or across molecular levels of the biological systems for analyzing their interaction, which leads to a function or effect.

This approach is now gaining ground in exploration of high-level physiologic functions. SB combines both “reductionist” and “integrationist” approaches to allow individual characteristics of different parts and their interaction with each other and their environment to be studied in more “holistic” manner. [2]

This approach tries to elucidate how parts and their interaction give rise to maintenance of the biological entity and to predict its physiologic behavior. This process requires advanced mathematical and computational modeling techniques. One of the main limitations of SB is the challenge in analysis and interpretation of the data. The analysis requires a sophisticated statistical method to estimate the significance of observed correlations and other dependencies.

System Biology and Study of Immune System

Utilization of SB strategies to immune system will enable comprehensive analysis of complex interactions involving host defense mechanism under normal and inflammatory/ infectious diseases. There are multiple interactions among different pathways that ultimately lead to T-cell activation. One way to study and analyze interaction among different and interconnected biomedical pathways is through SB. SB can further analyze the interaction of different components of immune system at different levels such as DNA sequencing, transcriptome analysis, signaling and proteomics. [3–5]

The most difficult challenge is SB is computation, analyzing and interpreting of the data. This needs people with different backgrounds from information technology, biomedical and computer science. SB approaches to study the immune system can range from measuring large number of changes in the hemostasis of immune system, to using quantitative methods to assess the relationship between different interconnected components such as cells and cytokines. [3–5]

SB has made significant contributions to our understanding of immunology. Although SB is far from the true and full description of all levels of immune response, the techniques and technologies of SM continue to improve at a rapid pace.

Tolerance as Ultimate Goal of Transplantation

Organ transplantation is not only considered as the last resort therapy but also as a treatment of choice for many patients with end-stage organ damage. Recipient-mediated acute or chronic immune response is the main challenge after transplant surgery. Nonspecific suppression of host immune system is currently the only method used to prevent organ rejection. Lifelong immunosuppression will cause significant side effects such as infections, malignancies, chronic kidney disease, hypertension and diabetes. [6] This is more relevant in children who have a longer life-expectancy so may receive longer period of immunosuppressive medications. Efforts to minimize or complete withdrawal of immunosuppression would improve the quality of life and long-term outcome of pediatric transplant recipients. [7]

Current Status of Transplantation in United States

There are currently more than 100, 000 patients on waiting list for solid organ transplantation in United States. The biggest challenge in organ transplantation is organ shortage.

Since 2004 number of living donor has decreased. [8] Decrease in number of living donors can be due to loss of income during post-operative recovery and expenses which may not be paid by insurance companies. [8] Interestingly number of organ transplantation has been increased about 50% in from 1988 to 2011 (OPTN). The number of deceased donor organs available for transplantation has decreased in recent years. [9] The number of wait listed patents has increased by 12 fold. However there is still a big gap between demand and supply of organs for transplantation.

The current challenges for continuous IS therapy are:

1 The side effects of chronic IS
2 Cost
3 Non-compliance

Side Effects of Immunosuppression

There are various studies which estimates the consequences of post-transplant non-specific IS, either by side effect of IS or as a complication of a defect in body`s defense mechanism. The type and dosage of most IS after transplantation have been changed during past decades. [10, 11] For example while usages of cyclosporine and azathoprine have been decreased from 1995 to 2005, usage of mycophenolate and tacrolimus have been increased. In 2005 92% of post-transplant patients received tacrolimus. [10, 11]

Immunosuppressive therapy can be divided to induction and maintenance therapy. Table 1 shows some of the medications that more commonly used for post-transplant IS along with their most common side effects (Table 1). [8, 12]

Two groups of medications that are well-known to cause direct side effects but yet are backbone of all IS therapies, are calcineurin inhibitors (CNI) and corticosteroids. However current efforts are minimizing their adverse effects by close drug monitoring and multiple drug combination, we cannot fully inhibit their side effects. [12]

Corticosteroids put the patients under the risk of developing a wide range of medical problems from poor wound healing, susceptibility to infections, cardiovascular risk factors such as hypertension (by up regulating of $\alpha 1$ receptors) and hyperlipidemia to growth retardation and even changing of appearance (Buffalo hump, moon face ,etc.). [13]

CNI are associated with renal toxicity both in renal and non-renal transplantation. [13–15] Rate of renal dysfunction in pediatric recipients of non-renal transplantation can be about 55% [12, 16, 17] and 3-6% of them can develop ESRD. [12, 18] Neurotoxicity, hypertension, hyperlipidemia, Diabetes mellitus and hyperkalemia are among other side effects of CNIs. [8]

Half of heart transplant recipients, 30% of liver recipients [12, 18] and 50% of pediatric kidney recipients also have hyperlipidemia. [12] Neurological disorders affect approximately 20% of liver recipients [12] and risk of developing *DE NOVO* malignancies are also 3-5 time higher than normal population.[12]

The risk of developing post-transplant lymphoproliferative disorder (PTLD) which has close relation with EBV infection is found to be higher in children comparing to adults as it is more likely that they be EBV seronegative at the time of transplantation. [13] Most pediatric patients receive transplantation at the age which they have naïve immune system and are seronegative to many viruses including EBV and HSV. [19] CMV infection, BK virus nephropathy and PCP are among other complications. [20] In general, Infection as the main cause of hospitalization after kidney transplantation is related to IS. [13, 21] About 85% of liver transplant patients contract infections (bacterial, viral or fungal) which is the cause of death in 28.4% of them. [12]

Table 1. Common side effects of immunosuppression medication

	MEDICATION	SIDE EFFECTS
Medication used for induction therapy	Anti-CD25 receptor antibodies (basiliximab, daclizumab)	Anaphylaxis, allergic reaction
	Anti-CD52 monoclonal antibody (alemtuzumab)	T-cell depletion, which increases the risk of infection, in particular CMV reactivation
	Anti-thymocyte globulin (ATG)	Lymphopenia, Serum sickness, anaphylactic reaction,shock,bronchospasm
Medication used for maintenance therapy	Corticosteroids	Cushinoid appearance, fluid retention, diabetes mellitus, hypertension, growth impairment, hyperlipidemia, osteopenia, impairment in wound healing12, Failure to thrive
	Calcineurin inhibitors (CNI)	Nephrotoxicity, neurotoxicity, hypertension, hyperlipidemia and hyperkalemia, diabetes mellitus, increased bone resorption, hirsutism, gingival hyperplasia, hearing impairment and cholestatic syndrome12
	Azathioprine	Hepatic nodular hyperplasia, Portal sclerosis, Myelosuppression
	Mycophenolate	GI disturbance, myelosuppression growth retardation

One report by SPLIT registry indicates that infants have the highest risk to develop infections after liver transplant among other subgroups. [6, 9]

Patient dependency on lifelong non-specific IS is an unsolved problem after transplantation. [22] Tolerance eliminates the complications of long-term IS use [13] which is a great challenge for pediatric transplant. It can also improve the patient's compliance which is the main problem in adolescents with chronic disease. [23] Adolescents are particularly prone to non-compliance with their medical regimen as a result of developing sense of authority and poor judgment at this age. [24]

Costs

Insurance coverage for long-term IS medication is a considering problem. Over 70% of kidney transplant programs report that their patients have serious problems paying their medication costs. [25] More than 68% of all programs report even deaths and graft loss because of cost-related IS non-adherence. However these problems are more significant in adults than pediatrics, but even children and their families are potentially at risk to face these problems. [25] In average the annual cost of IS is 10-15, 000 dollars.

Non-Adherance

Daily usage of IS medication may affect the mental health of patients particularly adolescents and their families. Both of these groups are prone for developing psychiatry problems such as depression, post-traumatic stress disorders (PTSD) and other anxiety disorder. [24] Adolescents have the worst outcome of graft survival mainly as a result of non-compliance. [20] Education about the potential risk of non-adherence is challenging in this group. Less complex medical regimen, medication with less side effects and cosmetic change can potentially be more successful. [24] Achieving the state of tolerance however seem to be the best possible solution to overcome adolescent non-adherence.

What is Tolerance?

Tolerance is the Holy Grail of transplantation. The concept of tolerance was first introduced in 1953 when Billingham et at showed that in-utero injection of bone-marrow cells to mice resulted the acceptance of skin graft from the same inbred donor while maintaining the ability to reject grafts from other breeds. [24, 25]

In "True tolerance" ,the organ transplanted, functions normally for a durable time and the recipient is also immunosuppression-free [23] and there is no detectable immune response to the donor antigens. [26] It is defined as a permanent and specific immunological acceptance of allograft antigens without the Immunosuppression usage. [27]

Operational tolerance is defined as the absence of rejection with normally functional allograft while the patients is immunosuppression-free but it does not necessarily mean lack of immune response towards the graft but rather the lack of destructive response. [27] In "Operational tolerance" there may be some immune response but it has no significant clinical presentation. [26] Operational tolerance usually results from elective or non-elective IS withdrawal. Prope or near tolerance is a term used when allograft functions normally and has normal histology but patients receive minimal immunosuppression. [27] Tolerance induction, theoretically is a method of modification of host immune system in a manner that it does not reject the organ transplanted, but being otherwise normally functional and competent, can result in long-term survival of both the organ and the patient.

Definition of tolerance in animal models differs as of clinical one. It is defined as acceptance of a graft without immunosuppression use ability to accept subsequent graft from the same donor while having the ability to reject an organ from a third party. [28]

Preclinical Experience for Tolerance Induction

Tolerance is in fact a process not a sudden event. [26] Interestingly most cases of reported operational tolerance after kidney transplantation are the result of noncompliance.

The usual result of discontinuation is graft rejection; nonetheless some patients do not reject the organ. [26] In 2007 Koyama et al. also Showed that CD8+ memory T cells can prevent mixed chimerism and tolerance induction in monkeys. [20, 29] While an immunosuppressive regimen had induced mixed chimerism when kidney transplantation and bone marrow transplantation had been performed simultaneously, the same regimen had failed to do so when bone marrow was transplanted after kidney transplantation. After finding that a numerous number of memory T cells remain even after that regimen, they added humanized anti-CD8 monoclonal antibody and depleting CD8+ memory Tcells and could achieved tolerance in nonhuman primates. [20, 29]

In 1995 Kawai et al. developed a nonmyeloablative regimen that could produce mixed chimerism and renal transplant tolerance in monkeys. [20, 30] Adams et al. showed that virally induced alloreactive memory response in mice, is a barrier to tolerance. They showed a base threshold of memory T cell is needed to induce rejection, with CD8+ memory T cell having the main role. [20, 31] Studies on animal and human models revealed several mechanisms which have role in tolerance.

Strategies of tolerance induction in general can be summarized as below:

1. *Using of Hematopoietic stem cells (HSC) for induction of activated T-cells apoptosis:*
   T-cell activation without sufficient support (other cytokines, co-stimulatory factors, etc.) leads to T-cell anergy on death. Blood transfusion is a well-known use of Hematopoietic cells for tolerance induction. [26] Recipients of blood transfusions have improved graft survival. [26]
2. *Using of (HSC for induction of Chimerism.*
   Homer`s Iliad is maybe the first literary which describes Chimera. A monster reared by Amisodorus with lion front, snake tail and a goat in the middle. [32] Full chimerism is total replacement of recipient bone marrow with donor`s bone marrow. This process has a high morbidity and mortality which can exceed that of IS therapy. [26] Mixed Chimerism is defined as making host immune system in a manner that it composed of both donor and recipient cells. [23] To achieve this, host bone marrow is largely preserved but partially replaced by that of donor. [26] By preventing the bone marrow rejection, the donor`s hematopoietic cells populate in the host bone marrow and thymus and cause central deletion of alloreactive cells. [26]
   In new strategies donors are administered granulocyte stimulating factor and then stem cells are collected by plasmapheresis and then injected to host peripheral or portal veins days before surgery. There are also some reports which indicate that after stem cell transplantation and development of chimerism, not only physicians were able to discontinue IS but also the primary genetic defect which had caused organ failure, was also cured. [33] Patients with primary immunodeficiency disorders usually are not suitable candidates for solid-organ transplantation as post-transplant IS can worsen their immune system condition. [34] These patients can benefit from chimerism approach using HSC. [34]
   Aplastic anemia can be developed in up to 28% of orthotropic liver recipients. [35] Bone marrow transplantation is treatment of choice in this situation. In 1991 Kawahara et al. reported a 6.5 year boy who received cadaveric liver transplantation after hepatic failure due to non-A-non-B hepatitis. The patient developed aplastic

anemia 3 weeks post-transplant. After about 2 years of treatment with immunosuppression, he underwent bone-marrow transplant from his HLA-identical sister and about 3 weeks later he developed full chimerism and ultimately weaned off all IS.

3. *Depletion.*

   In this strategy host lymphocyte depletion is the goal. However the ideal situation is to delete only graft-activated cells but aggressive lymphocyte depletion is unavoidable. [23]

   To achieve the best result depletion should be done before graft reperfusion to minimize the number of inflammatory signal producing cells. [23]

   Different agents are used for this procedure either in preclinical or clinical trials.CD3-dyphteria immunotoxin and deoxy spergualin either as immunotherapy or combined therapy and Campath-1H are some examples.

4. *Altering the co-stimulatory signal.*

   T cells normally require two signals to become activated. The first signal (antigen-specific signal), is provided through interactions between T cell receptor and MHC molecules on APCs. The second (co-stimulatory) signal is provided by the interaction between molecules expressed on the membrane T cell.

   There are several well-known co-stimulatory pathways that altering them can induce tolerance in animals; CD28-B7 co-stimulatory pathway in which ligation of CD28 on the surface of T-cell to the B7 on the surface of APCs can cause T-cell activation is one example. [23] CTLA4 (CD152)-B7 pathway which limits T-cell activation and CD154-CD40 pathway are other examples. [23]

5. *Thymic Transplantation.*

   Thymic transplantation is a way to achieve immunocompetence in children with thymic agenesis (Digeorge syndrome). [26, 36] T-cell maturation in donors transplanted thymus which presents alloantigen, can result in selective tolerance in rats. [26, 37, 38]

   It is important to note that while many Immunomodulatory strategies can induce tolerance in animals, they fail to achieve the same efficacy in clinic. [22] We cannot always extend preclinical experiments to clinical ones as most preclinical models are in rodents which have almost a naive immune system comparing to human. However studies on non-human primates seem to be more applicable to human due to their complex immune system. [23] Different organs have particular tolerance mechanism.[39] It means that we cannot extend the mechanism of tolerance in kidney to that of liver. Studies show that there is no common transcriptional or blood cell markers between kidney and liver transplant tolerance. [39, 40] In addition, it seems that mechanism of tolerance are age-related too. [39] However some blood gene set have been identified to be common in adult and pediatric liver transplants. [19]

Clinical Experience in Tolerance Induction

Tolerance induction has been attempted mostly in adult transplant patients. Studies conducted by Massachusetts general hospital (MGH) [41], Stanford University [42] and

Northwestern memorial hospital [43] show the results of tolerance induction in adults. There are only few studies regarding tolerance induction in pediatrics which we have mentioned in table 3. In a study by Kawai et al., at MGH, five patients (22-46 years old) with end stage renal disease received combined kidney and bone-marrow transplantation from their relatives after receiving a non-myeloablative perioperative regimen. While rejection happened in one of them, the other four patients could be withdrawn from all IS. [41]

Patients with multiple myeloma (MM) who develop ESRD are usually not eligible for routine treatment of either of these two diseases. [44] Stem-cell transplant which is the main treatment for multiple myeloma rarely indicated for ESRD patients and in other hand their malignancy does not let them to be a good candidate for transplantation. [44] Spitzer et al. from MGH reported a trial on seven patients with concomitant MM and ESRD who underwent combined HLA-matched kidney and bone-marrow transplant after being prepared with conditioning regimen.

All seven patients developed mixed chimerism. Three patients gained normal or near-normal renal function while being off-immunosuppression, two other patients also had normal renal function but IS therapy re-started because of chronic GVHD. Two patients died, one because of progressive MM and other because of developing therapy-related acute myeloid leukemia.

In another report by Stanford university group, 16 adult patients enrolled the study between 2005-2011. All of them received kidney transplant after being conditioned with total lymphoid irradiation and antithymocyte globulin.

They had a previous study at 2002 in which despite achieving chimerism in three out of four combined kidney and bone marrow transplant patients, none of them could be withdrawn from IS. In their latter study 8 patients could be withdrawn for 1-3 years, three patients were just withdrawn and one patient was at the midst of withdrawal. Four patients could not be withdrawn due to return of underlying disease and failure to achieve chimerism or due to rejection. [42]

And finally in a report by Northwestern memorial hospital, five out of eight HLA-mismatched kidney and hematopoietic stem cell recipients developed durable chimerism and could be weaned off all IS 1 year after transplant. [43]

Tolerance Biomarkers

During past decade many efforts have been developed to identify tolerant-related biomarkers. For example, to detect biomarkers for identifying pediatric patients who have more chance to benefit from IS minimization or withdrawal (tolerance) L. Li et al., conducted a study.

A set of 13 genes were identified to exist in all of their operational tolerant patients, suggesting a method to more specifically select patients for operational tolerance. In their study approximately 60% of stable pediatric liver transplant patients had high possibility for tolerance, This in turn suggest that in almost the same percentage of these patients IS can be minimize or withdrawn. [19]

Currently identified or under investigation biomarkers are mostly related to dendritic cell, Treg cells and gene polymorphism. [45, 46] A list of tolerance-related biomarkers that are suggested to have a role in withdrawal and/or rejection are listed in table 2. [45–48]

Mechanism of tolerance seems to be different in different organs for example while natural killer related transcripts have the main role in operational tolerance of the liver; B cell related ones have importance in operational tolerance after kidney transplants. [47]

Table 2. Biomarkers which have role in tolerance and/or rejections

Tolerance/Rejection	Increasing (↑frequency)	Decreasing(↓frequency)
Increase chance of tolerance	PD-L1/CD86 ratio pDC/mDC ratio Vδ1/ɤδ T cell ratio HLA-G CD4+CD25+ FOXP3+ IL-10	Vα24+Vβ11mNKT TNF-α
Increase risk of rejection	Cyclex, CD154+ Tc memory cells mDC:pDC ratio IL-23 IL-17	

What Is the Future of Tolerance?

With increasing the number of transplants and improvement in both surgical technologies and tolerance induction regimens, transplantation is entering a new promising era. Improvements in the efficiency of the donation process, public awareness and live donation, development of standardized operation and post-operation protocols, have made donation and organ transplantation more widespread.

Attempts must be conducted to achieve several goals:

First, to recognizing tolerance-vulnerable patients. There has been some progress in this regard and some potential biomarkers have been identified. [46] Patients should be selected carefully and the ones who fulfill the primary criteria must be enrolled. It might be interesting to point that in contrast to other solid-organs, acute rejection after liver transplant does not cause any long-term sequel. [19] Patients who develop graft rejection after IS withdrawal can turn back to normal graft function with IS reinitiating [21, 49] Tolerance after liver transplant is more frequent than other organs. [19] This can be related to its characteristic venous endothelium and vascular architecture and dual blood supply. [28, 50] Some articles report a tolerance rate of about 20% in adult liver transplantations [19, 51–54] and even higher rate in children. [19, 47, 55] Combination of above mentioned facts can suggest that liver can be the best organ for study of tolerance induction in pediatrics. In selected cases, IS withdrawal or minimization after liver transplantation can be achieved more frequently than other organs.[19]

Since the earliest days of solid organ transplantation, weaning or minimizing IS has been desirable. Identification of suitable patients who can enroll in this process is still a challenge. Identification of tolerance biomarker is a goal. However, there are still no reliable immunologic parameters to assess the feasibility of IS weaning. Better understanding of the factors involve in immune system activation and tolerance allow us to develop tailored IS treatments in which the medication dosage would be reduced and tolerogenic mechanism promoted. Increasing the frequency of reported successful operational tolerance or tolerance induction works like a driving force for development of new weaning protocols. Further studies must focus on development of gene expression profiling as a mean to detect candidates for IS weaning. Additionally, safer and less toxic protocols with more success rate must be designed.

System Biology and Tolerance

There are multiple interactions among different pathways which ultimately lead to T cell activation. One way to study and analyze interaction among different and interconnected biomedical pathways is through Systems Biology (SB). Lwin et al., [56] studied the SB approach to indentify serial genetic interaction to target T-cell activation. Using gene-gene interactions catalogued in the Bimolecular Interaction Network Database (BIND), Biological General Repository for Interaction Datasets (BioGRID), and Human Protein Reference Database (HPRD), they built a network and identified highly interconnected genes (ATP5C1, IL6ST, PRKCZ, MYC, MAPK1, FOS, and JUN), which were involved in cell proliferation, differentiation, and activation responses. They tried to knock down these genes using siRNA to cause T-cell hyporesponsiveness in Jurkat and HeLa Cells. In both cell lines, targeted gene knockdown using siRNA showed decreased cell proliferation and ATP production. However, there were inconsistencies between two cell lines in ATP production, perhaps due to different transfection efficiencies of siRNA. They achieved a transfection efficiency of 30%–50% in Jurkat cells and 90% in HeLa cells. Another limitation is the number of genes studied. The authors only studied a limited number of hub genes out of more than 1400 hub genes identified in the network. Despite these limitations, the authors should be congratulated on the use of a novel approach in studying T-cell activation. This approach could potentially lead to studies to prevent rejection or induce tolerance by indentifying genes and their interaction, which leads to T-cell activation.

SB could potentially lead to studies to prevent rejection or induce tolerance by indentifying genes and their interaction which leads to T cell activation.

References

[1] Hodgkin A. l., Huxley A. f. A quantitative description of membrane current and its application to conduction and excitation in nerve. *J. Physiol. (Lond.),* 1952;117 (4): 500–544.

[2] Kohl P., Crampin E. J., Quinn T. A., Noble D. Systems biology: an approach. *Clin. Pharmacol. Ther.,* 2010; 88 (1):25–33.

[3] Zak D. E., Aderem A. Systems biology of innate immunity. *Immunol. Rev.,* 2009; 227 (1): 264–282.

[4] Shapira S. D., Hacohen N. Systems biology approaches to dissect mammalian innate immunity. *Curr. Opin. Immunol.,* 2011; 23 (1):71–77.

[5] Gottschalk R. A., Martins A. J., Sjoelund V. H., Angermann B. R., Lin B., Germain R. N. Recent progress using systems biology approaches to better understand molecular mechanisms of immunity. *Semin. Immunol.,* 2012.

[6] Feng S. Long-term management of immunosuppression after pediatric liver transplantation: is minimization or withdrawal desirable or possible or both? *Curr. Opin. Organ Transplant.*, 2008;13(5):506–512.

[7] Londoño M-C., Danger R., Giral M., Soulillou J-P., Sánchez-Fueyo A., Brouard S. A need for biomarkers of operational tolerance in liver and kidney transplantation. *Am. J. Transplant.,* 2012; 12 (6):1370–1377.

[8] Saidi R. F. Current status of liver transplantation. *Arch. Iran. Med.*, 2012; 15 (12): 772–776.

[9] Saidi R. F., Markmann J. F., Jabbour N., et al., The faltering solid organ donor pool in the United States (2001-2010). *World J. Surg.*, 2012; 36 (12):2909–2913.

[10] Bucuvalas J. C., Alonso E., Magee J. C., Talwalkar J., Hanto D., Doo E. Improving long-term outcomes after liver transplantation in children. *Am. J. Transplant.,* 2008; 8 (12): 2506–2513.

[11] Horslen S., Barr M. L., Christensen L. L., Ettenger R., Magee J. C. Pediatric transplantation in the United States, 1996-2005. *Am. J. Transplant.,* 2007; 7 (5 Pt 2):1339–1358.

[12] Tredger J. M., Brown N. W., Dhawan A. Immunosuppression in pediatric solid organ transplantation: opportunities, risks, and management. *Pediatr. Transplant.*, 2006; 10 (8): 879–892.

[13] Traum A. Z., Kawai T., Vacanti J. P., Sachs D. H., Cosimi A. B., Madsen J. C. The need for tolerance in pediatric organ transplantation. *Pediatrics,* 2008; 121 (6): 1258–1260.

[14] Solez K., Vincenti F., Filo R. S. Histopathologic findings from 2-year protocol biopsies from a U.S. multicenter kidney transplant trial comparing tarolimus versus cyclosporine: a report of the FK506 Kidney Transplant Study Group. *Transplantation,* 1998; 66 (12):1736–1740.

[15] Ojo A. O., Held P. J., Port F. K., et al., Chronic renal failure after transplantation of a nonrenal organ. *N. Engl. J. Med.,* 2003; 349 (10):931–940.

[16] Arora-Gupta N., Davies P., McKiernan P., Kelly D. A. The effect of long-term calcineurin inhibitor therapy on renal function in children after liver transplantation. *Pediatr. Transplant.*, 2004;8(2):145–150.

[17] Fine R. N. Renal function following liver transplantation in children. *Pediatr. Transplant.*, 2005; 9 (5):680–684.

[18] Midtvedt K., Hjelmesaeth J., Hartmann A., et al., Insulin resistance after renal transplantation: the effect of steroid dose reduction and withdrawal. *J. Am. Soc. Nephrol.,* 2004; 15 (12):3233–3239.

[19] Li L., Wozniak L. J., Rodder S., et al., A common peripheral blood gene set for diagnosis of operational tolerance in pediatric and adult liver transplantation. *Am. J. Transplant.,* 2012; 12 (5):1218–1228.

[20] Traum A. Z., Ko D. S., Kawai T. The potential for tolerance in pediatric renal transplantation. *Curr. Opin. Organ Transplant.*, 2008; 13 (5):489–494.

[21] Koshiba T., Li Y., Takemura M., et al., Clinical, immunological, and pathological aspects of operational tolerance after pediatric living-donor liver transplantation. *Transpl. Immunol.,* 2007; 17 (2):94–97.

[22] Ohe H., Waki K., Yoshitomi M., et al., Factors affecting operational tolerance after pediatric living-donor liver transplantation: impact of early post-transplant events and HLA match. *Transpl. Int.,* 2012; 25 (1):97–106.

[23] Pearl J. P., Preston E., Kirk A. D. Tolerance: is it achievable in pediatric solid organ transplantation? *Pediatr. Clin. North Am.,* 2003;50(6):1261–1281, vii.

[24] Stuber M. L. Psychiatric issues in pediatric organ transplantation. *Child Adolesc. Psychiatr. Clin. N. Am.*, 2010;19(2):285–300, viii–ix.

[25] Evans R. W., Applegate W. H., Briscoe D. M., et al., Cost-related immunosuppressive medication nonadherence among kidney transplant recipients. *Clin. J. Am. Soc. Nephrol.*, 2010; 5 (12):2323–2328.

[26] Girlanda R., Kirk A. D. Frontiers in nephrology: immune tolerance to allografts in humans. *J. Am. Soc. Nephrol.,* 2007; 18 (8):2242–2251.

[27] Wieërs G., Gras J., Bourdeaux C., Truong D. Q., Latinne D., Reding R. Monitoring tolerance after human liver transplantation. *Transpl. Immunol.,* 2007; 17 (2):83–93.

[28] Alex Bishop G., Bertolino P. D., Bowen D. G., McCaughan G. W. Tolerance in liver transplantation. *Best Pract. Res. Clin. Gastroenterol.*, 2012;26(1):73–84.

[29] Koyama I., Nadazdin O., Boskovic S., et al., Depletion of CD8 memory T cells for induction of tolerance of a previously transplanted kidney allograft. *Am. J. Transplant.,* 2007; 7 (5):1055–1061.

[30] Kawai T., Cosimi A. B., Colvin R. B., et al., Mixed allogeneic chimerism and renal allograft tolerance in cynomolgus monkeys. *Transplantation,* 1995; 59 (2): 256–262.

[31] Adams A. B., Williams M. A., Jones T. R., et al., Heterologous immunity provides a potent barrier to transplantation tolerance. *J. Clin. Invest.,* 2003; 111 (12):1887–1895.

[32] Chimera (mythology). Wikipedia, the free encyclopedia. 2012. Available at: http://en.wikipedia.org/w/index.php?title=Chimera_(mythology)&oldid=525368151. Accessed January 3, 2013.

[33] Matthes-Martin S., Peters C., Königsrainer A., et al., Successful stem cell transplantation following orthotopic liver transplantation from the same haploidentical family donor in a girl with hemophagocytic lymphohistiocytosis. *Blood,* 2000;96 (12):3997–3999. Available at: http://bloodjournal.hematologylibrary.org/ content/96/12 /3997. Accessed December 23, 2012.

[34] Hadzić N., Pagliuca A., Rela M., et al., Correction of the hyper-IgM syndrome after liver and bone marrow transplantation. *N. Engl. J. Med.,* 2000; 342 (5): 320–324.

[35] Kawahara K., Storb R., Sanders J., Petersen F. B. Successful allogeneic bone marrow transplantation in a 6.5-year-old male for severe aplastic anemia complicating orthotopic liver transplantation for fulminant non-A-non-B hepatitis. *Blood,* 1991; 78 (4): 1140–1143.

[36] Rice H. E., Skinner M. A., Mahaffey S. M., et al., Thymic transplantation for complete DiGeorge syndrome: medical and surgical considerations. *J. Pediatr. Surg.,* 2004; 39 (11): 1607–1615.

[37] Posselt A. M., Barker C. F., Tomaszewski J. E., Markmann J. F., Choti M. A., Naji A. Induction of donor-specific unresponsiveness by intrathymic islet transplantation. *Science,* 1990; 249 (4974): 1293–1295.

[38] Odorico J. S., Barker C. F., Posselt A. M., Naji A. Induction of donor-specific tolerance to rat cardiac allografts by intrathymic inoculation of bone marrow. *Surgery,* 1992; 112 (2): 370–376; discussion 376–377.

[39] Roedder S., Gao X., Sarwal M. M. The pits and pearls in translating operational tolerance biomarkers into clinical practice. *Curr. Opin. Organ Transplant.*, 2012; 17(6): 655–662.

[40] Lozano J. J., Pallier A., Martinez-Llordella M., et al., Comparison of transcriptional and blood cell-phenotypic markers between operationally tolerant liver and kidney recipients. *Am. J. Transplant.,* 2011; 11(9): 1916–1926.

[41] Kawai T., Cosimi A. B., Spitzer T. R., et al., HLA-mismatched renal transplantation without maintenance immunosuppression. *N. Engl. J. Med.,* 2008; 358 (4): 353–361.

[42] Scandling J. D., Busque S., Dejbakhsh-Jones S., et al., Tolerance and withdrawal of immunosuppressive drugs in patients given kidney and hematopoietic cell transplants. *Am. J. Transplant.,* 2012; 12 (5): 1133–1145.

[43] Leventhal J., Abecassis M., Miller J., et al., Chimerism and tolerance without GVHD or engraftment syndrome in HLA-mismatched combined kidney and hematopoietic stem cell transplantation. *Sci. Transl. Med.*, 2012; 4 (124): 124ra28.

[44] Spitzer T. R., Sykes M., Tolkoff-Rubin N., et al., Long-term follow-up of recipients of combined human leukocyte antigen-matched bone marrow and kidney transplantation for multiple myeloma with end-stage renal disease. *Transplantation,* 2011; 91 (6): 672–676.

[45] Castellaneta A., Thomson A. W., Nayyar N., De Vera M., Mazariegos G. V. Monitoring the operationally tolerant liver allograft recipient. *Curr. Opin. Organ Transplant.*, 2010; 15 (1): 28–34.

[46] Mazariegos G. V. Immunosuppression withdrawal after liver transplantation: what are the next steps? *Transplantation,* 2011; 91 (7): 697–699.

[47] Talisetti A., Hurwitz M., Sarwal M., et al., Analysis of clinical variables associated with tolerance in pediatric liver transplant recipients. *Pediatr. Transplant.*, 2010; 14 (8): 976–979.

[48] Sánchez-Fueyo A. Identification of tolerant recipients following liver transplantation. *Int. Immunopharmacol.,* 2010; 10 (12): 1501–1504.

[49] Feng S., Ekong U. D., Lobritto S. J., et al., Complete immunosuppression withdrawal and subsequent allograft function among pediatric recipients of parental living donor liver transplants. *JAMA.*, 2012; 307 (3): 283–293.

[50] Gao Z., McAlister V. C., Williams G. M. Repopulation of liver endothelium by bone-marrow-derived cells. *Lancet,* 2001; 357 (9260): 932–933.

[51] Devlin J., Doherty D., Thomson L., et al., Defining the outcome of immunosuppression withdrawal after liver transplantation. *Hepatology,* 1998; 27 (4): 926–933.

[52] Mazariegos G. V., Sindhi R., Thomson A. W., Marcos A. Clinical tolerance following liver transplantation: long term results and future prospects. *Transpl. Immunol.,* 2007; 17 (2): 114–119.

[53] Girlanda R., Rela M., Williams R., O'Grady J. G., Heaton N. D. Long-term outcome of immunosuppression withdrawal after liver transplantation. *Transplant. Proc.,* 2005; 37 (4): 1708–1709.

[54] Koshiba T., Ito A., Li Y., Wu Y., Takemura M., Sakaguchi S. [Regulatory T cell based transplant tolerance--freedom from immunosuppression]. *Nippon. Rinsho*., 2007; 65 (3): 557–567.

[55] Mazariegos G. V., Reyes J., Marino I. R., et al., Weaning of immunosuppression in liver transplant recipients. *Transplantation,* 1997; 63 (2): 243–249.

[56] Lwin W. W., Park K., Wauson M., et al., Systems biology approach to transplant tolerance: proof of concept experiments using RNA interference (RNAi) to knock down hub genes in Jurkat and HeLa cells in vitro. *J. Surg. Res.,* 2012; 176 (1): e41–46.

Editors Contact Information

Dr. A. X. C. N. Valente
Center for Neuroscience and Cell Biology
Systems Biology Group
University of Coimbra
Biocant - Biotechnology Innovation Center,
Cantanhede, Portugal
Tel: (351) 231419040
andre.valente@biocant.pt

Prof. Abhijit Sarkar
Department of Physics and Vitreous State Laboratory,
Catholic University of America,
Washington DC, US

Prof. Yuan Gao
Lieber Institute for Brain Development,
Johns Hopkins Medical Campus,
855 N. Wolfe Stree, Suite 300, Baltimore, Maryland, US

Index

A

D

E

F

G

H

I

N

O

P

Q

R

S

T

U

V

W

X

Y